Integrated Endocrinology

Integrated Endocrinology

John Laycock and Karim Meeran
Both of Imperial College London

WILEY-BLACKWELL
A John Wiley & Sons, Ltd., Publication

This edition first published 2013 © 2013 by John Wiley & Sons, Ltd

Wiley-Blackwell is an imprint of John Wiley & Sons, formed by the merger of Wiley's global Scientific, Technical and Medical business with Blackwell Publishing.

Registered office: John Wiley & Sons, Ltd, The Atrium, Southern Gate, Chichester, West Sussex, PO19 8SQ, UK

Editorial offices: 9600 Garsington Road, Oxford, OX4 2DQ, UK
The Atrium, Southern Gate, Chichester, West Sussex, PO19 8SQ, UK
111 River Street, Hoboken, NJ 07030-5774, USA

For details of our global editorial offices, for customer services and for information about how to apply for permission to reuse the copyright material in this book please see our website at www.wiley.com/wiley-blackwell.

Library of Congress Cataloging-in-Publication Data has been applied for
ISBN 978-0-4706-8813-7 (hardback) – 978-0-4706-8812-0 (paperback)

A catalogue record for this book is available from the British Library.

Wiley also publishes its books in a variety of electronic formats. Some content that appears in print may not be available in electronic books.

Set in 9.5/13pt Meridien by Laserwords Private Limited, Chennai, India
Printed and bound in Singapore by Markono Print Media Pte Ltd

First Impression 2013

Contents

Preface

Both of us have always been inspired by the interest shown by our students in the subject of endocrinology, and we are grateful to them for enhancing our desire to produce a readable textbook covering the main aspects of the subject. Because the understanding of endocrinology is enhanced by a consideration of the clinical conditions associated with diminished or excessive production of individual hormones, we hope that the integration we have attempted between basic and clinical aspects is successful. In addition we are grateful to our wives and families, who have given us their continual support, and to our colleagues, in particular Waljit Dhillo and Gareth Leng who gave us many suggestions for improvement in those chapters which they read for us.

John Laycock and Karim Meeran

CHAPTER 1
The Molecular Basis of Hormones

Endocrine glands and their hormones

Introduction

In 1849 Claude Bernard postulated that the internal environment of the cells in the body (the *milieu intérieur*) is constantly regulated. In 1855 he also proposed that substances can be synthesised and secreted internally, within the body, by demonstrating the production and release of glucose from the liver. In these ways, perhaps he can be considered to be the 'father' of endocrinology even though we would not nowadays consider glucose to be a hormone.

While there are descriptions clearly relating to what we now know as endocrine glands going back at least 2000–3000 years in various parts of the world, it is interesting to note that endocrine glands were first classified as such only from the early 1900s. Initially, there were certain accepted methodologies which were used to ascertain whether or not tissues and organs had a true endocrine function. For example, since hormones are released into the bloodstream, certainly according to the original definition (see section 'What is a hormone?'), it is not surprising to find that endocrine glands in general have a very high blood flow per gram of tissue compared to most other organs. When assays were sufficiently developed to estimate, and later to measure, hormone levels in the blood, the concentration of a hormone would be expected to be greater in the venous effluent from an endocrine gland than in its arterial affluent. Nowadays, this general principle can be used to precisely locate the presence of an endocrine tumour.

Furthermore, removal of the endocrine gland being studied would be expected to result in observable changes to some aspect of the body's physiology. For example, one classic observation by Berthold, also in 1849, linking the testes (male sex glands) with a specific feature, was the disappearance of the comb on the head of a cockerel – a male characteristic in this species – when the testes were removed. Indeed, transplantation of the testes to another part of the body with an adequate blood circulation

---

Integrated Endocrinology, First Edition. John Laycock and Karim Meeran.

apparently restored the cock's comb. Likewise, the appropriate administration of an extract from the suspected endocrine gland should restore that physiological function, following the gland's removal. Clearly, such determining studies as these were essential in identifying what we can now call the 'classical' endocrine glands including the thyroid, the parathyroids, the gonads and the adrenals. However, nowadays we also recognise the much more disparate sources of hormones, from tissues not instinctively considered to have an endocrine function. For instance, the heart, lungs, kidneys and liver are better known for their other better described physiological functions than for the production of hormones, and yet this is undoubtedly one of their roles. Clearly the removal of such an organ would be problematic with regard to determining its endocrine function!

What is an endocrine gland (or tissue)?

An endocrine gland (or tissue) is usually defined as a group of cells which synthesises a chemical that is released into the surrounding medium, essentially into the blood. An endocrine gland is most easily recognised if it consists of cells with clearly identifiable, intracellular secretory machinery allowing for the expulsion of synthesised molecules out of that cell into the surrounding medium (blood). This is in contrast to an exocrine gland, such as a salivary gland or the main part of the pancreas, which secretes molecules into a duct leading to the exterior of the body. Nowadays, we know of various endocrine glands and tissues in the body which do not quite so readily fit the description given here; indeed, many of them are better known for other physiological actions. A general classification of endocrine glands would now have to include the following categories.

'Classic' endocrine glands

Various clearly defined glands have been identified as having an endocrine function and these can be classified as the 'classic' endocrine glands. They include the gonads, the thyroid, the adrenals, the parathyroids, the pancreatic islets of Langerhans and the pituitary. Each of these glands produces one or more different hormones. Only those glands producing amino acid–derived (e.g. amine, polypeptide and protein) hormones would contain secretory granules.

Gastrointestinal tract

Historically, the term hormone was actually used by two physiologists, William Bayliss and Ernest Starling, in 1909 to describe a molecule (secretin) produced by the gastrointestinal tract. This organ is extremely large with a clearly defined physiological function regarding digestion and absorption, and it is also the source of numerous hormones. While many of these hormones have clear gastrointestinal effects and fall within the domain of gastroenterologists, some of them do have more widespread effects including on the central nervous system (CNS). These hormones are of particular interest to endocrinologists currently trying to determine the mechanisms involved in the regulation of food intake, hunger and appetite (see Chapter 16).

Central nervous system

While the hypothalamus is a part of the brain with a well-defined endocrine function, releasing molecules from nerve endings either into a specific blood portal system linking it to the anterior pituitary or into the general circulation via the posterior pituitary (see Chapter 2), it is quite possible that other parts of the brain also produce molecules which could be secreted into brain fluids such as the intracerebroventricular (icv) or general brain extravascular fluids. Certainly, the dendritic release of molecules known to be hormones into the icv fluid in the third ventricle has opened up possibilities of communication between different parts of the brain using a fluid distinct from blood. Furthermore, within the brain is a small organ which has a more clearly defined endocrine function, called the pineal gland. It is an interesting gland with an afferent nerve pathway originating from the eyes; in Hinduism and Buddhism it is known as the third (or inner) eye, and is considered to be a symbol of enlightenment. It produces melatonin, a hormone which plays a role in regulating various functions related to the internal circadian 'clock', located in the suprachiasmatic nucleus in the hypothalamus. It is released during the night and its production is inhibited by daylight.

Placenta

Another tissue which has highly specific reproductive functions, and which has a major endocrine role, is the placenta. During pregnancy many hormones are produced by this tissue, often in conjunction with the developing fetus, so that the endocrine tissue as a whole is often referred to as the feto-placental unit (see Chapter 9).

Other endocrine tissues

More recently, and as mentioned earlier, tissues better known for other physiological functions have been shown to have endocrine roles. They include the liver, the kidneys, the heart, the blood and adipose tissue. Immune (e.g. lymphoid) tissue also produces molecules which have 'endocrine' effects on their distant target cells.

The cells of an endocrine gland produce molecules for export into the general circulation as a consequence of the integration of various signals reaching those cells. At any given time, these endocrine cells may receive numerous differing signals, some stimulatory and others inhibitory.

The manner in which each endocrine cell integrates these various signals is one area of endocrinology which is just beginning to be unravelled.

What is a hormone?

A hormone is that molecule produced by a certain cell or cells (the endocrine gland or tissue) which is exported out of the cells and is transported to its target cells by a circulating fluid medium, by definition (but not necessarily) the blood. Hormones influence their target cells to respond in a specific way, normally to the benefit of the organism. It is part of the homeostatic response to an altered environment, whether internal or external. The hormone conveys a message from one part of the body to another, and can therefore be considered as a messenger molecule. The response of the target cell to that first messenger, or hormone, is often produced in response to an intracellular cascade of activated or inhibited molecules which can be considered to be 'second' messenger systems.

Molecules or metabolites that are nutrients or excretory products would have to be excluded. For this reason, Claude Bernard's original discovery that the liver releases an internal secretion which is the energy substrate glucose means that it does not actually conform to the definition of an endocrine gland on this basis; this secretory product is not a hormone. However, the liver does produce other molecules which are not simple nutrients or excretory products, but which are true messengers ('first' messengers) secreted into the bloodstream which carries them to their distant target tissues. Consequently, it can truly be considered to be an endocrine tissue (see Chapter 3). Likewise, the definition has to exclude those molecules secreted from nerve terminals and which traverse the tiny synaptic gaps between neurones to act as neurotransmitters or neuromodulators. Of course, this does not exclude the possibility that neurones, for instance in the CNS, can be endocrine cells producing true hormones. Indeed, the study of these neurones and their neurosecretions is such a rapidly expanding research area that it is now a well-established entity in its own right, called neuroendocrinology.

It is apparent that, despite the definition given here, it is not always easy to appreciate whether a molecule is a hormone or not. For example, consider the group of molecules called cytokines. These are proteins produced by cells of the immune system which clearly have a communicating function between different cells. They are transported by the blood and can have diffuse effects in the body, including the CNS, and therefore act as hormones. This group of molecules includes the interleukins (IL1-18) and various other 'factors' such as the tumour necrosis factors and interferons, all components of the immune system. Not surprisingly, endocrinologists and immunologists show much interest in them.

One way we could establish whether a molecule is a hormone or not, is to determine whether it has specific receptors on its target cells. This would certainly allow us to exclude simple nutrients and excretory products which do not have any as such. Indeed, there are occasions when a gene product leads to the discovery of a new protein, which may be a receptor protein, for instance. Such a molecule is sometimes called an orphan receptor until a ligand (a molecule that binds to a receptor, such as a hormone) for it has been identified.

As indicated earlier, hormones, by definition, are chemicals which are produced by specific cells and released into the bloodstream to exert their effects on distant target cells. However, it is quite likely that they can be released into (or enter) other circulating fluids such as the cerebrospinal fluid, seminal fluid, amniotic fluid and lymph. All these fluids are essentially made up of water containing a variety of solutes and ions, maybe cells and cell fragments. As chemicals, some of the hormones will be hydrophilic (i.e. 'water-loving') molecules such as amino acids, peptides and proteins. They can similarly be considered to be lipophobic ('lipid hating'). Other hormones, such as steroids, will be lipophilic ('lipid-loving') molecules; they can also be described as hydrophobic ('water hating'). Not surprisingly, the chemical nature of any hormone in question will have an important bearing not only on its synthesis but also on its storage, its release from the endocrine cell, its transport in the fluid medium and its mechanism of action. A relatively new consideration is that some molecules, generally considered to be gaseous, have also been shown to have an endocrine role, although because they are very short-lived in the general circulation they probably have an effect only on cells near their sites of production. Nitric oxide is the clearest example of such a molecule; it is ubiquitous, being produced in many different tissues and exerting its effects locally.

While it is clear that most hormones exert their *endocrine* effect on distant target cells, some, such as nitric oxide, have an effect on nearby adjacent cells. They are described as having a *paracrine* effect. Also, we now know that a few hormones may actually have an immediate effect on their own cells of production, influencing their own processes of synthesis, storage and release. This is called an *autocrine* effect (see Figure 1.1).

There is another term that has been used to describe the location of action for some hormones: *cryptocrine*. The term describes the actions of molecules (such as hormones) which a cell produces, and which act within a closed space associated with its cell of production. One good example of such a cryptocrine activity is that of the Sertoli cell in the testis producing factors which act on developing spermatids within that cell's enclosed environment (see Chapter 6).

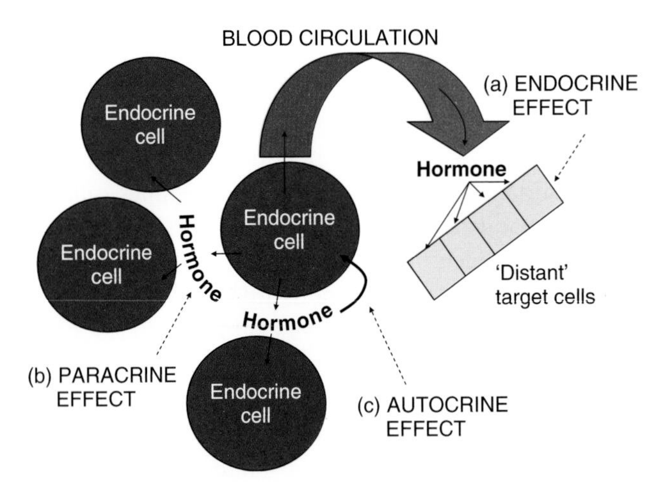

Figure 1.1 Diagram illustrating the release of a hormone from its endocrine cell and having its effect (a) on distant target cells (endocrine effect), (b) on cells nearby (paracrine effect) or (c) on its own endocrine cell of production (autocrine effect).

Interestingly, adjacent cells can have specific physical relationships between each other which permit them to 'converse'. One example is provided by gap junctions where the membranes of two cells are in contact. They are membrane proteins called connexins which contain pores (connexons) allowing small molecules such as ions and nutrients to cross from one cell to the other. Another example is the dynamic (labile) formation of tight junctions which fuse adjacent cells to each other. The tight junctions, consisting of transmembrane proteins, can actually temporarily trap extracellular fluid between the cells. Any molecule secreted by one cell into this tiny extracellular space would reach a high concentration locally, and if receptors for that molecule are present on the second cell, then an effect can be exerted which would not occur if the hormone was simply released into the general circulation to reach the second cell via a longer route involving considerable dilution (Figure 1.2).

The main hormones, their chemical nature and their main sites of synthesis are given in Table 1.1. Many of the hormones, and some of the endocrine glands, have alternative names, and they can be used interchangeably. Details of the various differing nomenclatures will be given in the relevant chapters of this text dealing with the individual glands and their hormones.

Hormones versus neurotransmitters

It will be apparent that hormones form a regulatory system which functions to maintain the body's homeostasis in response to perturbing

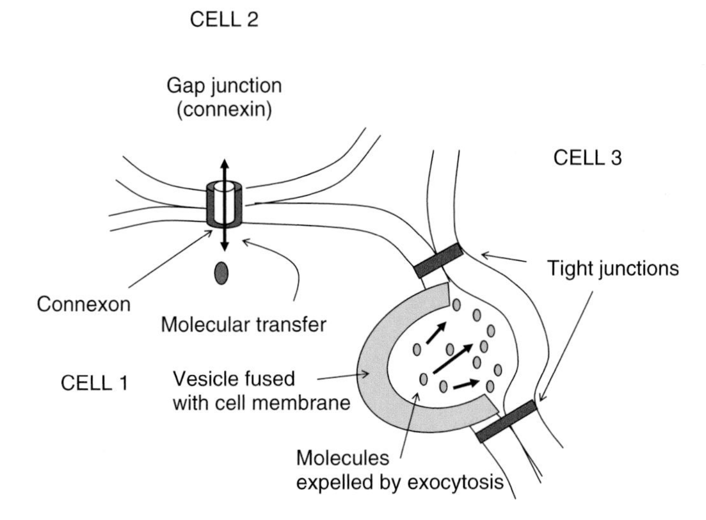

Figure 1.2 Diagram illustrating adjacent cells with gap and tight junctions, allowing paracrine communication between them. Molecules can pass from cell 1 to cell 2 via the pore (connexon) in the gap junction (connexin). They can also be released (e.g. by exocytosis) into an enclosed space defined by the presence of dynamic tight junctions, such as between cells 1 and 3 (shown) where they can reach a high concentration.

influences (stimuli) from within, as well as from outside, the body. The other major control system is the nervous system which is comprised of neurones which generally contact other neurones within the CNS via synaptic gaps, or target cells such as skeletal muscle fibres via neuromuscular junctions, for example. Both these control systems involve the use of chemical 'messengers' and have certain similarities, but there are also clear differences between them.

Neurones, when stimulated, produce neurosecretions which are released usually at the nerve terminals. Most of the time these neurosecretions are released across the synaptic gaps between neurones, and therefore they act as neurotransmitters, or at neuromuscular junctions. However, some neurones release neurosecretions into the blood, in which case they are clearly hormones. Furthermore, it is increasingly clear that some of these neurosecretory molecules could be transported to more distant target cells by the cerebrospinal fluid, or by the more general brain extracellular fluid, in which case they can also justifiably be considered to be hormones.

One important part of the brain which plays an essential role in regulating our internal environment is the hypothalamus. This part of the brain exerts many of its effects by controlling a number of peripheral endocrine glands. Some of the hypothalamic neurones release their neurosecretions into a specialised blood system. They are transported by the blood down to their target cells which comprise a 'mediating' endocrine gland called

Table 1.1 The principal hormones, together with their chemical group and main sites of synthesis.

Hormone type	Examples	Main endocrine gland
Amino acid or amino acid derived	Thyroxine (T4) and tri-iodothyronine (T3)	Thyroid
	Adrenaline and noradrenaline	Adrenal medulla
	Dopamine	Hypothalamus
Polypeptide	Insulin	Pancreas
	Glucagon	Pancreas
	Vasopressin	Neurohypophysis
	Oxytocin	Neurohypophysis
	Corticotrophin	Adenohypophysis
	Calcitonin	Thyroid parafollicular cells
	Parathormone	Parathyroid glands
	Somatostatin	Hypothalamus
	Corticotrophin-releasing hormone	Hypothalamus
	Thyrotrophin-releasing hormone	Hypothalamus
	Gonadotrophin-releasing hormone	Hypothalamus
	Inhibin	Testis, ovary
	Activin	Ovary
	Angiotensin II	Blood
	Atrial natriuretic peptide	Heart
Protein	Somatrotrophin	Adenohypophysis
	Prolactin	Adenohypophysis
	Erythropoietin	Kidney
	Cytokines (various, e.g. interleukins)	Immune system cells
	Leptin	Adipose tissue
	Ghrelin	Stomach
Glycoproteins	Thyrotrophin	Adenohypophysis
	Luteinising hormone	Adenohypophysis
	Follicle-stimulating hormone	Adenohypophysis
Steroids	Aldosterone	Adrenal
	Cortisol	Adrenal
	Testosterone	Testis
	17β-oestradiol	Ovary
	Oestrone	Ovary
	Progesterone	Ovary
	Calcitriol	Kidney
Prostaglandins	PGA1, PGA2, PGE1, PGE2, PGF1, etc.	Various
Thromboxanes and prostacyclins	TXA, TXB and PGI	Various
Gaseous molecules	Nitric oxide	Various (e.g. endothelial cells)
	Carbon monoxide	Various (e.g. hypothalamic neurones)

the pituitary (or hypophysis). Thus the hypothalamus can quite correctly be considered not only as part of the CNS but also as an endocrine gland in its own right. The hypothalamus together with the pituitary is generally called the hypothalamo-pituitary (or hypothalamo-hypophysial) system, or axis, and this will be considered in some detail in Chapters 2, 3 and 4. Indeed, the endocrine system has, rather poetically, been likened to an orchestra regulating many functions of the body, in which case the pituitary gland acts as the leader (principal first violin) while the hypothalamus is the conductor. As for all generalities, there are plenty of exceptions to this important control pathway involving the hypothalamus, and some endocrine glands have no obvious links, direct or indirect, with the hypothalamo-hypophysial system.

Both neural and endocrine systems are essential in regulating the various differing activities of the body, and both are dependent on the release of specific chemicals as either neurotransmitters or hormones, so what are their distinct characteristics?

There are many similarities between them. For instance:

- Neurotransmitters have been generally considered to be small molecules such as acetylcholine, amino acids (e.g. glutamine) or amino acid–derived molecules (e.g. gamma-amino butyric acid, or GABA), while hormones can also be amino acid derived, but also include polypeptides, steroids and larger proteins and glycoproteins. This difference now seems to be much less clear-cut: neurones can also produce polypeptides, and even steroids (neurosteroids) which have direct or modifying effects on adjacent or other postsynaptic neurones. Some hormones, initially thought to be synthesised only in peripheral endocrine cells, are now known to be produced in the CNS, by either neurones or other cells such as glia. The opposite is also true, with some molecules originally being clearly identified as neurotransmitters (e.g. in the CNS) also shown to be produced by typical secretory cells in peripheral tissues.
- Neurotransmitters and many hormones are essentially released from vesicles into the surrounding fluid by very similar mechanisms, involving calcium ions and an expulsion system called exocytosis which involves intracellular microtubules and filaments.
- Their mechanisms of action are generally similar, essentially involving either ion channels or G protein–related receptors.

While there are similarities between neurotransmitters and hormones, there are also some crucial differences. The more obvious differences are given in Table 1.2.

Table 1.2 Some differences between neurotransmitters and hormones.

	Neurotransmitters	Hormones
Transmission	Across synaptic cleft	By blood (or other fluid)
Target cells	Direct to specific neurones or other cells	Can be some distance from source cells
Speed of action	Milliseconds	From seconds up to days

Synthesis of hormones

Many cells synthesise molecules which are necessary for the proper functioning of those cells some of which are intracellular (e.g. enzymes), while others may be molecules (e.g. glucose) destined for export to other cells in the body. Intracellular molecules can be carbohydrates (e.g. energy substrates), lipids (e.g. for membranes) or amino acid–derived polypeptides and larger proteins (e.g. enzymes), for example. One group of molecules synthesised for export to other cells consists of hormones, and these can be amines, polypeptides, proteins or steroids. Some of the proteins may well have carbohydrate components attached to them, and they are then called glycoproteins.

Most hormones fall into one of two groups because of the chemical nature of their structures: they are either 'water-loving, lipid-hating' (hydrophilic, lipophobic) molecules, or are the opposite and are 'water-hating, lipid-loving' (hydrophobic, lipophilic). The former group of molecules generally consists of the polypeptide and protein hormones, while the steroid hormones mainly comprise the latter. This difference between the two types of molecules has much relevance regarding their synthesis, storage and release and also influences their transport in the blood or other fluids, their binding to receptors and their mechanisms of action. However, there are other molecules which are very small, can have very short lives and do not readily fall into either of these two main categories. This third group includes gaseous molecules such as nitric oxide and carbon monoxide, and amines which are amino acid–derived hormones such as the iodothyronines and catecholamines which share some of the properties of both of the other two main groups.

Polypeptide and protein hormones

These hormones consist of chains of amino acids. Polypeptides are often, arbitrarily, described as being chains of 2–100 amino acids, while proteins are generally taken to be longer chains of 100 or more amino acids. Some proteins are glycosylated, having carbohydrate residues attached to them (glycoproteins), and can also comprise more than one chain linked together. All these hormones are often synthesised initially as larger precursor molecules called prohormones which are cleaved to form more than one product, including the known hormone(s).

The endocrine cells which synthesise and export polypeptide and protein molecules contain a number of well-developed intracellular organelles which are involved in the processing and packaging of these molecules. These are the endoplasmic reticulum, the Golgi complex and the secretory granules. The synthesis process begins with the activation (or de-repression) of a specific gene on a chromosome within the nucleus of the endocrine cell by a transcription-stimulating (or -inhibiting) factor. The human genome contains approximately 30,000 genes, each one containing the code for a specific protein. When a specific gene is activated, resulting in a new protein molecule being synthesised, we talk about the gene having been 'expressed' (i.e. it is a process of gene expression). The gene, which consists of a segment of deoxyribonucleic acid (DNA), contains the nuclear code for a chain of amino acids. Nuclear chromosomes are made up of long strands of DNA. Each strand of DNA is made up of two chains of nucleotides linked together into the shape of a double helix. Each nucleotide consists of a base which projects towards the centre of the double helix, and an associated pentose sugar and phosphate which together form the backbone of the structure. Other molecules are also associated with the helix. The genetic code for the protein to be synthesised is provided by a series of base triplets, each one coding for an amino acid. Transcription of the gene code sequence into a corresponding molecule of messenger ribonucleic acid (mRNA) is initiated by an enzyme called RNA polymerase. Only one of the two DNA strands is used as a template for any given protein, so unzipping of the strands is an initial step. The start of the relevant segment of DNA is identified by the presence of a special nucleotide sequence called the promoter which is where the RNA polymerase attaches. Another sequence called a terminator identifies the end of the gene sequence. There are four DNA bases – adenine, cytosine, guanine and thymine – each pairing with a different base in a complementary manner. During transcription, each cytosine, guanine and thymine in the DNA template pairs with a guanine, cytosine and adenine respectively in the RNA strand, while adenine pairs with uracil (not thymine, as in the other DNA strand in the double helix). Each triplet of transcription nucleotides, called a codon, is associated with a specific amino acid.

Not all the DNA along a gene is transcribed into protein. Each gene is composed of exons which do transcribe into proteins, and introns which do not. As the complementary mRNA strand is synthesised within the nucleus, the introns are removed by enzymes called small nuclear ribonucleoproteins (snRNPs, called 'snurps'). The product of this process is a functional strand of mRNA which passes through a pore in the double membrane comprising the nuclear envelope. The transport process is active, and selective, for that mRNA molecule. The mRNA molecule that is synthesised initially from a gene's DNA can then be split (or 'spliced') into more than one mRNA component, each of which can be translated into a different protein. This process is called alternative splicing.

To get from the mRNA molecule to a new protein, the coded information provided in the nucleotide sequence needs to be translated into a complementary sequence of amino acids, linked by peptide bonds. The translation process occurs in ribosomes which are either present free in the cytoplasm or bound to the membrane of the rough endoplasmic reticulum (RER) which encircles, and is attached to, the nucleus. For proteins due for export such as hormones, the translation process is usually associated with the ribosomes on the endoplasmic reticulum. Each ribosome consists of two subunits. The small subunit has a binding site for the mRNA, while the larger subunit has two binding sites for other small RNA molecules which bind specific amino acids and *transfer* them to the ribosome (transfer RNA, or tRNA).

The translation process begins with an mRNA molecule binding to the small ribosomal subunit at the mRNA-binding site. An initiator tRNA then binds to the promoter codon on the mRNA strand and this is where translation begins. The initiator tRNA is complementary to the mRNA codon, and is called the anticodon. The tRNA molecule is also bound to the larger ribosomal subunit where the anticodon is translated into methionine amino acid, which is always the starter sequence for the subsequent growing chain of amino acids. The next mRNA codon pairs with the subsequent tRNA molecule anticodon, and its attached amino acid then binds with the previous amino acid by a peptide bond, and so on (Figure 1.3).

The peptide bonding is catalysed by an enzyme component of the ribosome larger subunit. The subsequent arrival of other tRNA molecules ensures that peptides are linked to each other according to the code provided by the mRNA. This polymerising process of amino acids results in the formation of the new protein molecule. The synthesis of the protein is completed when a stop codon is reached; the protein molecule then separates from the final tRNA and the ribosome itself separates into its large and small subunits. More complete details of this process can be found in specific textbooks on molecular biology.

The synthesis of a polypeptide hormone generally begins with the formation of an initial larger protein called a pre-prohormone. This molecule moves from its ribosome through the membrane of the RER. The transfer

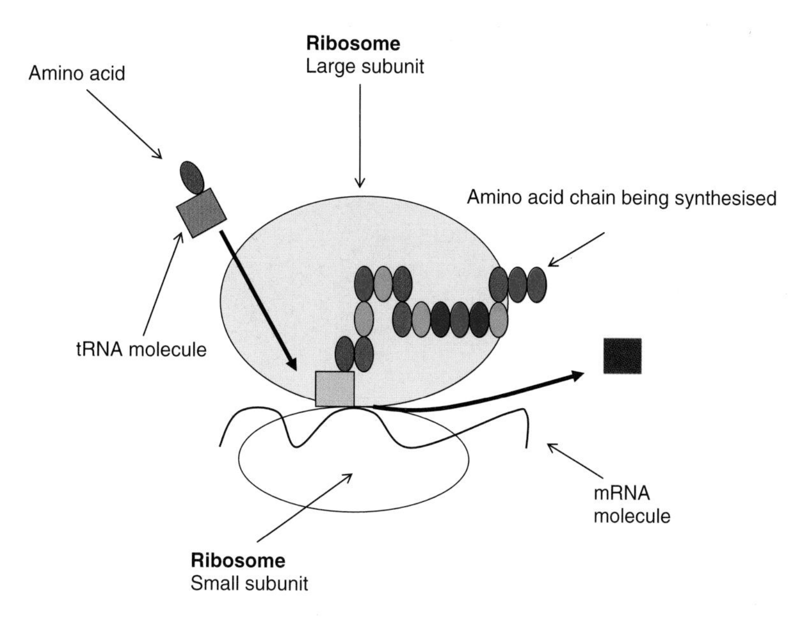

Figure 1.3 Diagram illustrating the arrival of a tRNA molecule, with its specific amino acid, at its binding site on the large ribosomal subunit where the amino acid links to the previous amino acid in the sequence by a peptide bond.

into the RER follows recognition of the initial segment of the molecule which is called the signal peptide. This is subsequently cleaved off the main molecule on entry into the sacs, or tubules, comprising the RER, producing the prohormone precursor. Subsequently, this precursor passes from the RER into the adjacent Golgi complex where it can be further modified. For example, here a prohormone can be enzymatically cleaved to form more than one protein or polypeptide, or it can be glycosylated, or otherwise modified. The final stage in the process of peptide hormone synthesis is the incorporation of the prohormone breakdown products into vesicles, which bud off from the Golgi membrane. There may be various peptide breakdown products from the same prohormone precursor, including the known biologically active hormone itself. In many cases, the other peptides have no known function although as research progresses they are likely to be shown to have some role, or biological activity, somewhere in the body.

The vesicles, invariably present in protein hormone–secreting endocrine cells, are an important storage source of the hormone (Figure 1.4).

Interestingly, because all eukaryotic cells within an individual contain the same genetic material, with relevant genes normally expressed only in specific tissues, it is apparent that abnormal polypeptide hormone

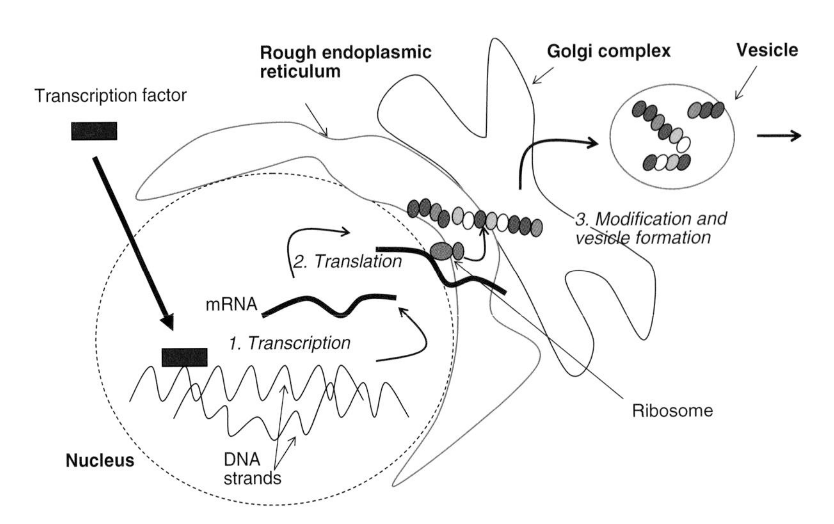

Figure 1.4 Diagram illustrating the three major steps involved in the synthesis of a hormone following the binding of the transcription factor to its site on a DNA strand: (1) Transcription of the protein 'code' from DNA to mRNA, (2) translation from mRNA to protein and (3) processing (chemical modification) and packaging of protein hormone into vesicles ready for export from the endocrine cell.

production is possible by non-endocrine tissue should that gene expression become activated. This is what can happen when a cell becomes abnormally stimulated by an appropriate chemical (carcinogen) such that the gene becomes 'switched on' inappropriately. The consequence is that some tumours of non-endocrine tissue (e.g. lung) can become abnormal (ectopic) sources of protein hormones. In addition to the direct effects of the tumour growth itself, problems can also arise due to the inappropriate, unregulated and very often exceedingly high circulating levels of the bioactive hormone being produced.

Steroid hormones

The other major group of hormones consists of steroids synthesised from the same initial precursor, cholesterol. Steroids are lipid soluble (lipophilic), so when they are synthesised by an endocrine cell they are immediately capable of moving out of that cell. Until recently it was believed that they simply move out of the cell by diffusion through the lipid membrane, but increasingly there is evidence for specific transporters in the cell membrane which would certainly aid the process. Not surprisingly, very little steroid hormone is found within its cell of synthesis since it would simply pass out of the cell immediately, so it is generally produced on demand (i.e. when the endocrine cell is stimulated appropriately). Consequently, the synthesis process involves the activation of specific intracellular enzymes which

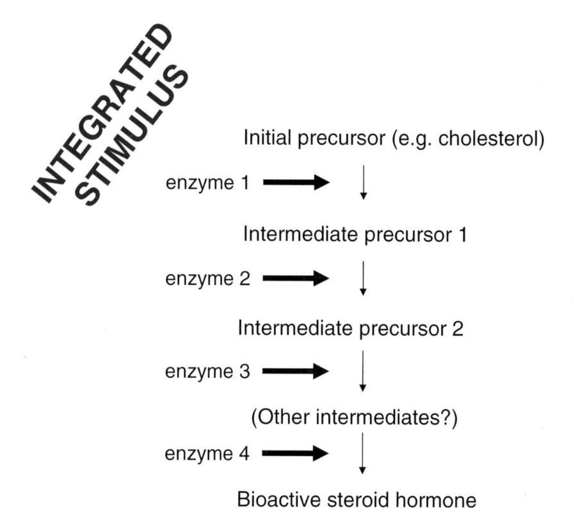

Figure 1.5 Diagram illustrating how the integrated stimulus to the endocrine cell, via intracellular mechanisms, activates enzymes which are involved in the formation of the biologically active steroid hormone from precursor molecules. It is quite possible that more than one bioactive hormone is produced when the synthesis pathway is stimulated.

catalyse chemical conversions such as hydroxylation and aromatisation, from precursors to the final bioactive hormone molecules.

Just as with peptide- and protein-producing endocrine cells, the cells producing steroid hormones receive constant stimulatory and inhibitory signals about the internal environment from a variety of sources, and these are somehow integrated in order to produce the final endocrine response. In these cells the various enzymes converting earlier molecular stages to the final bioactive molecule are the targets for the signalling pathways (Figure 1.5).

Amino acid–derived hormones

A few important hormones, such as those of the thyroid gland (iodothyronines) and the adrenal medulla (catecholamines), have very specific synthesis pathways and these will be considered in detail elsewhere in the relevant chapters of this volume. Essentially, the initial precursor molecule is an amino acid, and this is enzymatically altered to produce the final bioactive molecule. Their general properties vary, and are specific for each type of hormone. Thus, they are usually stored either in the endocrine cells or in follicles, they are transported in the blood either freely or protein bound and they bind to their receptors which are located on the plasma membranes of their target cells.

The eicosanoids: Prostaglandins, thromboxanes, prostacyclins and leukotrienes

There are other non-steroidal molecules which can be considered to be 'hormones' (see section 'What is a hormone?') and which also readily pass through plasma membranes by means of specific transporters, and therefore are only synthesised on receipt of appropriate stimuli. For example, the prostaglandins are lipids derived from 20-carbon essential fatty acids, called eicosanoids, found in most cells in the body. There are three series of prostaglandins, each derived from specific precursors: series 1 derived from gamma-linolenic acid, series 2 from arachidonic acid and series 3 from eicosapentaenoic acid. The best described are the molecules derived from arachidonic acid, which include the thromboxanes, prostacyclins and leukotrienes. All these groups of molecules together form the prostanoids, ubiquitous lipids which have physiological (and pathological) effects within the cells in which they are formed (autocrine) or on adjacent cells (paracrine). They have very short half-lives, are not transported to distant cells and therefore differ from the 'classic' hormones described here.

The precursor substrates are released from the cell membrane phospholipids by the action of specific enzymes. Thus, for example, arachidonic acid is released from cell membranes by the action of phospholipase A_2 and is then acted upon by other enzymes called cyclooxygenases (COX1 and COX2) to form intermediate prostaglandins PGG_2 and the unstable intermediate PGH_2. These prostaglandins are rapidly converted to other prostaglandins of the series by specific prostaglandin synthases, forming PGE_2, $PGF_{2\alpha}$ and so on (Figure 1.6). PGH_2 is also the precursor for prostacyclin (PGI_2) and thromboxanes by the action of yet other enzymes, prostacyclin and thromboxane synthases.

The gaseous molecules

The gaseous molecules that have been shown to act like locally acting hormones are nitric oxide and, it has been suggested, carbon monoxide. These molecules have very short half-lives, measured in seconds, and therefore act only on cells in close proximity to the cells producing them.

The free radical nitric oxide (NO) was first shown to be produced in living cells by Moncada and colleagues in the late 1980s (Palmer *et al.* 1987). They showed that the previously sought endothelium-derived relaxing factor (EDRF) was in fact this essentially toxic NO molecule. It was subsequently identified as having numerous physiological effects in maintaining homeostasis and is synthesised in many tissues in the body. It is particularly important as a regulator of the vasculature. There are three nitric oxide synthases (NOS): inducible (iNOS) and two calcium-dependent constitutive enzymes found in either vascular endothelial cells (eNOS) or nervous tissue (nNOS). In the vasculature, for example, NO

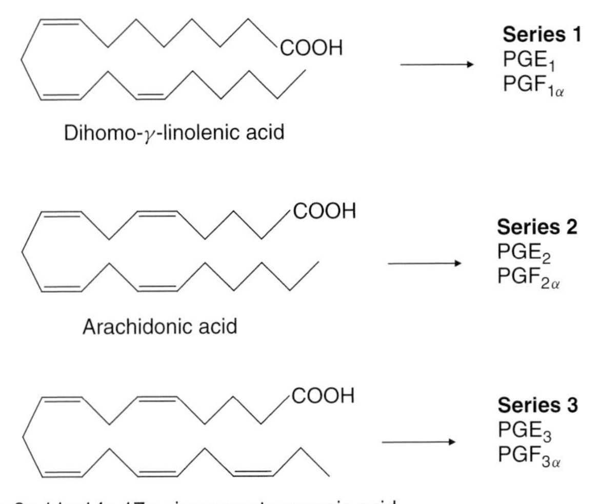

Figure 1.6 The precursor substrates linolenic, arachidonic and eicosapentanaenoic acids, derived from cell phospholipids, for series 1, 2 and 3 prostaglandins PGE and PGF.

is synthesised from the amino acid L-arginine by the action of eNOS or iNOS, as shown in Figure 1.7. The NO molecule diffuses from the endothelial cell into the adjacent outlying vascular smooth muscle cells where it stimulates the conversion of guanosine triphosphate (GTP) to cyclic guanosine monophosphate (cGMP).

The stimuli for eNOS include a direct effect of shear stress produced by blood flow against the endothelial cells, and a receptor-activated pathway, both of which stimulate the release of calcium ions from intracellular storage sites. The increase in the intracellular free calcium ion concentration results in the stimulation of the constitutive eNOS. The iNOS is activated by a longer term inflammatory response. The production of NO then results in the activation of guanyl cyclase in the adjacent smooth muscle and the subsequent synthesis of cGMP. The cGMP induces muscle relaxation partly by activating potassium channels, inducing membrane hyperpolarisation which results in the inhibition of calcium entry into the cell through voltage-dependent calcium channels. The consequent decrease in the intracellular calcium ion concentration decreases calcium–calmodulin formation which in turn decreases myosin–actin binding (the contractile protein 'machinery') resulting in muscle relaxation. It also stimulates myosin light chain phosphatase which dephosphorylates myosin light chains, contributing to the smooth muscle relaxation.

Nitric oxide has a short half-life of only a few seconds because it is rapidly inactivated by superoxides, these being reactive oxygen species (ROS) produced within the cell.

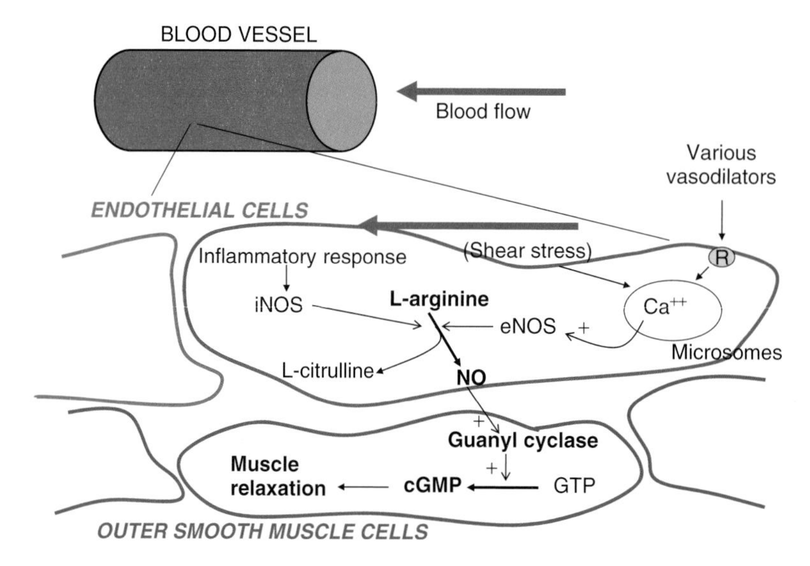

Figure 1.7 Diagram illustrating how the generation of nitric oxide (NO) by vascular endothelial cells results in the relaxation of the underlying smooth muscle cells by cGMP. R = receptor.

Storage of hormones

The chemical structure of a hormone will play an important part in determining whether a hormone is stored prior to its release or not. As with the synthesis pathways which differentiate protein and polypeptide hormones from the others, there is also a general difference between these two main groups of hormones with respect to whether or not they are stored.

Protein and polypeptide hormones

Proteins and polypeptides are generally stored in intracellular vesicles (see section 'Polypeptide and Protein Hormones'), these being formed as part of their synthesis pathway. Thus, when a protein or polypeptide hormone-producing cell is stimulated, not only will the genomic synthesis pathway be activated, but also the vesicles containing stored hormone can be mobilised and shifted to the cell's plasma membrane. This is important to appreciate: that the endocrine cell can be stimulated to not only synthesise new protein and polypeptide molecules for longer term use, but also release previously synthesised hormone molecules stored in the vesicles. Differing intracellular (second messenger) pathways for each of these activities can be induced separately or together, depending on the incoming stimulus.

However, another source of preformed protein and polypeptide hormones can be available within the blood. Some of these hormones when released bind to plasma proteins, and then circulate in a bound form which is in dynamic equilibrium with any free (unbound) hormone. One example is growth hormone (somatotrophin) which has a number of binding proteins in the blood. The binding of protein and polypeptide hormones to plasma proteins is not the norm, however. In general, these hormones are stored in intracellular vesicles, to be released when the endocrine cell is stimulated appropriately. Other possible sources of stored protein and polypeptide hormones are cellular components in the blood, such as platelets. For example, the polypeptide hormone vasopressin is present within platelets at quite high concentrations, so it can be released from these circulating 'containers' when they are disrupted. This may well be physiologically appropriate in certain circumstances such as at a site of haemorrhage, since one action of this hormone is to cause vasoconstriction.

Amino acid–derived hormones

These hormones, such as the iodothyronines and catecholamines, are present in the endocrine glands in storage form. The catecholamines adrenaline and noradrenaline are stored in vesicles like the larger polypeptide and protein hormones, and are similarly released by exocytosis. The iodothyronines are stored in a more unusual, extracellular form within follicles lined by the thyroid follicular cells which synthesise them. When the follicular cells are stimulated appropriately, the iodothyronines are taken back into the cells and ultimately released into the general circulation. While the catecholamines are transported in the free unbound state, the iodothyronines are transported mostly bound to plasma proteins. For further details about the catecholamines and iodothyronines, see Chapters 11 and 14 respectively.

Steroid hormones

Steroids, being lipophilic, are not generally stored within the endocrine cells in bioactive form, but are mainly synthesised on demand. All cells will contain some cholesterol (e.g. in the form of esters), so since cholesterol is the parent precursor molecule for steroid hormones, one could just about get away with saying that there is a tiny amount that could be considered to be present in storage form as the precursor.

Does this mean that such hormones are not present in the circulation until the endocrine cells are stimulated to synthesise and release them? The answer is no, because steroid hormones are transported in the blood mainly bound to plasma proteins. A dynamic equilibrium exists between free (biologically active) hormone molecules and those bound to plasma proteins. In probably simplistic terms, this dynamic equilibrium between free and bound forms of hormone ensures that much of the hormone at any moment is 'stored' in the blood as the inactive, bound component.

The equilibrium can be expressed in terms of concentrations of the three components:

$$[\text{free hormone H}] + [\text{plasma protein P}] \leftrightarrow [\text{protein-bound hormone HP}]$$

At equilibrium, the equation reaction can be expressed as:

$$K_a = [HP]/[H][P]$$

where K_a is the association constant.

Not only would this equilibrium reaction provide a constant source of hormone in the circulation in 'storage form' as the protein-bound hormone component, but also it means that if the equilibrium is upset, then the components automatically rearrange themselves in order to restore that equilibrium. In theory, at least, in regions of the body where there may be an increased presence of hormone receptors, for instance, then the bioactive free component would bind to its receptors (having the higher affinity for binding) resulting in a local disequilibrium state. The physical forces in operation would result in an unloading of hormone from the protein-bound to the free hormone state in order to restore the balance (Figure 1.8). The same equilibrium reaction can also be used to describe the relationship between the hormone and its (protein) receptor.

The plasma proteins involved in the transport of hormones fall into two main categories: those that are carriers for specific hormones, and those that are not particularly specific but are present in high concentrations and can 'mop up' large amounts of hormones. The defining characteristic for a plasma protein is whether it has a high or a low affinity for binding hormones, and what capacity it has for binding to them. Those plasma proteins which have a high affinity for specific hormones, such as for glucocorticoids (e.g. cortisol), can transport relatively large amounts of these hormones. They are usually globulins (e.g. cortisol-binding globulin). Because these globulins are not present in particularly high concentrations in the blood, they may have a high affinity for specific hormones, but they have a low capacity for hormones generally. In contrast, albumins form the largest plasma protein component of the blood. While this fraction does not have a high affinity for any hormone in particular, because there is so much of it, it will have a relatively high capacity for hormone transport and can carry a significant proportion of many hormones in the circulation.

While the globulin proteins in the blood are relatively specific regarding which hormones they can bind, there is some overlap between them. For instance, sex hormone–binding globulin as its name suggests can bind not only androgens such as testosterone but also oestrogens, while cortisol-binding globulin (or transcortin) binds not only the glucocorticoid cortisol

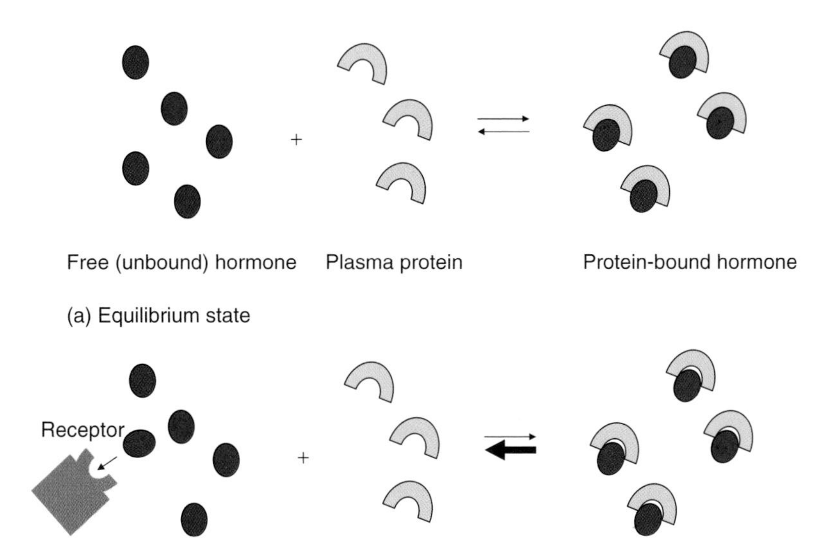

Figure 1.8 Diagram illustrating (a) the equilibrium state existing between the free and bound hormone concentrations and the concentration of relevant binding plasma proteins, and (b) the theoretical unloading of hormone from the bound component whenever there is a decrease in free hormone as exemplified by the presence of receptors with a greater affinity for the hormone than the plasma protein molecules.

but also the mineralocorticoid aldosterone, both being steroid hormones produced by the adrenal cortex.

Another point to bear in mind clinically is that plasma proteins are synthesised by the liver, and their synthesis can be influenced by specific hormones such as oestrogens, so conditions which affect liver function or the production of oestrogens can influence the protein-binding capacity of the blood.

Eicosanoids and gaseous hormones

These hormones tend to be lipophilic and are not stored to any significant extent either within their cells of production or elsewhere. They have short half-lives, so they are synthesised only on demand (i.e. on receipt of specific stimuli) and rapidly removed.

Release of hormones

The release of a hormone is yet another stage at which hormone production can be regulated. As will be already appreciated, different categories of hormone are either released from a pre-stored form or released on demand.

Protein and polypeptide hormones

Since protein and polypeptide hormones are generally stored in vesicles within the cells of their synthesis, their release requires that (i) the vesicles be moved to the outer plasma membrane, and (ii) the vesicle membranes fuse with that plasma membrane, allowing the vesicle contents to be expelled to the exterior of the cell. The latter process is called exocytosis.

Within each eukaryotic cell exists a dynamic structure called the cytoskeleton, which consists of a continually changing lattice-like framework of protein filaments which not only provide structural support to cell organelles but are also involved in various intracellular activities such as cell division and the movement of vesicles to the outer plasma membrane. The protein filaments which comprise this cytoskeletal network are actin filaments, intermediate filaments and microtubules. These filaments are held together and linked to the other intracellular structures by a variety of accessory proteins.

Actin is a globulin protein (G-actin) which can polymerise to a filamentous form (F-actin) known as a microfilament, in the presence of adenosine triphosphate (ATP). Each 8 nm long microfilament can link an intracellular protein to the plasma membrane. The actin filaments are particularly concentrated underneath the plasma membrane. Intermediate filaments, so called because their length is intermediate between the actin and microtubule at approximately 10 nm, consist of various proteins with specific characteristics, generally providing a supporting network within the cell. Microtubules are made up of the protein tubulin which is a dimer of α- and β-tubulin polypeptide molecules which bind to guanosine triphosphate (GTP). Each microtubule is a dynamic hollow structure made of polymers of tubulin, with a length of approximately 25 nm. Hydrolysis of the GTP bound to β-tubulin destabilises the microtubule allowing it to depolymerise, which accounts for the dynamic nature of its structure. Microtubules have various functions within the cell, including an involvement in regulating the movement of vesicles from within the cytoplasm towards the plasma membrane.

The movement of vesicles from the Golgi complex to the outer plasma membrane is sometimes called vesicle 'trafficking'. This process is believed to involve the movement of vesicles along microtubules towards the plasma membrane, with the process of exocytosis occurring only when the actin filaments below the membrane depolymerise, allowing the vesicles to penetrate the actin layer and reach the membrane (Figure 1.9). Various other molecules, such as recognition and anchoring proteins, play an important role in the intracellular transport process of vesicles. Clathrin is a surface protein on the vesicle which 'recognises' specific molecules to be

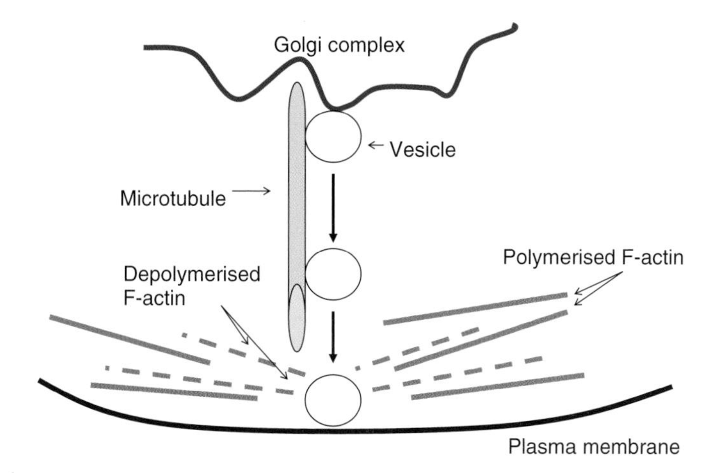

Figure 1.9 Diagram illustrating the current view that the vesicle containing hormone molecules is 'trafficked' down the microtubule towards the plasma membrane which is reached following the depolymerisation of the F-actin mesh lying underneath it.

transported within the vesicle, while other proteins are important for the docking of vesicles to the target membranes (e.g. the plasma membrane). These are called soluble NSF attachment protein receptors (SNAREs, where NSF is *N*-ethylmaleimide-sensitive fusion protein). Clearly, there are various signalling pathways which activate the various components of the intracellular vesicle-trafficking system within the cell. Our current knowledge of this process is still relatively rudimentary, but more details can be found in standard cell biology textbooks.

The actual process of exocytosis culminates in the fusion of the vesicle membrane with the plasma membrane resulting in expulsion of vesicle contents to the exterior of the cell. Exocytosis of hormones expressly results in the hormone molecules entering the bloodstream through the endothelial lining of the blood vessels (capillaries) vascularising the endocrine gland. This may be transcellular or through the gaps between the endothelial cells.

Amino acid–derived hormones

As indicated throughout this chapter, the hormones which fall within this category have similarities and differences with the protein and polypeptide hormones. Catecholamines are stored in vesicles within the cells of the adrenal medulla, so the secretory process is similar to the one described here. The synthesis, storage and release of the iodothyronines from the thyroid are all processes which differ somewhat from the standard, and will be discussed further in Chapter 14.

Steroids and other lipophilic hormones

These hormones, being lipophilic, are not stored in vesicles but are synthesised on demand. Their release into the bloodstream has generally been considered to be the consequence of the passive process of diffusion through the lipid component of the cell membrane. However, there is the possibility that at least some of these molecules may actually be secreted by a process involving membrane transporters, but at present this is generally speculative.

Transport of hormones in the circulation

It should already be apparent that a hormone, once released into the general circulation, can be transported free (i.e. unbound), be bound to a plasma protein or indeed be transported within a cell (e.g. an erythrocyte or leukocyte) or cell derivative (e.g. platelet) in the blood. It is generally believed that it is the free, unbound hormone which represents the bioactive fraction. The relationship between free and bound components has already been considered in section 'Steroid hormones', and while the importance of plasma proteins is particularly relevant for steroids and other small hormones, it should be noted that protein and polypeptide hormones can also be transported partly within the protein-bound component. Growth hormone, also known as somatotrophin (see Chapter 3), is one such hormone: there are a number of binding proteins for this hormone, with one at least being derived from the extracellular component of the growth hormone receptor found in target cell membranes.

Transport of hormones within the cells and cellular components of blood may also occur. For example, the gaseous hormone nitric oxide can interact with haemoglobin within erythrocytes, and it is feasible that preservation of the molecule within the red blood cell could allow it to be transported for release elsewhere in the circulation. Also, a significant fraction of the polypeptide hormone vasopressin in the circulation is transported within the platelets.

Hormone receptors

Once a hormone reaches its target cell, it must first of all recognise it (and be recognised), and then influence it in some way by a process called signal transduction. The first step is therefore the recognition process between hormone and target cell, and this requires the presence of some specific molecule associated with that target cell to which the hormone can bind. This molecule is known as the receptor. It can be located in the target cell's plasma membrane with an extracellular domain projecting outside the membrane into the circulating medium, or it may be located within the cell, usually in the cytoplasm or nucleus. Hundreds of receptor proteins have so far been identified. In principle, a receptor should be specific for a given hormone molecule. In practice, the receptor usually is capable of actually binding to other chemically similar molecules, or ligands, in much the same way as a 'specific' globulin in the blood can actually bind more than one chemically similar hormone (see section 'Steroid hormones').

There are five specific points to appreciate about hormones and their receptors:

1. Any one hormone may actually have a number of different receptors to which it can bind. The distribution of those subtypes of receptors can be tissue specific, allowing that hormone to have varying effects on different target cells and tissues.
2. The converse is also true: that any receptor is not necessarily specific for only one hormone, but can bind other ligands with differing affinities.
3. Receptor subtypes are likely to be linked to different intracellular second messenger pathways which, when activated, can produce varied responses in different target cells.
4. A very important concept in endocrinology is that a hormone's effect in a target tissue can be influenced not only by the concentration of bioactive hormone in the close vicinity of the target cell, but also by the presence of varying numbers of receptors. Thus an important feature of endocrinology will be the relationship between hormone concentration and receptor number. It also becomes apparent that a clinical disorder arising from the loss of a hormone's activity can be caused not only by a deficiency in its production but also by a deficiency (or abnormality) of its receptors.
5. If a hormone remained bound to its receptor, then that particular stimulated intracellular pathway could remain active indefinitely. It is clearly important to have cellular regulatory mechanisms which disrupt the hormone receptor interaction and inactivate the hormone (see the following section).

Protein and polypeptide (and certain other) hormone receptors

Not surprisingly, protein and polypeptide hormones being lipophobic are unlikely to penetrate target cell membranes, certainly without the assistance of specific transport systems to get them across. Consequently, it is not surprising to find that these hormones generally have their receptors embedded in the plasma membranes of their target cells. Some eicosanoids such as prostaglandins also have cell surface receptors.

A typical receptor of this type would be a transmembrane protein having an extracellular domain to which the hormone molecule can attach, a section of the receptor spanning the membrane and finally an intracellular domain which would be associated with a specific transduction process. One large, ubiquitous family of such receptors is the seven-transmembrane receptor group, so named because all the receptors in this group have seven membrane-spanning helices connecting the extracellular and intracellular domains. The intracellular domain is connected to a specific intracellular guanine nucleotide-binding protein, called a G protein, which is the initial part of the intracellular transducing mechanism activating the cell in a specific fashion. These receptors are therefore called G protein–coupled receptors (GPCRs). It is clear that these GPCRs have complex regulatory mechanisms, and at least 30 intracellular molecules which can influence them have so far been identified, called regulators of G protein signalling (RGSs). Furthermore, other novel intracellular molecules which can activate the G proteins have been identified, and they are called activators of G protein signalling (AGS). Clearly, the intricate intracellular regulation of the G proteins adds a whole new dimension to the cellular transduction of the initial hormone-induced process.

Some membrane receptors may have a single strand spanning the membrane, with an extracellular domain with hormone-binding capability and an intracellular domain having intrinsic enzyme activity when activated. Such receptors are identified by the nature of the enzymes on their intracellular domains (e.g. tyrosine kinase receptors). A variant of this type of receptor would be the homodimer which is essentially a double single-spanning receptor (see Figure 1.10). There is also evidence to suggest that heterodimers can exist also, where two different receptors can link together. These receptors do not have to be single-spanning membrane receptors but can be GPCRs. This suggests some intriguing possibilities, particularly when each receptor is linked to different intracellular second messenger systems. (The molecules which transduce the hormone's response within the cell are called the 'second messengers'.)

When the hormone, or other ligand, binds to the extracellular domain the receptor conformation changes, resulting in activation of the enzyme associated with the intracellular domain or G protein. Activation of this intracellular enzyme is associated with the transduction of the hormone's 'message' to the interior of the target cell.

Hormone receptor removal

The binding of the hormone to its 'recognition' site leads to the desensitisation of the receptor in response to a further hormonal challenge by a series of mechanisms that include a mere decrease in its affinity for the agonist, an internalisation with or without recycling to the membrane, an increased proteolytic degradation or ultimately a blockade of its synthesis at a pre- or post-transcriptional level. Internalisation refers to the loss of

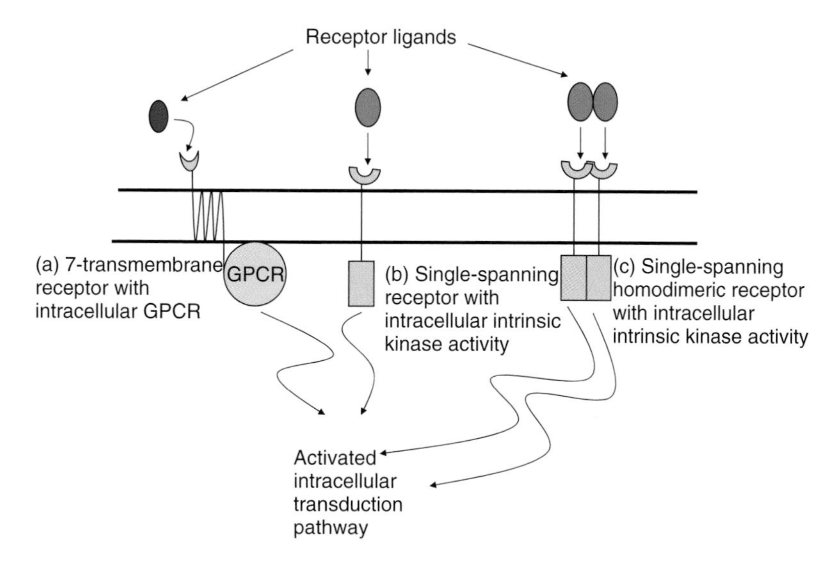

Figure 1.10 Diagram illustrating types of membrane-spanning receptors: (a) the seven-transmembrane receptor with the extracellular domain to which the ligand (e.g. hormone) binds, the seven transmembrane helices and the intracellular domain linked to the G protein–coupled receptor (GPCR); (b) the single membrane-spanning receptor with the extracellular ligand-binding domain and the intracellular domain with intrinsic enzyme (kinase) activity and (c) the same as for (b) but in a homodimeric form.

membrane receptors and results from the combined effects of receptor uptake by endocytosis, and recycling.

As mentioned earlier, much of the magnitude of a hormone's action is due to the concentration of specific receptors in its target cells. The down-regulation of receptors is one mechanism by which the prolonged effect of a specific hormone is reduced. It can be a global phenomenon, where the ligand-induced internalisation process is compounded with irreversible sequestration and intracellular proteolysis. The mechanisms responsible for these events involve two main families of proteins: the G protein–coupled receptor kinases (GRKs), and the arrestins. After binding the hormone, the receptor undergoes a conformational change and binds to a GRK which then phosphorylates it on serine or threonine residues located on the intracellular loops or on the C-terminal tail. This phosphorylation diminishes the receptor's affinity for the ligand and mainly leads to its subsequent binding to an arrestin molecule. This physically prevents coupling of the receptor to its G protein and therefore to the transduction of the signal. Phosphorylation of the receptor can also be caused by kinases other than the GRKs, such as protein kinases A and C (PKA and PKC, respectively), thus sustaining the desensitisation process. In other cases, the tightly bound arrestin dissociates from the receptor and accompanies it into the cell where the complex stays for extended periods of time within endosomal vesicles before being directed to lysosomes or being recycled.

The reuptake of receptors is by the process of endocytosis, mediated by various proteins including clathrin. Clathrin can be recruited to membranes by a variety of proteins. One of these, AP180 (an adaptor protein), is a very efficient recruitment protein that binds to phosphoinositol 4,5 bisphosphate (PIP2) in the plasma membrane. It also induces the polymerisation of clathrin into a lattice within the cytoplasm in the region of the cell membrane. Clathrin plays a key role in the formation of vesicles for internalisation, along with other proteins. The arrestins target the areas of the cell membrane where clathrin is located, called clathrin-coated pits. The arrestins can behave as adaptor proteins facilitating the recruitment of receptors to the plasma membrane domains where the clathrin-coated pits develop, prior to the endocytosis process. Arrestins play this role not by binding to clathrin itself but by interacting with other endocytic elements, including the adaptor protein AP2 (Figure 1.11).

Steroid hormone receptors

Receptors for steroids are also proteins, but because these lipophilic hormones (and a few others) can enter their target cells with ease, their receptors are found intracellularly, most commonly in the cytoplasm as well as in the nucleus. As with protein and polypeptide hormones, while each receptor may recognise specific molecules, they are not usually

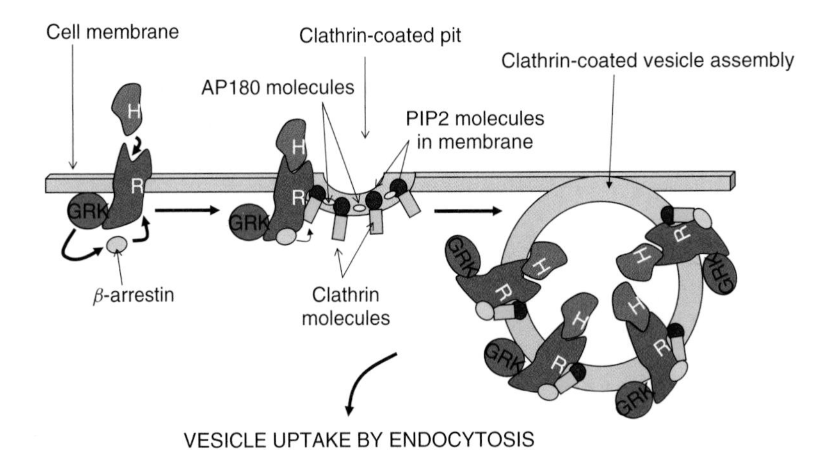

Figure 1.11 Diagram illustrating the uptake of a hormone receptor complex by endocytosis. This process is initiated when the hormone molecule first binds to its receptor in the cell membrane. It elicits the transduction of the hormonal effect via induction of its second messenger system and, at the same time, the GRK associated with the receptor is activated. See section 'Hormone receptor removal' for further details.

totally specific for the one hormone and can bind other chemically similar molecules with varying affinities. Once bound to their receptors, these hormone receptor complexes can influence the target cell genome to induce changes in protein synthesis. These new proteins, often enzymes, then mediate the actions of the hormone within the target cells (see the following section). Furthermore, there is some evidence (mainly indirect) suggesting that some steroids may also exert rapid non-genomic effects, perhaps through cell membrane receptors. This remains a controversial issue and an area of much research interest.

Mechanisms of action

Once a hormone has bound to its receptor on a target cell, the 'message' of that hormone molecule needs to be somehow passed on within the target cell so that it can respond in such a way that the body's general homeostasis is ultimately restored. The process by which that message is passed on to the target cell is known as transduction. One important feature of this process is that one hormone molecule, by binding to its receptor, can 'switch on' that cell in such a way that the message is multiplied intracellularly by the production of a whole cascade of molecules. Thus the signal provided by any one hormone molecule binding to its receptor can be amplified many times. Another feature, discussed in a previous section, is the establishment of an internal regulatory process whereby hormone receptor activation can be terminated.

Protein and polypeptide hormones

As explained elsewhere, the hormone molecules released into the bloodstream act as (first) messengers carrying information from one part of the body to another in order to induce a specific response. In general, once the hormone binds to its receptor, it can instigate the activation of specific pathways involving other intracellular (second) messenger molecules. For proteins and polypeptides, and other hormones which bind to the extracellular binding sites of their receptors in the plasma membranes of target cells, there are various second messengers which can be produced, each one associated with one or more specific intracellular pathways. These pathways are usually cascades of activated enzymes, in a sequence culminating in the production of the appropriate response to the initial stimulus, which can vary from the opening of an ion channel to the synthesis of a new protein molecule.

The mechanisms of action are related in part to the types of membrane receptor to which the hormones bind. The two main receptor types relevant to most hormones are the enzyme-coupled receptors and the more numerous G protein–coupled receptors.

Mechanisms linked to ligand-gated channel receptors

Some receptors are part of the chemical structure of ion channels in the plasma membrane. They are often referred to as ligand-gated channels to distinguish them from other ion channels such as voltage-gated and stretch-activated channels. The best known examples are neurotransmitters such as acetylcholine and gamma aminobutyric acid (GABA). When the messenger molecule binds to the extracellular receptor portion of the ion channel, it can open to specific ions, allowing them to flow down electrochemical gradients across the cell membrane.

Mechanisms linked to enzyme-coupled receptors

There may be ligand-linked receptors in the cell membrane which combine target cell recognition to a specific, and direct, response. For example, some transmembrane receptors have an intracellular enzyme as part of their structure, so that when the hormone binds to such a receptor, then the intracellular enzyme component is activated. Examples would include receptor tyrosine kinases, receptor serine and threonine kinases and receptor guanyl cyclases.

Once activated by the binding of ligand to an extracellular binding site on the transmembrane receptor, the intracellular enzyme component becomes capable of activating specific amino acids in associated proteins, by phosphorylating them. Some of these receptors are single-strand transmembrane proteins, while others may be paired to form homo- (or hetero-) dimers with adjacent receptor molecules. For example, growth factors activate tyrosine kinase–coupled receptors (TKCR) resulting in adjacent identical receptors forming homodimers. Consequently the activated tyrosine kinases phosphorylate tyrosyl residues in the intracellular (tail) components of the TKCRs and can result in (i) autophosphorylation, prolonging and magnifying the activity associated with the activation of the receptors, and/or (ii) activation of tyrosyl residues in other, intracellular, proteins (Figure 1.12).

There are many intracellular proteins which can be activated by phosphorylation. The result is that the activation of different intracellular 'second messenger' pathways involving various intracellular proteins can occur, culminating in more than one biological response to the original hormonal signal. Furthermore, the activation of a single receptor kinase can result in the phosphorylation of a stream of intracellular proteins, providing an internal amplification of the initial hormonal stimulus.

Hormones such as insulin and various growth factors bind to tyrosine kinase receptors of this type.

Mechanisms linked to G protein–coupled receptors (GPCR)

This is a very large group of transmembrane receptors with a common general structure, comprising an extracellular ligand-binding domain, seven

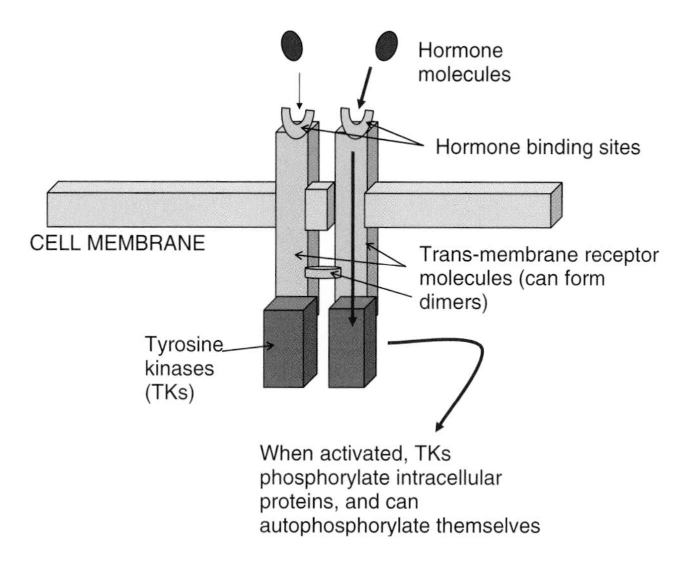

Figure 1.12 Diagram illustrating a membrane tyrosine kinase receptor, here shown as a homodimer. Binding of the hormone to the extracellular domain activates the receptor proper, resulting in the activation of tyrosine kinases present in the intracellular component. The activation of these tyrosine kinases can result in autophosphorylation of the receptor itself, or the phosphorylation of nearby receptors. However, the main effect will be to phosphorylate intracellular tyrosyl-containing proteins which can then act as intracellular second messengers, transducing the initial hormonal signal ultimately into the cellular biological response.

intramembrane loops and an intracellular enzyme-linked region which is activated when the ligand (hormone) binds to the extracellular domain. Many hormones bind to receptors belonging to this group, including adrenaline, glucagon, vasopressin and parathormone.

The intracellular domain of a GPCR is indeed a guanine nucleotide protein which is comprised of a triad of subunits called α, β and γ. Both α and γ subunits are linked to the cell membrane. Initially, the unstimulated Gα unit is inhibited by the Gβγ subunits, and is linked to a guanosine diphosphate (GDP) molecule. When a hormone molecule binds to the extracellular domain of its receptor, the subsequent conformational change of the receptor activates the Gα subunit which then preferentially binds to intracellular guanosine triphosphate (GTP) instead of GDP. Binding of GTP to Gα overcomes the inhibitory influence of the Gβγ subunits which then dissociate from it. The Gα–GTP activates a specific intracellular enzyme, which then initiates formation of a second messenger. The intracellular second messenger can then, in turn, induce a cascade of intracellular steps which culminate in the production of the appropriate cellular response to the initial stimulus. The α subunit has GTPase activity, so it then hydrolyses GTP back to GDP restoring the G protein to its initial state. There are various

Gα subunits: for example, one (Gαs) can stimulate a membrane-bound enzyme, adenyl cyclase, while another (Gαi) can inhibit it.

Adenyl cyclase is an enzyme which is part of one of the principal intracellular second messenger systems in cells. When activated, this enzyme catalyses the conversion of adenosine triphosphate (ATP) to the intracellular second messenger cyclic adenosine monophosphate (cAMP) with the release of two phosphates. There are 10 known isomers of adenyl cyclase, each one having distinguishing properties which identify specific cAMP-activated pathways. An important intracellular protein which is activated by cAMP is PKA which then phosphorylates other intracellular proteins.

Another membrane enzyme which, when activated, catalyses the formation of another important intracellular second messenger system is phospholipase C which is located in the plasma cell membrane. When stimulated, it catalyses the breakdown of a membrane phospholipid called phosphatidylinositol-4,5-bisphosphate (PIP2) into two components, both of which act as intracellular second messengers: inositol triphosphate (IP3) and diacylglycerol (DAG). These second messengers, in turn, switch on cascades of further intracellular events. IP3 readily diffuses into the cytosol and, through its binding to calcium channels on the endoplasmic reticulum, allows the release of calcium from this intracellular storage site into the cytoplasm. DAG, in the presence of calcium, directly activates PKC which then phosphorylates a series of protein substrates which may, or may not, differ from those of PKA. DAG can also be cleaved into arachidonic acid, the precursor for prostaglandin synthesis (see Figure 1.13).

One important intracellular event which can stimulate various subsequent reactions within the cell involves second messenger–induced mechanisms which raise the intracellular free calcium ion concentration. Calcium ions play an important role in triggering various intracellular chemical reactions, so regulation of their intracellular (cytoplasmic) concentration is often an important component in controlling a cell's response to the initial stimulus. The cytoplasmic free calcium ion concentration can be regulated through two different mechanisms: by controlling (i) calcium ion channels in plasma cell membranes, which link the intracellular compartment to the exterior of the cell, and (ii) calcium ion channels in the membranes of internal organelles, such as the endoplasmic reticulum, the sarcoplasmic reticulum (in muscle) and microsomes, which act as intracellular calcium stores. Some hormones can directly influence calcium ion channels located in the plasma membrane (see ligand-gated receptors in earlier section), while others do so by initiating the formation of second messenger transduction pathways which then influence calcium channels either in the plasma cell membrane or in the membranes of intracellular organelles (see Figure 1.13). The intracellular calcium ion concentration

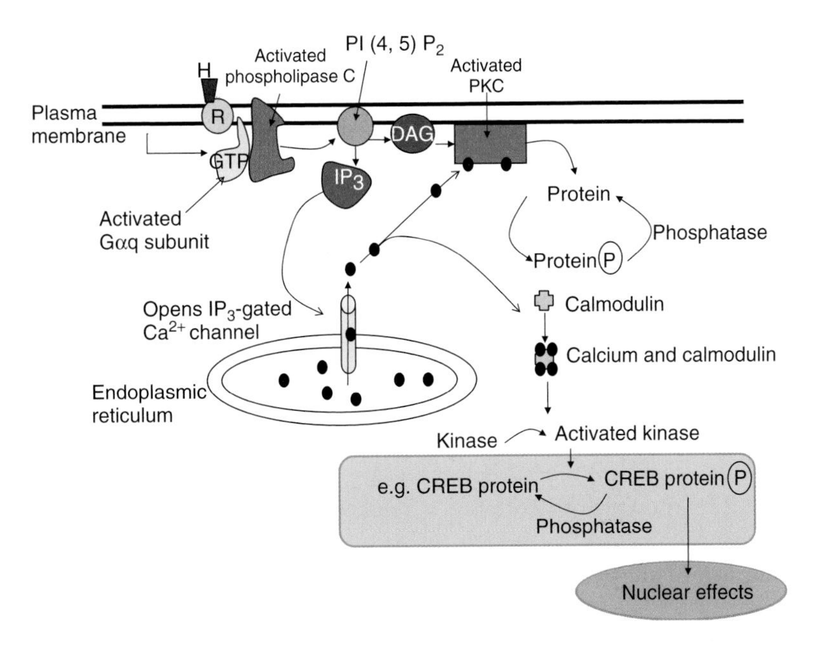

Figure 1.13 Diagram illustrating how a hormone (H) binding to the extracellular domain of its transmembrane receptor (R) can activate the α subunit of the associated G protein, resulting in the activation of the membrane-located enzyme phospholipase C. This enzyme then catalyses the formation of two different second messengers, diacylglycerol (DAG) and inositol triphosphate (IP3), which in turn switch on cascades of intracellular events which can include the release of calcium ions from intracellular stores such as the endoplasmic reticulum, or the activation of protein kinases such as protein kinase C (PKC). Longer term nuclear effects can also be induced by the activation of cyclic adenosine monophosphate (cAMP) response element binding protein (CREB).

is considerably lower (10 000-fold) than its extracellular concentration (approximately 1.15 mM of free ions), so the opening of cell membrane calcium channels results in an inward movement of these ions into the cytoplasm, temporarily increasing its intracellular concentration. Likewise, the opening of calcium channels in intracellular organelle membranes such as in the sarcoplasmic or endoplasmic reticulum will also result in the movement of these ions from the intracellular storage site (with their higher concentration) to the cytoplasm. The concentration of free cytoplasmic calcium ions is normally very low (approximately 0.1μM), so any increase in its concentration or in the frequency of its oscillations can have a marked signalling impact. In particular, calcium can bind to calmodulin and activate the calcium- and calmodulin-dependent kinases which interestingly can also phosphorylate cAMP response element binding protein (CREB) which acts as a DNA transcription factor (see Figure 1.13).

Thus, a ligand (e.g. hormone) can bind to its GPCR and influence cytoplasmic events such as the direct regulation of different enzymes and/or the intracellular free calcium ion concentration, and these actions will tend to produce rapid effects within seconds to minutes. However, that same ligand could also influence longer term (hours or days) events such as the synthesis of new proteins, these being either new intracellular enzymes or peptides and proteins for export from the cell. As can be seen from Figure 1.14, cAMP molecules cause the dissociation of inhibitory regulatory units attached to PKA, thus activating the PKA molecules which then traverse the nuclear membrane through nuclear pores (Figure 1.14).

Steroid hormones

The intracellular hormone receptor complex acts as a transcription factor, binding to a location along the target chromosome acting as the DNA-binding site known as the promoter, which is likely to be a protein such as a histone associated with the DNA. Indeed, it is known that the beaded appearance of the chromosome is due to the presence of nucleosomes

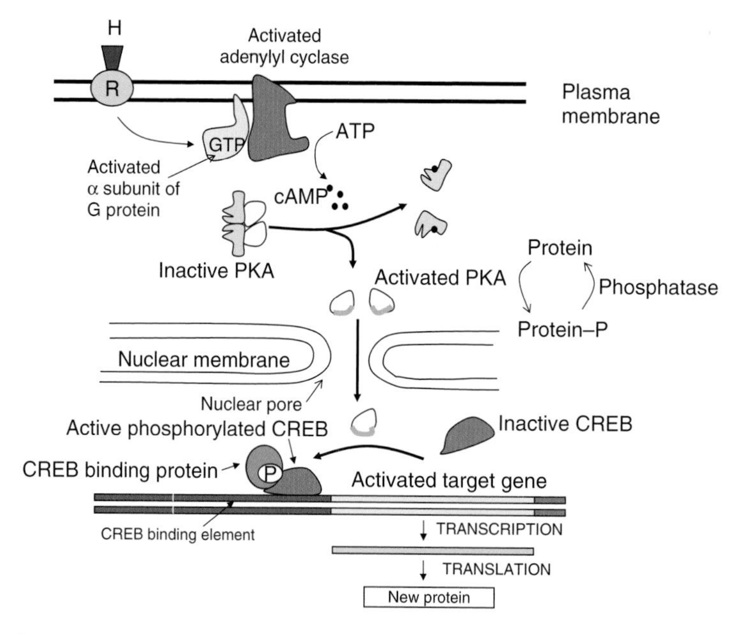

Figure 1.14 Diagram illustrating how a hormone (H) can bind to the extracellular domain of its transmembrane receptor (R) and activate a second messenger pathway such as is indicated here, involving cyclic adenosine monophosphate (cAMP). This second messenger, in turn, activates a protein kinase (here, PKA) which can traverse the nuclear membrane of the cell where it phosphorylates a cAMP response element binding protein (CREB). This molecule then acts as a transcription factor, binding to a CREB binding element (site) on the target gene and switching it on or off.

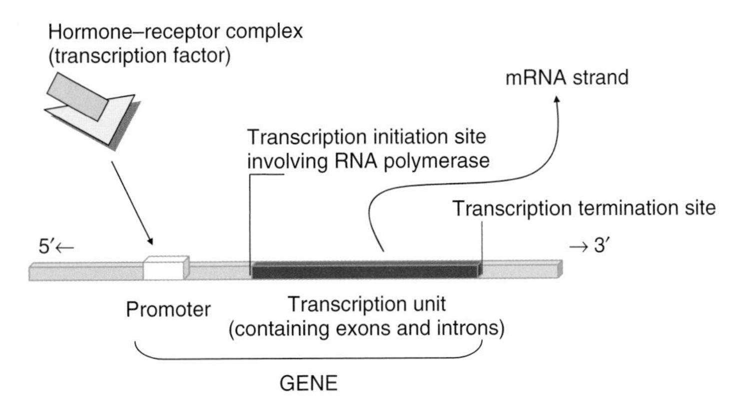

Figure 1.15 Diagram illustrating how the hormone receptor complex, acting as a transcription factor, influences the transcription of a gene's DNA into the complementary mRNA strand which then gets translated into a new protein.

which are tightly wound regions of DNA around a variety of proteins such as histones. The actual transcription unit, which is the DNA segment which carries the genetic information which will be transcribed into the mRNA for the end product protein, consists of introns and exons of which only the exons are transcribed (see section 'Polypeptide and protein hormones'). There will also be the promoter (or regulatory) region located up-stream to the transcription unit in the 5′ direction of the DNA strand, to which the transcription factor (in this case the hormone receptor complex) will bind, and a specific transcription termination site which determines where that particular gene transcription ends (see Figure 1.15).

As mentioned earlier, as with all protein synthesis the transcribed mRNA will ultimately be translated into the new protein which then mediates the cellular response to the hormone. It is also notable that lipophobic hormones such as proteins and polypeptides can have indirect genomic effects (as discussed earlier).

Feedback control

If a released hormone is allowed to act in an uncontrolled manner, then the effect that is generated can be prolonged to a point when that effect becomes disadvantageous to the organism. Indeed, some disorders are actually caused by control systems malfunctioning, an obvious example being tumours which can be defined as abnormal (i.e. uncontrolled) growths of cells, where the abnormality is usually genetic.

Not surprisingly, therefore, control systems exist in order to regulate endocrine function and thus maintain homeostasis. When the integrated signal to an endocrine cell is such that more hormone is required by the body, then that hormone is released in increased amounts (and maybe synthesised for longer term use) until the necessary action of that hormone has reduced the necessity for further increased production of it as indicated by a reduction in stimulus strength, upon which the endocrine cell returns to its basal activity state. This control mechanism is called negative feedback.

Negative feedback

Direct negative feedback

By far the most common form of feedback regulating an endocrine cell is direct, relating the endocrine cell to the function being controlled. An example of such a direct negative feedback is provided by the regulation of insulin and a function it controls, the blood glucose concentration (Figure 1.16).

When the internal environment is disturbed in a particular manner, that disturbance acts as a stimulus which activates an endocrine cell to release its hormone into the general circulation. The hormone will be transported to its target tissues where it will exert its physiological effect in order to reduce that stimulus and restore the body's internal environment to its undisturbed state. Using the β-cells of the pancreatic islets of Langerhans as an example of an endocrine gland, imagine that the blood glucose

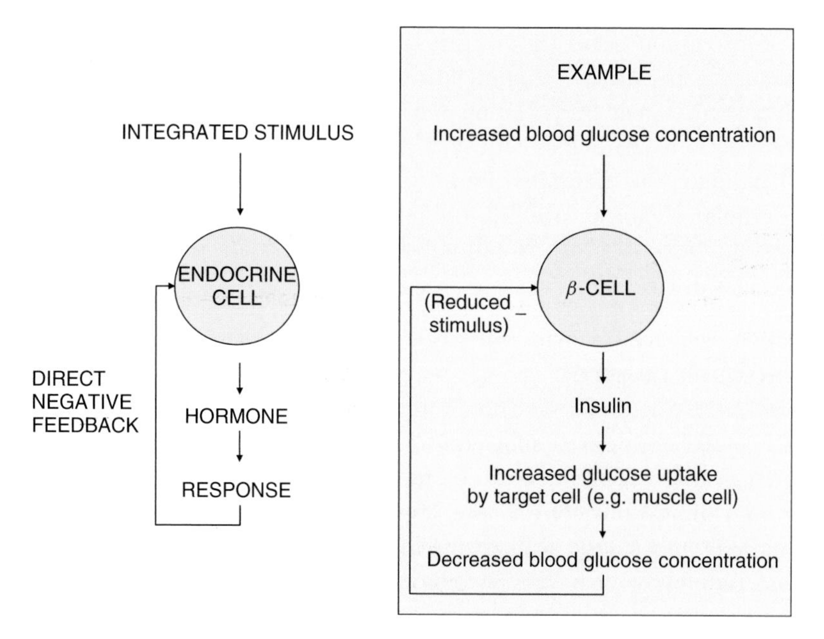

Figure 1.16 Diagram illustrating (left) the general principle of a direct negative feedback loop, and (right) an example provided by the hormone insulin.

concentration rises (e.g. following a meal). The internal environment is disturbed, so that the increased blood glucose concentration acts as a powerful stimulus on the β-cells and causes them to release their hormone, insulin. Insulin is then transported by the general circulation to its target cells (e.g. in muscle) where it stimulates the uptake of glucose. The stimulus acting on the β-cells is consequently reduced, the β-cells are no longer stimulated, insulin release is reduced and the internal environment is restored to its initial undisturbed state. Thus the 'response' to the hormone has exerted a direct negative feedback on the hormone's site of production, the β-cell. See Figure 1.16 and also Chapter 15 for further details.

Interestingly, some endocrine glands when stimulated exert their effects on target cells which may themselves also be endocrine cells. In this case, the production of hormone *b* from endocrine gland *B* acts on endocrine gland *C* to produce hormone *c*. In this case hormone *c* can exert a direct negative feedback on gland *B* (see Figure 1.17). The best examples of this situation are provided by hormones produced by cells of the anterior pituitary gland. Corticotrophin (hormone *b*) is produced by the anterior pituitary (endocrine gland *B*). It acts on the adrenal cortex (endocrine gland *C*) to produce cortisol (hormone *c*). Cortisol exerts a direct negative feedback on the anterior pituitary's production of corticotrophin (also known as ACTH; see Chapter 3).

Indirect negative feedback

The anterior pituitary gland provides us with another level of complexity in terms of negative feedback. In addition to the direct negative feedback loops shown in Figure 1.17, between the anterior pituitary (endocrine gland *B*) and other endocrine glands (e.g. adrenal cortex, C), indirect negative feedback loops can also be present. This is because the anterior pituitary is also influenced by the CNS, via a particular region of the brain called the hypothalamus. While this is described more fully elsewhere (see Chapters 2 and 3), we can stick with our example relating the anterior pituitary to the adrenal cortex in order to demonstrate indirect negative feedback involving the CNS. As will be discussed in Chapter 2, the hypothalamus, which is made up of nerve cells, or neurones, actually has an endocrine function in addition to its other roles. Various groups of neurones, called hypothalamic nuclei, send their axons down towards the pituitary gland which is attached to the base of the hypothalamus, and their neurosecretions are actually released into a special vascular system which links that region of the hypothalamus to the anterior lobe of the pituitary. Thus, the hypothalamus can in this regard be considered as another endocrine gland in its own right.

In this case, there will be not only the direct negative feedback loop operating between endocrine gland *C* (in the example, the adrenal cortex

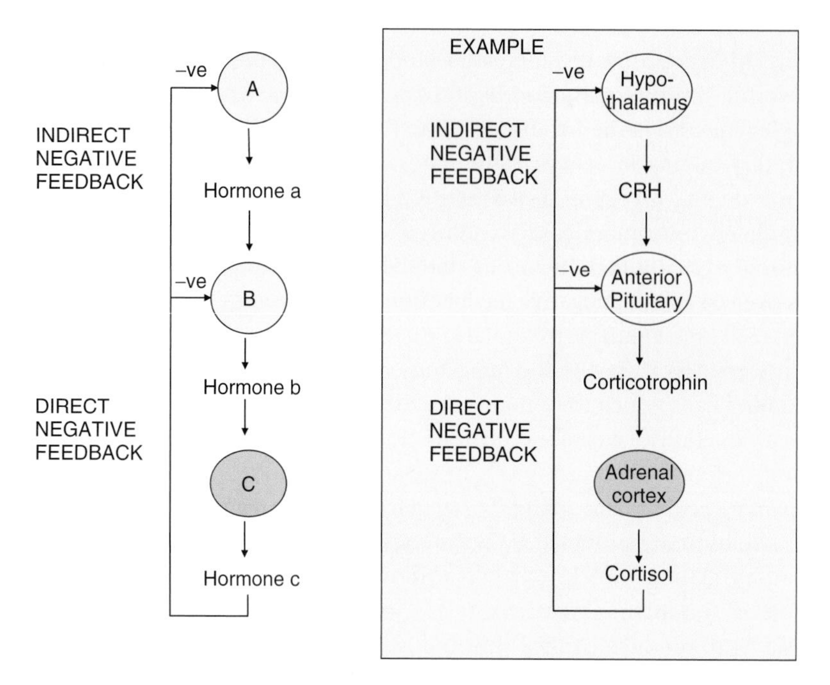

Figure 1.17 Diagram illustrating (left) the general principle of a direct negative feedback loop involving two endocrine glands (*C* and *B*) together with their hormones (*c* and *b*), and (right) an example provided by the adrenal cortex and the anterior pituitary and their hormones (cortisol and corticotrophin, respectively).

producing cortisol) and endocrine gland *B* (the anterior pituitary, influencing the production of corticotrophin) but also an indirect negative feedback loop operating back on endocrine gland *A* (the hypothalamus, influencing the production of corticotrophin-releasing hormone, or CRH). Other examples of direct and indirect negative feedback pathways include the regulation of the iodothyronines from the thyroid (Chapter 13), and the gonadal steroids from the testes or ovaries in males and females respectively (see Chapters 6 and 7). The indirect negative feedback in these cases links the peripheral endocrine gland (*C*) to the CNS (hypothalamus, *A*).

Short (or auto) negative feedback

When one examines the negative feedback loops which operate for various endocrine systems involving the hypothalamo-pituitary axis as indicated in Figure 1.17, it becomes apparent that the production of the anterior pituitary hormones is at least partly in response to the drive provided by specific hypothalamic hormones. Thus it would be quite feasible for the anterior pituitary hormones to influence their own production by feedback loops to the hypothalamic level. Indeed, not surprisingly, there

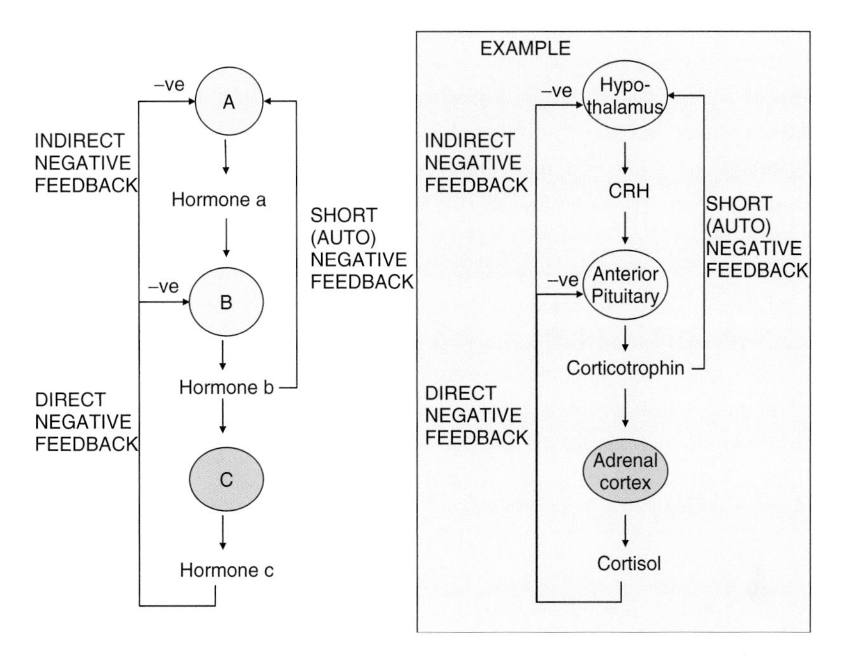

Figure 1.18 Diagram illustrating (left) the general principle of direct, indirect and short (or auto) negative feedback loops involving three endocrine glands (*A*, *B* and *C*) together with their hormones (*a*, *b* and *c*), and (right) an example provided by the hypothalamus, the anterior pituitary gland and the adrenal cortex and their hormones (corticotrophin-releasing hormone (CRH), corticotrophin and cortisol, respectively).

is plenty of evidence in support of this form of feedback control, which is called short (or auto) negative feedback (see Figure 1.18).

In the example given in Figure 1.18, the hypothalamus receives an indirect negative feedback from circulating cortisol and a short (auto) negative feedback from corticotrophin, while the anterior pituitary simply receives a direct negative feedback influence from cortisol. In fact, the situation relating the hormones from the hypothalamo-pituitary-adrenal axis described here is more complex than this, with other hormones and factors also exerting regulatory influences upon it. The point to appreciate is that the control exerted on any endocrine system is multifactorial, and can be at a number of different levels as described here. This allows for a quite sensitive, specific and precise degree of control over these important physiological regulatory systems.

Positive feedback

While it may seem counter-intuitive to learn that positive feedback loops do occur in nature, perhaps we should not be too surprised at their existence. Certainly, under general circumstances positive feedback *per se* would lead ultimately to chaos and dysfunction because of the loss of control over a physiological system. However, when different signals influence that system, then a positive feedback loop can arise but only under certain specific conditions. When those conditions are altered, then the positive feedback is no longer capable of being maintained and the initial status quo is restored. Various positive feedback effects are known to exist in nature, and in mammalian physiology there are various examples, which can be either between different parts of the body (different cells) or within cells. In all cases, under normal conditions, the induced change in environment is such that conditions allowing the positive feedback to develop in the first place are ultimately cancelled, and the effect is lost.

The best example of a positive feedback influence in endocrinology is the one which develops during the female menstrual cycle between a steroid hormone from the ovaries called oestradiol (more precisely,

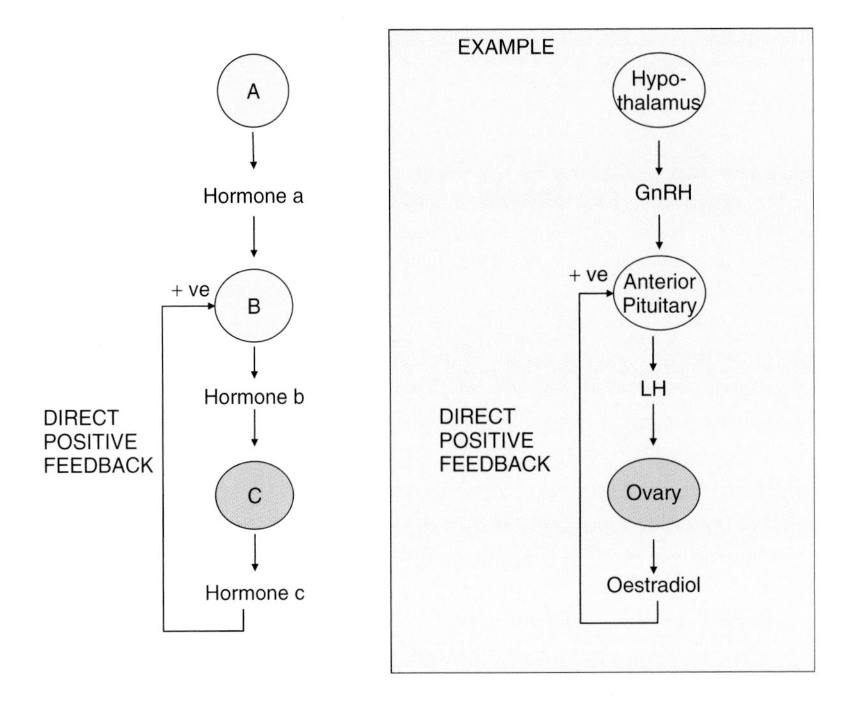

Figure 1.19 Diagram illustrating (left) the principle of positive feedback between two endocrine glands (*C* on *B*) by which hormone *c* sensitises endocrine gland *B* to the effect of stimulatory hormone *a* from endocrine gland *A*, and (right) the example of positive feedback by ovarian oestradiol on the anterior pituitary. The oestradiol effect is one of sensitising the anterior pituitary cells which produce luteinising hormone (LH) to the stimulatory effect of gonadotrophin-releasing hormone (GnRH) from the hypothalamus. The effect of the positive feedback is to stimulate a surge in LH which promotes the process of ovulation. The endocrine background is altered, the conditions for positive feedback are lost and the hypothalamo-pituitary-ovarian axis returns to its 'normal', or basal, state once again by negative feedback.

17β-oestradiol), and the hypothalamo-pituitary axis which plays a vital role in regulating the release of a ripe egg, or ovum, a process called ovulation (see Figure 1.19). For a more detailed discussion of ovulation, see Chapter 7.

Internal feedback

In addition to the control loops linking 'effect' to hormone via the general circulation, there is considerable regulation within the cells of an endocrine gland. We have already seen how the endocrine cell itself acts as an integrator of probably many different signals arriving at that cell at any given moment. This implies considerable intracellular regulation at various levels, including synthesis, storage, removal or release of the hormone. Furthermore, the target cell has many ways of regulating its response to the hormone, including receptor synthesis, recycling or removal, as well as the intracellular control of the activated second messenger systems.

Summary

1. Endocrinology is the study of endocrine glands which are composed of specialised cells synthesising hormones, and releasing them into the general circulation which transports them to their target cells.
2. Hormones are important not only because of their communication role regulating metabolic and other activities in response to perturbations to the body's homeostasis, but also because it is now appreciated that they are integrated with other 'systems' in the body such as the nervous and immune systems.
3. Hormones can be generally classified according to their chemical structures. For example, the two main groups (proteins and polypeptides as well as steroids) differ in their lipid solubilities making them either lipophilic or lipophobic, and this can define processes such as their synthesis, storage, release, receptor site and mechanism of action.
4. Hormones bind to receptor molecules in their target cells and induce actions either directly on cellular processes, or indirectly via the mediation of intracellular second messenger systems.
5. Simple and complex mechanisms exist to closely regulate each endocrine system, relating hormone production to the hormone-induced effect, in order to maintain homeostasis.

Conclusion

The field of endocrinology has widened considerably since the concept was first proposed, and the definitions are no longer as clear-cut as they once were. Indeed, the study of the various endocrine systems, and the roles they play, is now an essential and integral component in our current understanding of how the body functions.

Reference

Palmer, R.M., Ferrige, A.G. & Moncada, S. (1987) Nitric oxide release accounts for the biological activity of endothelium-derived relaxing factor. *Nature*, 327 (6122), 524–6.

CHAPTER 2
The Hypothalamus and the Concept of Neurosecretion

The hypothalamus

The hypothalamus is a part of the brain which plays an important role in the regulation of many vital physiological systems and in the maintenance of homeostasis. In addition to its clear role as an integrating centre mediating many neural responses, it also has an endocrine role; indeed, it can be considered an endocrine gland in its own right. This endocrine role is one which depends on its close relationship with the pituitary gland, a small endocrine gland attached to the base of the hypothalamus. Given the vital importance of the hypothalamus to the overall function of many physiological systems essential to life, it is perhaps surprising to appreciate that this part of the brain has an approximate volume of only 4 cm^3, and weighs about 4 g given that the total weight of an adult brain is of the order of 1300 g.

Structure

The hypothalamus is the part of the brain that lies immediately below the thalamus, together forming the major part of the ventral region of the diencephalon. The fetal hypothalamus develops from the sixth to twelfth weeks *in utero*. It is that part of the brain which is connected to an important endocrine gland called the pituitary, or hypophysis. Other structures in the vicinity of the hypothalamus include the third ventricle, which it surrounds, the posteriorly located mammillary bodies and the anteriorly located optic chiasma. The hypothalamus is mainly comprised of neurones, some of which project to distant regions elsewhere in the brain while others project down to the pituitary. The lower part of the hypothalamus, where it abuts the pituitary, is a particularly vascular region called the median eminence. The capillaries in this region of the brain have fenestrations (i.e. gaps or windows, from the Latin word *fenestra*) between adjoining endothelial cells, which allow for the movement of molecules between the blood and the brain extracellular fluid to occur in either direction. The median eminence therefore lies outside the blood–brain

---

Integrated Endocrinology, First Edition. John Laycock and Karim Meeran.

barrier which is provided by the tight, unfenestrated endothelial cells elsewhere in the brain vasculature, and which protects most of the brain from toxins and other potentially disruptive molecules in the blood.

There are many afferent nerve fibres which reach the hypothalamus from other parts of the brain, including an input from the eye (retina) and another from the olfactory cortex. It also receives sensory information (e.g. skin temperature) from the spinal cord via the reticular formation in the brainstem, via the internal organs (viscera) and from receptors in parts of the brain which lie outside the blood–brain barrier (in what are known as the circumventricular organs). It also receives information regarding the processing of other neural activities, for instance relating to behaviours and mentation. Thus the hypothalamus receives sensory and other information from all over the body (Figure 2.1), and consequently functions as an important integration centre which plays an essential role in regulating the internal environment (i.e. in homeostasis).

The third ventricle divides the hypothalamus into left and right halves. In each half, the efferent neurones can be loosely grouped into what are called hypothalamic nuclei which are paired, and located in the anterior (supraoptic), middle (tuberal) and posterior (mammillary) regions of the hypothalamus. The anterior hypothalamus includes the supraoptic, medial preoptic, paraventricular and suprachiasmatic nuclei. The middle

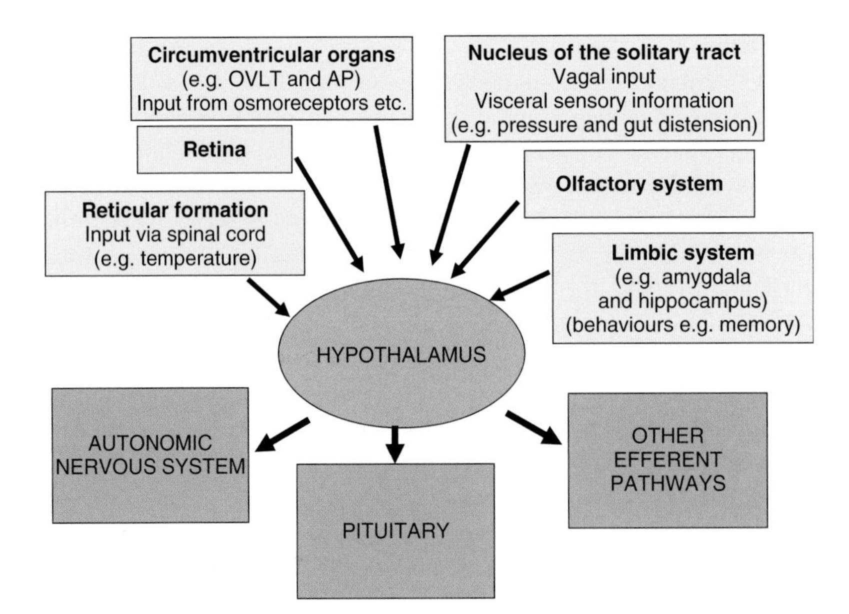

Figure 2.1 Diagram illustrating the main afferent pathways to, and efferent pathways from, the hypothalamus. AP = area postrema, OVLT = organum vasculosum of the lamina terminalis.

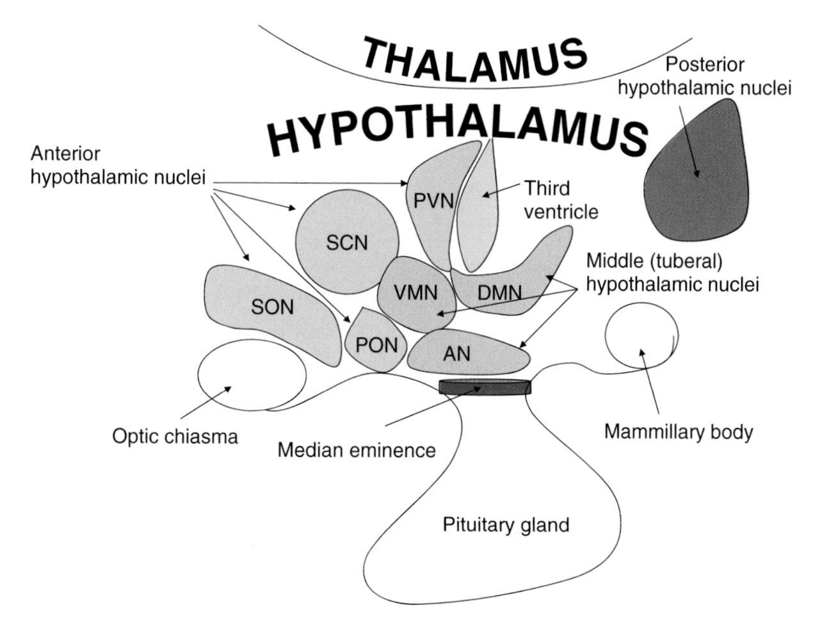

Figure 2.2 Diagram illustrating the principal regions of the hypothalamus, the hypothalamic nuclei and other markers of the region. AN = arcuate nucleus; DMN = dorsomedial nucleus; PON = preoptic nucleus; PVN = paraventricular nucleus; SCN = suprachiasmatic nucleus; SON = supraoptic nucleus and VMN = ventromedial nucleus.

hypothalamus includes the arcuate, ventromedial and dorsomedial nuclei, while the posterior hypothalamus remains fairly indistinct in terms of specific neuronal groupings (see Figure 2.2).

There are three main efferent neuronal pathways by which the hypothalamus exerts its physiological effects:

1. One important efferent pathway originates in the anterior (particularly paraventricular) hypothalamic nuclei, and connects the hypothalamus to the autonomic nervous system (ANS) via the dorsal longitudinal fasciculus and the dorsal tegmental tract of the brainstem. These neurones pass (i) to the midbrain central periaqueductal grey matter, for pain modulation; (ii) to the medullary autonomic cardiovascular and respiratory centres; (iii) to the ventral tegmental area involved in reward, motivation and cognition behaviours; (iv) to the brainstem parasympathetic nuclei (e.g. the dorsal motor nucleus of the vagus) and (v) down the spinal cord as thoraco-lumbar preganglionic sympathetic, and lumbo-sacral preganglionic parasympathetic, neurones.
2. The second efferent pathway links the hypothalamus to other parts of the brain, and can be part of various neuronal circuits. For example, afferent neurones from the hippocampus, which is an area of the brain involved with conscious memory, synapse with efferent nerve fibres

from the hypothalamus. These efferent fibres pass to the thalamus where they may synapse with yet other neurones which ultimately terminate in specific areas of the cerebral cortex.

3. The third efferent pathway is specifically directed down to the median eminence which is the highly vascular region at the base of the hypothalamus adjacent to the pituitary gland (see Figure 2.2). Most neurones have their nerve terminals in the median eminence. Some neurones pass through the median eminence down through the pituitary stalk to terminate in the posterior part of the pituitary gland. All these neurones making up the third efferent pathway have their cell bodies in the discrete hypothalamic nuclei described in this section, such as the arcuate, supraoptic and paraventricular nuclei.

Blood supply to the hypothalamus

The hypothalamus is partly supplied with arterial blood by vessels coming from the circle of Willis. However, most of the blood arises from vessels coming off the internal carotid arteries which include the suprachiasmatic, tuberal and hypophysial arteries. The suprachiasmatic arteries provide blood to the anterior nuclei, the tuberal arteries to the dorsomedial and ventromedial nuclei, and the hypophysial arteries to the median eminence and pituitary, as well as to the arcuate nuclei. From the circle of Willis, branches from the anterior communicating arteries supply the preoptic area while the anterior, medial and posterior cerebral arteries also contribute to the blood supply of many of the nuclei. While the density of the capillary beds in the hypothalamus is comparable to that found in most brain grey matter, the density is particularly high in the nuclei associated with the magnocellular neurones, these being the paraventricular and supraoptic nuclei. Other richly dense capillary networks are also associated with circumventricular organs linked to the hypothalamus such as the organum vasculosum of the lamina terminalis (OVLT) and the median eminence itself. The venous outflow is mainly via the anterior cerebral, basilar and interpeduncular veins. The basilar veins provide the largest part of the drainage from the hypothalamus and they also collect blood draining from the hypophysial arteries. Blood from the basilar vein enters the great cerebral vein of Galen, which ultimately drains into the jugular vein.

Of particular interest is the drainage of blood from the capillary network in the median eminence, because it is part of a special blood portal system linking the hypothalamus to the anterior lobe of the pituitary gland (the adenohypophysis). Blood mainly from the superior hypophysial artery (a branch of the internal carotid artery) enters the capillary network in the median eminence (primary network) from which blood passes down through the pituitary stalk via special venous portal vessels. The blood then enters a second capillary network within the anterior pituitary

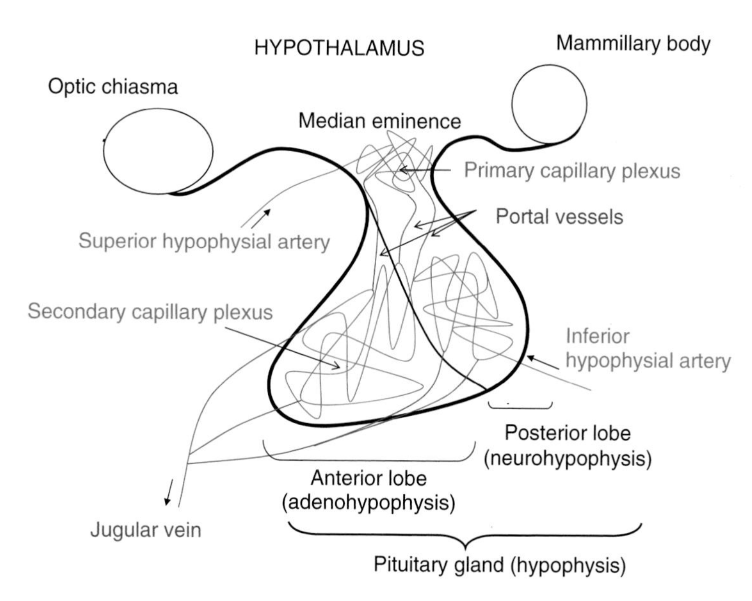

Figure 2.3 Diagram illustrating the hypothalamo-adenohypophysial portal system linking the primary capillary plexus in the median eminence to the secondary capillary plexus within the anterior lobe of the pituitary, and the blood flow to the neurohypophysis.

before draining into the cavernous sinus, ultimately entering the internal jugular vein. This specialised blood system is called the hypothalamo-adenohypophysial portal system. The posterior lobe of the pituitary (the neurohypophysis) receives arterial blood mainly from the inferior hypophysial artery, the blood entering a capillary network from which venous blood also passes into the cavernous sinus (see Figure 2.3).

Hypothalamic neurosecretions: Neurotransmitters and hormones

The nerve fibres of the hypothalamus are like neurones anywhere else in the body: they have dendrites, cell bodies and axons. The dendrites are the 'sensory' components of the neurone, receiving information from other neurones, while the cell bodies contain the cell nuclei and the main machinery necessary for synthesising molecules, particularly those which will be released when the cell is stimulated appropriately. These molecules are transported down the nerve axons, ultimately reaching the nerve terminals where most of them will be released, again by an appropriate stimulus. All molecules secreted by neurones are called neurosecretions. If they are released across the synaptic gaps between a neurone terminal and the dendrites of another neurone, then they are called neurotransmitters. However, the hypothalamic neurones which comprise the efferent pathway passing down to the median eminence, and those which pass through the median eminence to terminate in the posterior pituitary, actually release their neurosecretions into blood, and therefore they are true hormones.

Those hypothalamic neurones which originate within specific nuclei and whose axons terminate close to the capillaries in the median eminence act as if they were endocrine cells. In the median eminence the capillaries are fenestrated, allowing molecules to enter the blood. When the neurones are stimulated, action potentials travel down the nerve axons and depolarise the nerve terminals. As with all nerve fibres, exocytosis of granule contents results in the consequent release of the neurosecretory molecules which in this case enter the bloodstream through the gaps of the capillary endothelial cells. These neurosecretions, as discussed here, are therefore true hormones which pass down the portal vessels to the secondary capillary plexus in the anterior pituitary, hence reaching their target cells in this part of the gland. These hormones act on specific anterior pituitary, or adenohypophysial cells, which they either stimulate or inhibit and so they are generally called releasing hormones or inhibiting hormones. Their concentrations in the blood, while high enough in the portal system to exert their effects in the anterior pituitary, are too low to be usefully measured once the blood draining the pituitary reaches the general circulation. The close relationship between hypothalamus and anterior pituitary is often called the hypothalamo–anterior pituitary axis, also known as the hypothalamo-adenohypophysial axis.

Other hypothalamic neurones, originating in the paraventricular and supraoptic nuclei, pass down through the pituitary stalk to terminate close to capillaries in the posterior lobe of the pituitary, also called the neurohypophysis. These neurosecretions are also hormones but they actually reach the general circulation, and unlike the hypothalamic-releasing or -inhibiting hormones, they act on distant target cells in the body. As for the anterior lobe, the relationship between the hypothalamus and the posterior pituitary is called the hypothalamo-posterior pituitary axis, or rather the hypothalamo-neurohypophysial axis.

All these hormones – synthesised in neuronal cell bodies in the hypothalamus, and released from the nerve terminals either into the special hypothalamo-adenohypophysial portal system to the adenohypophysis, or directly into the general circulation in the neurohypophysis – will be considered in much more detail later in Chapters 3 and 4, which deal with the anterior and posterior lobes of the pituitary respectively.

One important characteristic of hypothalamic hormones is that their release is pulsatile. This is important with respect to their actions on target cells in the adenohypophysis, for the adenohypophysial hormones are also released in pulses. The importance of this pulsatility is demonstrated by the administration of an analogue of a hypothalamic hormone called gonadotrophin-releasing hormone (GnRH) which, if administered as a continuous infusion, inhibits the release of gonadotrophins which are hormones from the anterior pituitary (see Chapters 6 and 7). In contrast, when the same analogue is given in pulses every 10 minutes, it induces a normal pulsatile pattern of gonadotrophin release.

Physiological role of the hypothalamus

The hypothalamus has an important role to play as an integration and control centre, mediating input from various parts of the body, including other areas of the central nervous system (CNS), and activating specific efferent pathways in response. As can be seen from Figure 2.4, the main efferent pathways activated from the hypothalamus are the autonomic nervous system and the endocrine system via the pituitary gland. Any perturbation of the internal environment will tend to bring about a normal restorative response via these two systems. Indeed stimuli from the external environment also influence these two systems and the release of the hormones associated with them, acting via the CNS and specifically the hypothalamus.

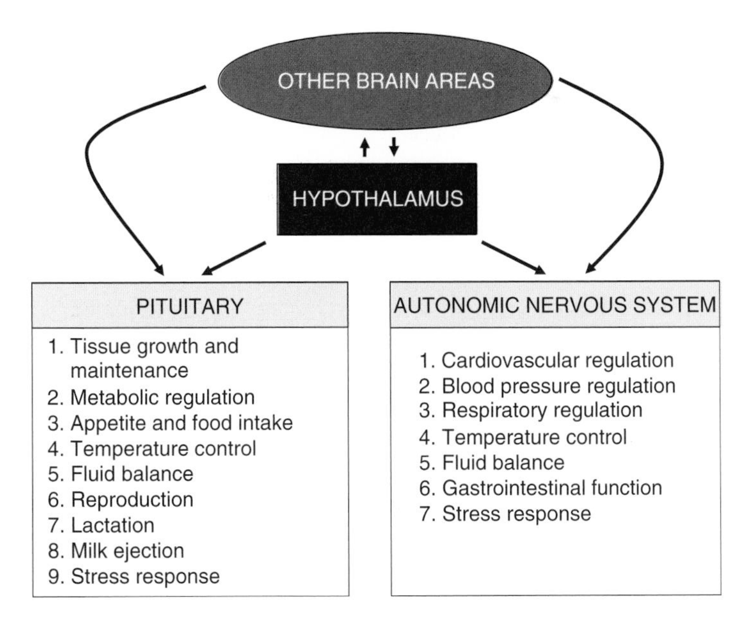

Figure 2.4 Diagram illustrating some of the key effects brought about following the hypothalamic activation of specific hormonal pathways via the pituitary gland, and the autonomic nervous system.

While the effects listed in Figure 2.4 are by no means the only ones affected by these two pathways, they do illustrate how they can work together in order to control certain physiological systems. For example, reproduction in the male involves not only the production of mature gametes, a process involving hormones from the pituitary gland, but also erection of the penis which is the gamete delivery system under vascular control by the autonomic nervous system. Another example is fluid balance, which involves the endocrine regulation of the kidneys by a pituitary hormone and the process of sweat regulation by the autonomic nervous system.

Not included in Figure 2.4 are the effects exerted by the hypothalamo-pituitary axes and the autonomic nervous system on the CNS. However, there are clear behavioural and other neural effects which are associated with the activation of these two efferent systems, and the limbic system is clearly involved in some of them.

CHAPTER 3

The Pituitary Gland (1): The Anterior Lobe (Adenohypophysis)

Introduction

The pituitary gland is an important endocrine gland because it is the source of a number of hormones of which some have, as their major effect, the regulation of hormone production by other endocrine glands elsewhere in the body. It has been described as the 'leader of the endocrine orchestra', with the hypothalamus conceivably acting as the overall conductor.

Embryological derivation

The pituitary gland, or hypophysis, has an interesting embryological development. It is formed as the result of a fusion between an upward-growing ectodermal extension from the roof of the primitive buccal cavity, called Rathke's pouch, which can be seen from the fourth week of development, and a downward ectodermal extension from the developing diencephalon region of the brain, to which it connects during the fifth week. The extension from the buccal cavity then separates from the developing pharyngeal region by the 12th week, remaining attached to the downward neural growth from the brain called the infundibulum, which it surrounds at the upper end forming the pituitary gland. That part of the gland which originated as the upward growth from the buccal cavity forms the anterior lobe of the pituitary, while the downward growth from the developing brain forms the posterior lobe. Cellular differentiation of the anterior lobe has begun by around the sixth week, with some early hormone production beginning by the seventh week.

General structure

The pituitary gland lies within a depression in the sphenoid bone beneath it, which is characteristically shaped liked a Turkish saddle and hence is

Integrated Endocrinology, First Edition. John Laycock and Karim Meeran.

known by its Latin name, *sella turcica*. In the adult, the two lobes of the gland are quite distinct, with the anterior lobe being made up of various types of secretory cells; it is also called the adenohypophysis. The posterior lobe is composed in large part of nerve axons from the hypothalamus, and is called the neurohypophysis.

The main bulk of the anterior lobe of the pituitary is called the *pars distalis*, while the upper part which wraps around the neural stalk (called the *infundibulum*) is known as the *pars tuberalis* (see Figure 3.1). The adenohypophysis consists of various cell types which can be distinguished using specific histological stains and other differentiating techniques such as immunofluorescence. The first stage of differentiation is to identify the cells as being acidophilic (40% of total, responding to acidic dyes such as eosin), basophilic (10% of total, responding to basic dyes such as Schiff blue) or chromophobic (not responding to colour dyes). The second stage of differentiation is to identify the acidophilic cells as those producing specific hormones, and they are the somatotrophs and lactotrophs which produce somatotrophin and prolactin respectively. Likewise, the basophilic cells comprise the thyrotrophs, corticotrophs and gonadotrophs which produce thyrotrophin, corticotrophin and the gonadotrophins respectively. The chromophobes include many which are probably immature acidophils and basophils, and cells which have released their hormones and are not responding to the histological dyes. While the acidophilic and basophilic cells are typical secretory cells containing numerous characteristic granules, the chromophobic cells are essentially devoid of granules.

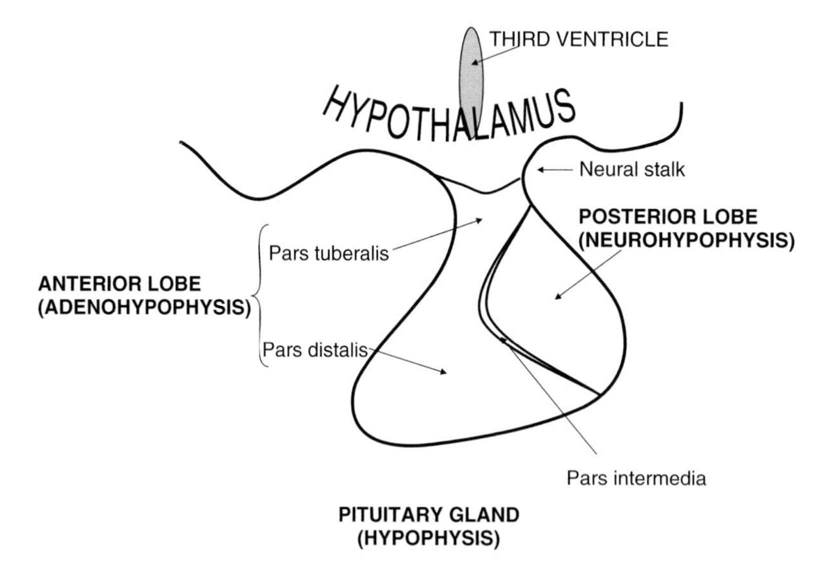

Figure 3.1 Diagram illustrating the different parts of the pituitary gland.

During pregnancy, the anterior pituitary increases in size, as do other endocrine glands, and at parturition can be over 25% greater than previously. In contrast, the posterior lobe, or neurohypophysis (also called the pars nervosa), is comprised of the many unmyelinated axons of larger than normal nerve fibres originating in the hypothalamus called magnocellular neurones. These large axons are distinguishable from those of other neurones not only by their size but also by the presence of swellings at regular intervals, particularly those at their distal end which are called Herring bodies after Percy Herring, the scientist who first described them in 1908. The nerve terminals and Herring bodies (and dendrites, even) are full of secretory granules. In between these nerve axons are many other cells: specialised glial cells called pituicytes. Originally, these cells were believed to be there simply to provide structural support for the nerve axons. In fact they are now known to be closely associated with the nerve axons and have various actions which are important for their normal function.

In between these two lobes is another very thin cellular area known as the *pars intermedia*. This area is much more developed in other vertebrates (e.g. amphibians and reptiles) than in the human where, for most of the time, it remains small and with little known function. Interestingly, during pregnancy, this intermediate lobe can expand but whether it has some function at this time is unclear. One hormone produced by the intermediate lobe cells in other vertebrates is melanocyte-stimulating hormone (MSH) which stimulates melanocytes to produce the black pigment melanin. In the skin, melanocyte stimulation results in the movement of pigment outwards from the cell centre along melanophores, expanding the region and producing the pigmentation associated with this molecule. In humans, MSH is a by-product from the cleavage of a large precursor molecule produced by certain cells of the anterior pituitary, called pro-opiomelanocorticotrophin (POMC). However, MSH is also produced elsewhere, and it is known to have effects in addition to melanocyte stimulation (e.g. in the brain).

The anterior pituitary and its link to the hypothalamus

For a long time, it was known that destruction of the pituitary gland resulted in the loss of many functions in the body, and that its total loss had severe consequences which could be fatal. It was also clear that its position, attached to the hypothalamus, was crucial for its own function, and while a nervous connection to the posterior lobe was clearly present, the link between the hypothalamus and the anterior lobe was unclear.

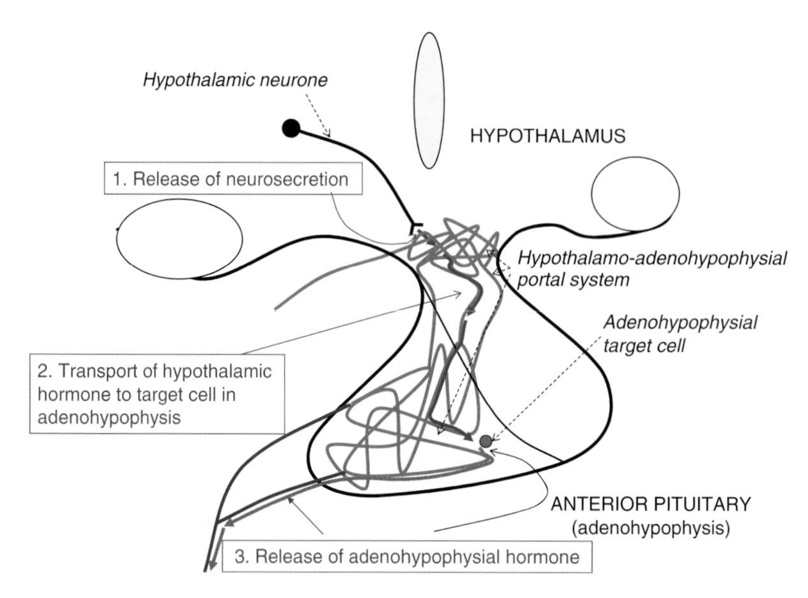

Figure 3.2 Diagram illustrating the way by which the hypothalamus exerts a controlling influence over the release of a specific adenohypophysial hormone from its target cell. The three main steps are the actual neurosecretory stage at the hypothalamic nerve endings in the median eminence region (i), the transport of the hypothalamic hormone to the adenohypophysis via the hypothalamo-adenohypophysial portal system (2) and the release of the adenohypophysial hormone (3) which ultimately enters the general circulation.

There is no appreciable innervation of the adenohypophysis, although a few peptidergic nerve fibres do terminate within it and could have a controlling influence. There is also evidence for some innervation by sympathetic fibres, specifically to the vasculature in this region. It is quite possible that regulation of adenohypophysial blood flow could have some influence on hormone release into the general circulation. However, it was the crucial observation that a special hypothalamo-adenohypophysial portal system is present, as described in Chapter 2, which provided the necessary clue to the manner by which the hypothalamus regulates adenohypophysial function (Figure 3.2). Around the middle of the 20th century, Geoffrey Harris performed a series of elegant experiments demonstrating that precise electrical stimulation of specific areas of the hypothalamus elicited ovulation in rabbits, an effect which could be abolished if the pituitary was separated from the hypothalamus. His description of these studies was published in 1955, and to many he is considered to be the 'father' of neuroendocrinology. By then it was generally accepted that the main direction of blood flow along the portal system was from the median eminence down to the anterior pituitary. It was not long before it became clear that hypothalamic control over the adenohypophysis was exerted by molecules released from nerve terminals in close proximity to the capillary walls of the primary capillary plexus. The subsequent identification of the first of these molecules (thyrotrophin-releasing hormone (TRH) and gonadotrophin-releasing hormone (GnRH); see the following section) by Roger Guillemin and Andrzej Schally jointly won them the Nobel Prize in Medicine in 1977.

These molecules are hypothalamic hormones, and most of them stimulate specific endocrine cells in the adenohypophysis and are therefore called releasing hormones. A few actually exert inhibitory control over the release of specific adenohypophysial hormones, so they are called release-inhibiting (or, simply, inhibiting) hypothalamic hormones. They are generally named according to their principal actions.

Hypothalamic neurones and their hormones

The cell bodies of the hypothalamic neurones are generally loosely grouped into nuclei as described in Chapter 2. They have dendrites which receive chemical information from other neurones usually originating in other parts of the central nervous system (CNS). The axons of many of these hypothalamic neurones terminate adjacent to the capillaries in the median eminence (see Figure 3.3).

As described in Chapter 1, the neurosecretory molecules are initially transcribed from nuclear DNA and translated from the consequent RNA strands to the protein precursor molecules. The synthesis process involves modification of the molecules in the rough endoplasmic reticulum, and their packaging into granules in the Golgi complex. The granules then move along the axons, to accumulate in the nerve terminals. The neurones function like other neurones when stimulated appropriately.

Action potentials generated by the integration of incoming stimulatory and inhibitory signals travel down the length of the axon to cause depolarisation of the nerve terminal membranes. This is associated with an influx of calcium ions which triggers the movement of intracellular granules containing the neurosecretory molecules to, and their fusion with, the nerve terminal membrane. The neurosecretion is then released into the surrounding interstitial fluid from which the molecules enter the blood flowing through the primary capillary plexus in the median eminence.

Each neurone releases its hormone into the hypothalamo-adenohypophysial portal system which transports it to its target cell in the anterior pituitary. Indeed, most (if not all) of the hypothalamic neurones release more than one type of neurosecretory molecule from their nerve endings. It is quite likely that subtle differences in the incoming signals

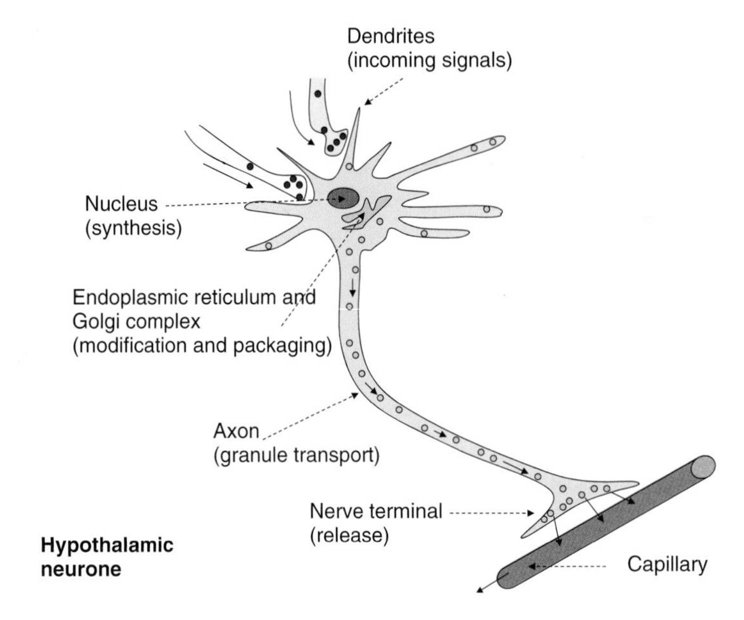

Figure 3.3 Diagram illustrating a typical hypothalamic neurone abutting a capillary in the median eminence, and the various intracellular events which are involved in the manufacture of the hypothalamic hormone. Also shown are the nerve terminals from other neurones close to dendrites which can release molecules which affect the hypothalamic neurone.

are integrated such that they can result in differing patterns of secretion and molecular constituents. It is important to remember the pulsatile nature of hypothalamic hormone release into the median eminence, as this is essential in determining not only the nature, but also the amount, of anterior pituitary hormone to be released.

As mentioned earlier, the anterior pituitary contains various different cell types which can be classified according to their principal hormonal secretions. Thus we have thyrotrophs which secrete thyrotrophin, corticotrophs which produce corticotrophin and so on. There are six principal adenohypophysial hormones which are well characterised, and these will be considered in some detail along with their principal target cells. The release of each one is associated with one or more hypothalamic hormones (see Figure 3.4).

Prolactin, produced by the lactotrophs, is the only adenohypophysial hormone under dominant inhibitory control, exerted by dopamine from the hypothalamus. Dopamine, derived from the amino acid tyrosine, is better known as a neurotransmitter within the CNS. However there are additional, stimulatory, hypothalamic influences on prolactin release, namely, TRH, and other potentially stimulatory molecules such as copeptin

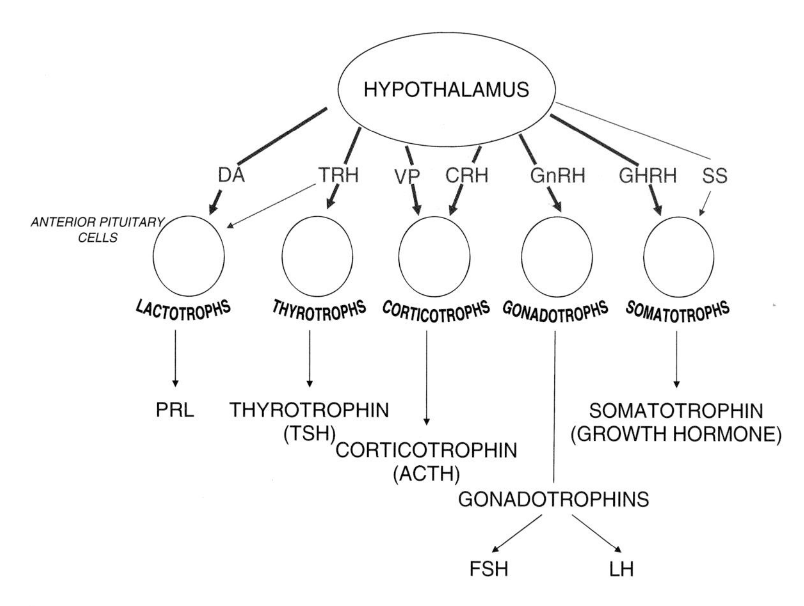

Figure 3.4 Diagram illustrating the principal hypothalamic hormonal influences on the anterior pituitary cell types. The purple and blue lines represent the stimulatory and inhibitory influences on each cell type respectively. CRH = corticotrophin-releasing hormone; DA = dopamine; GHRH = growth hormone–releasing hormone; GnRH = gonadotrophin-releasing hormone; SS = somatotstatin; TRH = thyrotrophin-releasing hormone; PRL = prolactin; FSH = follicle-stimulating hormone; LH = luteinising hormone and VP = vasopressin. TSH and ACTH are thyroid-stimulating hormone and adrenocorticotrophic hormone, respectively (see section 'Hypothalamic neurones and their hormones' for details).

(released with vasopressin; see Chapter 4) and RFamide-related peptide-1 (RFRP-1). TRH, a tripeptide from the hypothalamus derived from a much larger precursor, mainly exerts stimulatory control of thyrotrophin production from the thyrotrophs, hence its name. Thyrotrophin is sometimes also called thyroid-stimulating hormone (TSH). TRH is mainly produced by medial neurones located in the paraventricular nuclei.

Corticotrophin-releasing hormone (CRH) is a 41–amino acid polypeptide derived from a larger prohormone molecule of 191 amino acids, synthesised in specific hypothalamic neurones also in the paraventricular nuclei. It acts in conjunction with a nonapeptide, vasopressin (VP), which is released from some of the same neurones which also release CRH in the median eminence, as well as from other neurones. This dual hormonal influence is probably important in regulating corticotrophin release from the adenohypophysial corticotrophs under different circumstances, for instance in response to different stimuli, and this will be considered elsewhere (see Chapter 4). Corticotrophin is also known as adrenocorticotrophic hormone (ACTH).

Another hypothalamic hormone is the decapeptide GnRH which is also initially synthesised as a larger prohormone molecule, particularly in the preoptic nuclei. Interestingly, it somehow controls the release of two different gonadotrophins, luteinising hormone (LH) and follicle-stimulating hormone (FSH), from the adenohypophysial gonadotroph cells. However, it is quite possible that additional, separate populations of gonadotrophs exist which produce only one or the other of the two gonadotrophins on stimulation. And of course, differential release from the same source cell can be produced depending on other influencing factors acting in addition to GnRH. One recently identified hypothalamic hormone having an inhibitory effect on gonadotrophin release is called gonadotrophin inhibitory hormone (GnIH). Also called RFamide-related peptide-3 (RFRP-3), its physiological significance is currently under investigation.

Two hypothalamic hormones (at least) are known to regulate the production of another hormone from the anterior pituitary, the protein somatotrophin, or growth hormone (GH, sometimes specified as human growth hormone (hGH)). The dominant hypothalamic influence on its release is exerted by growth hormone–releasing hormone (GHRH), but there is a lesser inhibitory influence exerted by a 14–amino acid polypeptide called somatostatin. This molecule, like other molecules associated with the hypothalamus, can be synthesised by other tissues in the body. Interestingly, somatostatin exerts an inhibitory influence on all its target cells wherever they are. It is also produced by neurones in other parts of the CNS (as a neurotransmitter), by cells of the small intestine and by specific cells of the endocrine pancreas. In the latter two sites, processing of the precursor molecule actually produces a larger 28–amino acid version of somatostatin.

Pituitary tumours and the optic chiasma

As mentioned earlier, the pituitary gland sits in the *sella turcica* just below the median eminence region of the brain, and relatively close to the anteriorly situated optic chiasma. This anatomical region comprises afferent nerve fibres arising from the retinae of the eyes. The fibres from the retinae leave the eyes as the optic nerves; those fibres from the nasal parts of the retinae actually cross over to the opposite (contralateral) side of the brain, whilst those originating from the retinal cells in the outer (temporal) parts of the retinae pass to the brain on the same side (ipsilateral). Consequently, light from the same part of the visual fields of both eyes is directed to the same side of the brain (see Figure 3.5).

If a pituitary tumour develops and grows to such a size that it pushes up into the hypothalamus above, it can disrupt the optic nerve fibres

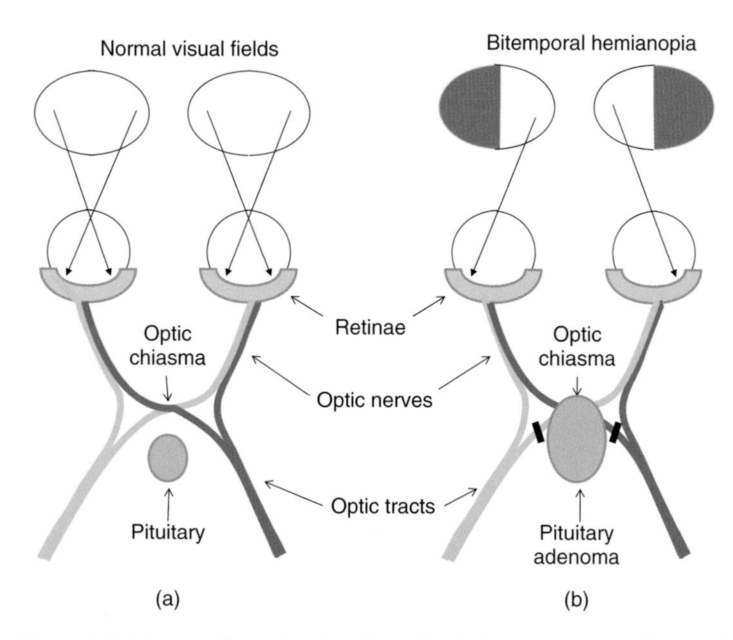

Figure 3.5 Diagram illustrating the effect of a pituitary tumour on the visual pathways from the eye. (a) Light reaching inner nasal and outer temporal retinae of each eye from opposite sides of the visual fields. The optic nerve from each eye comprises nerve fibres originating from cells in the two parts of each retina. Fibres from the inner nasal parts of each retina cross over at the optic chiasma in the brain, which lies close to the pituitary gland. (b) The effect of a pituitary tumour which has grown upwards to destroy the cross-over nerve fibres in the optic chiasma, producing a loss in temporal visual fields (i.e. loss of peripheral vision) known as bitemporal hemianopia.

crossing over in the optic chiasma. This disruption would affect the fibres originating in the nasal parts of the retinae of each eye, which receive light from the temporal sections of the visual fields. Occasionally, the loss of vision from the outer, temporal, visual fields is an early indicator of the presence of a growing pituitary tumour, and it can first be noted by an optician at an eye check using perimetry (see Chapter 5). The loss of peripheral sight affecting the temporal visual fields is known as bitemporal hemianopia (half-blindness affecting the two temporal visual fields). It is also sometimes called heteronymous hemianopia, meaning half-blindness on opposite sides of the visual fields for the two eyes.

The adenohypophysial hormones

Of the various types of cells in the anterior pituitary, those mainly associated with hormones are typical secretory cells with many intracellular granules. Most of these cells actually synthesise and release various

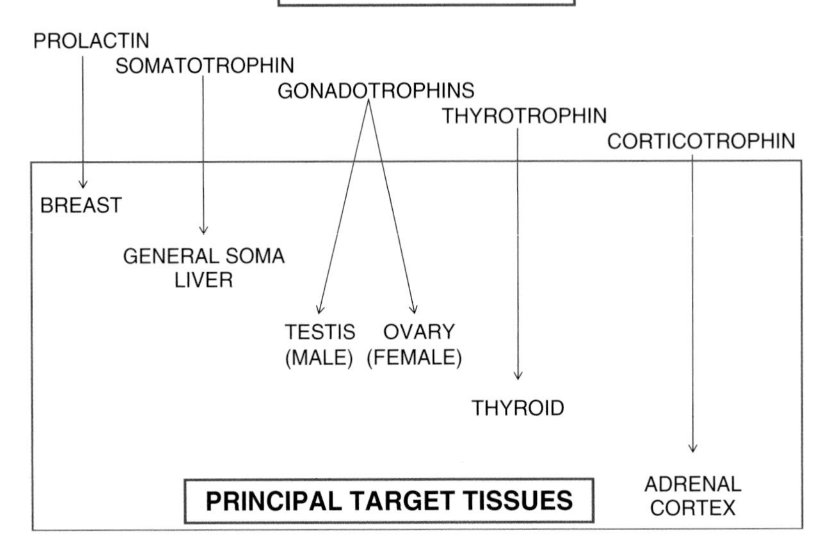

Figure 3.6 Diagram illustrating the main adenohypophysial hormones and their principal target tissues.

molecules, but their principal known roles are associated with the hormones mentioned in this chapter, and identified in Figure 3.6. In general, the adenohypophysial hormones stimulate the growth and maintenance of their target tissues, and this nutritive or 'trophic' role gives them their names. Thus, somatotrophin stimulates the growth and development of the 'soma', or body (Greek origin), hence its alternative name growth hormone. Likewise, thyrotrophin stimulates the growth and development of the thyroid gland, corticotrophin the outer parts (cortices) of the adrenal glands and gonadotrophins the gonads. Prolactin may appear to be the exception, but it does have an alternative (albeit rarely used) name, lactotrophin, which also relates the name to its main target tissue, the lactating mammary gland.

What is of particular interest regarding the adenohypophysial hormones is that (i) they are greatly influenced by the CNS through the hypothalamus and its hormones, and (ii) most of them actually have other endocrine glands as their principal targets. As mentioned in Chapter 2, one important feature of adenohypophysial hormone release is that it is pulsatile as a consequence of the hypothalamic influence. Each of these hormones will be discussed here, although many of them will also be considered in the chapters dealing with their target endocrine glands.

The adenohypophysial protein hormones

The two anterior pituitary hormones which are classified as proteins are growth hormone (somatotrophin) and prolactin. Each is produced by specific somatotrophic or lactotrophic cells respectively, and in addition by a small population of lactosomatotrophic cells which appear to produce both molecules. All three cell types are classified as acidophilic.

Somatotrophin (GH)

Synthesis, storage, release and transport

Somatotrophin is a large single-chain protein of 191 amino acids, initially synthesised as an even larger precursor molecule (pro-somatotrophin) mainly by the acidophilic somatotroph cells of the adenohypophysis. There are five hGH genes located on chromosome 17, one of which (GH-N) encodes for the 22 kd, 191 amino acid, version of GH. The other genes are associated with other variants of GH, in particular a GH version produced by the placenta (GH-V) which produces the bulk of circulating hGH during pregnancy. The hormone is stored in intracellular granules from which it is secreted into the blood by exocytosis. Its principal role is to stimulate the growth and maintenance of the general tissues of the body. The somatotrophs constitute approximately 40% of the anterior pituitary cells, and are generally located in the lateral parts of the adenohypophysis. There is some homology between the GH, prolactin and human placental lactogen molecular structures, which accounts for their overlapping physiological effects. Growth is a complex process involving increased synthesis of many molecules in all tissues, so not surprisingly somatotrophin stimulates many metabolic processes including protein synthesis and glucose metabolism. It plays a particularly important role in stimulating the growth and development which take place in the early years of life, and the expression of linear growth is the most obvious manifestation of its actions at this stage.

GH is released in pulses throughout the day, with levels being low to undetectable at the bottom of each trough. Obesity is associated with decreased pulse frequency and decreased GH release in response to various secretory stimuli, in contrast with the increased pulse frequency and responsiveness to stimuli seen with fasting. With increasing age there is also a decrease in pulse amplitude, so that by middle age the circulating peak levels are considerably lower than those measured during puberty, with unchanged pulsatility. Interestingly, the pulsatility reflects the alternating changes in GHRH and somatostatin secretory patterns, which also demonstrate gender differences.

Of the different sizes of GH secreted, the main component is the 22 kd form. Once released into the general circulation, it is either transported as a free (unbound) molecule or transported bound to other proteins, called binding proteins. One of these binding proteins has a high affinity for somatotrophin and is identical to the extracellular domain of the growth hormone receptor. Raised levels of this high-affinity binding protein in the circulation decrease the binding of the hormone to its target cell receptors by competing for the ligand.

The GH receptor and mechanism of action

The growth hormone receptor has extracellular and intracellular domains linked by a single membrane-spanning domain. Growth hormone binds to the extracellular components of two of these receptors, which dimerise together. The intracellular messenger system activated by the GH-induced receptor dimerisation involves JAK2 tyrosine kinase-induced autophosphorylation, and the phosphorylation of various intracellular signalling molecules including three signal-transducing activators of transcription proteins (STATs 1, 2 and 3) and mitogen activated protein kinase (MAPK). These migrate to the target cell nucleus where they stimulate specific target gene transduction and new protein synthesis. These proteins then induce the effects associated with GH activity (Figure 3.7). The intracellular system activated by GH stimulates its own self-regulation by the activation of another set of intracellular proteins called suppressor of cytokine signalling (SOCS) proteins which disrupt the JAK2 pathway.

Somatotrophin receptors are found in most tissues and organs of the body, with a particularly high concentration of them found on liver cells (hepatocytes). In fact, while the hormone can exert some of its effects directly in target tissues, many of them are actually produced indirectly, by other hormones synthesised in target cells, particularly the hepatocytes, in response to GH. Thus, the liver is an important target organ for GH and is an endocrine gland in its own right. The hormones it produces in response to GH are called insulin-like growth factors I and 2 (IGF1 and IGF2). They are mediators of some of growth hormone's actions, and for this reason they used to be called somatomedins. The principal mediator of many of GH's actions is the IGF1 molecule. The IGF2 molecule has more similarity with the insulin molecule, and has a greater affinity for the insulin receptor than IGFI.

The insulin-like growth factors IGF1 and IGF2

The insulin-like growth factors IGF1 and IGF2 are named because they have structures similar to that of pro-insulin, the precursor of the pancreatic hormone insulin, and have some insulin-like actions. Insulin is an important metabolic hormone which is considered in detail in Chapter 14.

IGF1 and IGF2 are polypeptides of 70 and 67 amino acids respectively, and both share approximately 50% homology with the insulin molecule. All three molecules have A and B chains linked by disulphide bonds, but their connecting C chains differ. The structural similarities between the IGF and insulin molecules account for the affinities of these hormones for each other's receptors.

Gene knock-out studies in experimental animal models indicate that both IGF1 and IGF2 are important in promoting fetal growth, with IGF2 being the major factor, while IGF1 plays the more important role as a growth promoter during post-natal development, along with GH. While GH appears to be the main regulator of the IGF1 gene, control over the IGF2 gene is less clear.

The IGFs are synthesised by a wide variety of cells in the body, including cardiocytes, where they can have local paracrine (and autocrine) effects, or be released into the general circulation, as from their main source the hepatocytes. There are at least six IGF binding proteins (IGFBPs 1–6) in the blood, and the hormones are virtually all transported as complexes with them. It would appear that the affinity of these IGFBPs is generally greater than for the IGF receptors. This suggests that they can inhibit IGF activity unless the hormones are dissociated from their binding proteins. Indeed there are various proteases in the circulation which appear to be able to stimulate this dissociation process. Furthermore, at least some of the IGFBPs may have direct effects of their own, independent of IGF binding, and certain cell surface molecules appear to act as their receptors.

There are three receptors to which IGF molecules can bind. Two of these receptors, IGF receptor 1 (IGFR1) and IGF receptor 2 (IGFR2), are specific for the IGF molecules, the former having an equally high affinity for both IGF1 and IGF2, while the latter has a greater affinity for the IGF2 molecule, and appears to mainly direct this hormone to an intracellular degradation pathway. The IGFR1 is by far the more important receptor regarding the mediation of physiological activities associated with both IGF1 and IGF2, and is found in virtually all tissues (except the liver). The third receptor which also binds IGF molecules is the insulin receptor which has a 100-fold lower affinity for IGF1 than for insulin itself. It has a close structural similarity to the IGFR1 receptor.

Physiological actions of GH

Growth hormone is released from the anterior pituitary somatotrophs largely in response to the integration of the hypothalamic signals mediated by GHRH and somatostatin. It acts on its target tissues directly via its own receptors, and indirectly by stimulating the release of IGF1 from the hepatocytes of the liver, which is an important target tissue (Figure 3.7).

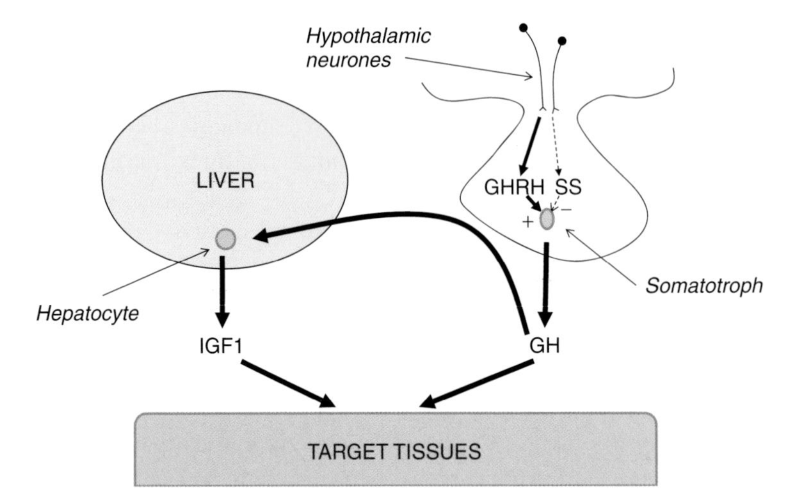

Figure 3.7 Growth hormone (GH) release from the anterior pituitary acts directly on its target tissues which include the liver, which in turn produces IGF1 which also acts as a growth promoter. GHRH = growth hormone–releasing hormone; SS = somatostatin and +/- = a stimulatory or inhibitory influence.

The physiological actions of GH and IGF1 can be considered generally to be metabolic and growth promoting, but other actions have also been described.

Metabolic and growth-promoting actions

1. Protein metabolism: GH stimulates protein synthesis by genomic and non-genomic mechanisms. It is generally associated with an increased muscle mass. The genomic action is concerned with the influence exerted by intermediate intracellular second messengers such as MAPK and STATs on specific genes resulting in the synthesis of proteins such as intracellular enzymes (see Figure 3.8). The other mechanism by which GH increases protein synthesis is indirect and non-genomic, and it involves membrane transporters for amino acids. There are at least three transporters, one each for basic, acidic and neutral amino acids, and their activities are increased by GH. IGF1 also stimulates amino acid uptake and protein synthesis in muscle and other tissues. GH stimulates differentiation of chondrocytes (cartilage cells) from precursors resulting in stimulation of bone growth directly, and also at least partly acts by stimulating hepatic IGF1 production, which stimulates chondrocyte proliferation. GH also stimulates local IGF1 production, for example from chondrocytes in the growth plates (epiphyses) of the long bones. In addition, GH appears to stimulate the hepatic production of IGF-BP3, and this molecule, in association with an acid-labile subunit, binds

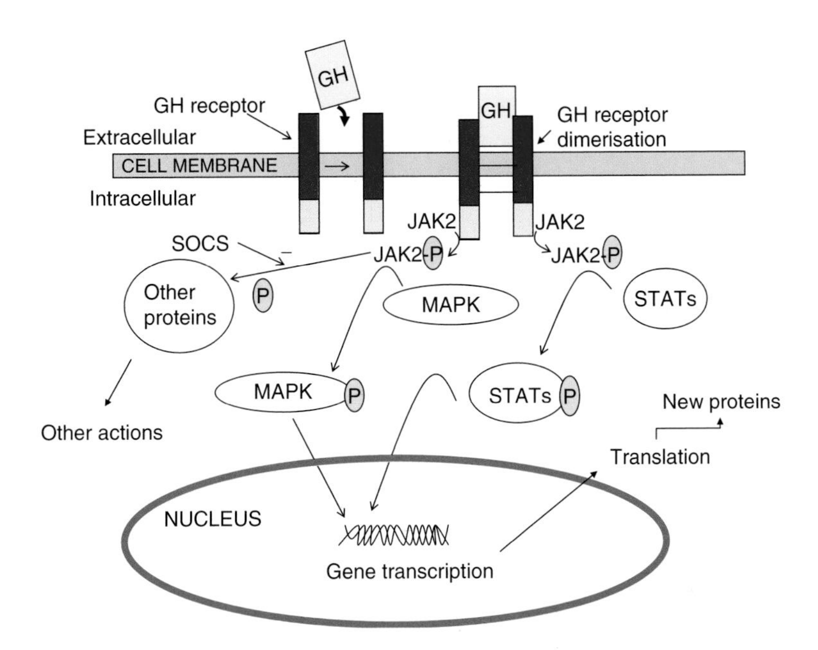

Figure 3.8 Diagram illustrating the mechanism of action of growth hormone (GH) following its binding to two GH receptors which dimerise resulting in phosphorylation of JAK2 kinases which, in turn, phosphorylate other intracellular proteins such as the signal transduction activating transcription proteins (STATs) and mitogen-activated protein kinase (MAPK). Gene transcription is a key target of many of these intracellular phosphorylated proteins. SOCS = suppressor of cytokine signalling.

IGF1 and transports it to the cells of the growth plate where protease-induced cleavage liberates IGF1. IGFI also appears to stimulate both the differentiation and proliferation of myoblasts in muscle.

2. Carbohydrate metabolism: The overall chronic effect of GH on carbohydrate metabolism is to raise the blood glucose concentration. It does this in various ways. For instance, it enhances hepatic glucose production by increasing gluconeogenesis (e.g. using the lipid breakdown product glycerol and certain glucogenic amino acids). However, its main influence on carbohydrate metabolism is indirect and is associated with the development of insulin resistance by peripheral tissues; it reduces the insulin-stimulated glucose uptake by peripheral cells such as in muscle and adipose tissue, and it decreases tissue glucose utilization. This may be a physiologically protective mechanism for the brain, since the increase in circulating glucose concentration enhances the amount of this major nutrient reaching the neurones of the CNS.
3. Fat metabolism: GH stimulates lipolysis in adipocytes by activating hormone-sensitive lipase, increasing the breakdown of triglycerides. The consequence is that circulating levels of non-esterified fatty acids

and glycerol increase, the latter being utilized by hepatocyctes for gluconeogenesis. GH also decreases fat deposition. The increase in fatty acid production is believed to be a major cause for the increase in insulin resistance which can develop when circulating GH levels are raised, for instance as a consequence of a GH-producing pituitary tumour. This would explain why excess circulating GH in acromegaly (and gigantism) is associated with a raised blood glucose concentration and an increased incidence of secondary diabetes mellitus.

The overall specific metabolic effects of GH combine with those of IGF1 to promote the growth and maintenance of tissues. During puberty, when GH and IGF1 circulating levels are highest, these effects combine with the growth-promoting actions of the androgenic and oestrogenic steroids to induce the growth spurt (see Chapter 8). Furthermore, the overall actions of GH on protein and fat metabolism are associated with the maintenance of a lean body mass. Indeed, the term 'middle-age spread' may well be at least partly attributed to the decline in circulating GH levels with age since the exogenous administration of GH is accompanied by some degree of restoration of a lean body mass.

Other GH actions

1. Immunoregulation: Involution of the thymus with ageing appears to be associated with the decreasing circulating concentrations of GH and IGF1 since the process can be reversed following administration of these hormones. Indeed, the thymus, the spleen and peripheral blood all appear to be capable of producing GH; furthermore, the GH receptor is expressed on different subpopulations of lymphocytes. Various studies indicate that GH stimulates the proliferation of T and B cells, modulates the migration of developing T cells from the thymus, stimulates immunoglobulin synthesis and may modulate cytokine responses to specific stimuli. However, in humans GH deficiency is not usually associated with immunodeficiency and only minor abnormalities of immune function have been reported, so the clinical significance of these actions is unclear.
2. Prolactin-like effects: GH can have some prolactin-like effects, particularly regarding lactogenesis, and it also promotes ductal elongation as well as the differentiation of ductal epithelia into terminal end buds in the mammary glands (particularly in females) during puberty.
3. Central effects: The sense of well-being reported by elderly subjects given exogenous 'replacement' GH suggests effects within the brain.

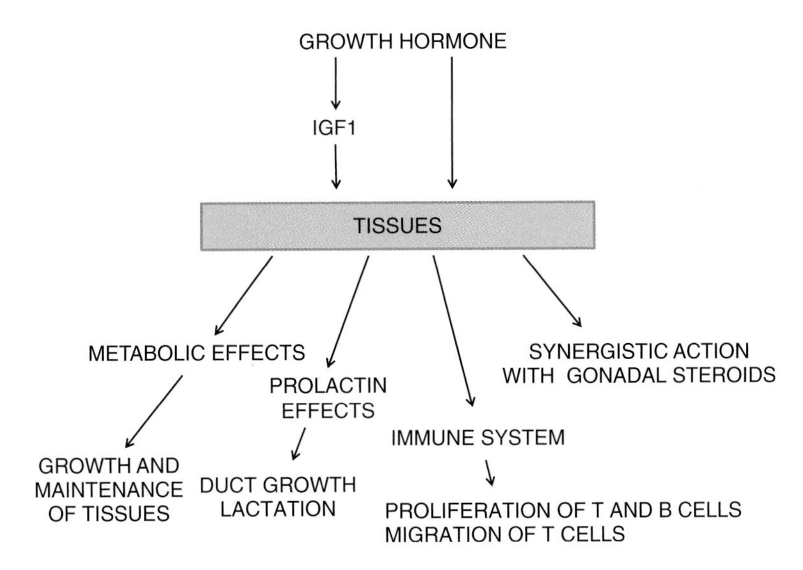

Figure 3.9 Growth hormone effects, which are at least partly mediated by insulin-like growth factor 1 (IGF1).

Relatively little is known about the precise pathways influenced by, and specific effects of, GH here.

4. Reproduction: GH is clearly implicated in the growth spurt of puberty, and it is likely that it acts synergistically with the gonadal steroids which are released in increasing quantities at this stage of development, to promote the growth and development of secondary sexual characteristics, for example (see Figure 3.9).

Clinical presentations of hypo- and hyper-production of GH

The direct and indirect growth-promoting effects of GH manifest themselves most clearly in the growing child: lack of GH results in short stature (pituitary dwarfism) while an excessive production (e.g. by a pituitary tumour) results in gigantism. The metabolic effects associated with these conditions in childhood, and in adults, are as one would expect from the description of them given above. Thus in hyposecretion of GH there is a tendency for increased adiposity and decreased muscle mass, while in the hypersecretion state (called acromegaly in adults) increased insulin resistance can manifest itself ultimately as secondary diabetes mellitus, together with the excessive growth of soft tissues including the visceral organs. Indeed, cardiovascular and respiratory problems in gigantism and acromegaly, if unsuccessfully treated, can result in increased morbidity and mortality (see Chapter 5).

Control of GH secretion

Central influence

Control of GH secretion is chiefly from the hypothalamus via GH-releasing hormone (GHRH) and, to a lesser extent, somatostatin (see Figure 3.9). GHRH is a 40–amino acid polypeptide synthesised in the cell bodies of neurones mainly in the arcuate nucleus while somatostatin neurones originate in the supraventricular nucleus (see Chapter 2). There is an interesting reciprocal relationship between the amplitudes of the pulses of GHRH and somatostatin; when GHRH is at its peak within a pulse, somatostatin is in its trough. Furthermore, there is clearly a gender difference such that the circulating concentrations of GH tend to be higher in females than in males. While greater peak amplitudes of GH pulses are observed in males, females have a higher, but more disorganised frequency of pulses resulting in a raised baseline circulating level. This reflects the gender difference believed to exist for GHRH and somatostatin release patterns; somatostatin release and negative feedback effects probably contribute to the determination of the gender-specific pulsatile release of GH.

The importance of the central influence over GH release is also apparent when considering the circadian variation, with greater pulses during the night. This is particularly relevant during puberty when considering the large pulses generated during sleep, with stage 4 of slow-wave sleep particularly associated with them. It has been proposed that the decreased amount of stage 4 slow-wave sleep with age correlates with the corresponding decreased production of nocturnal GH release. Certainly, by middle age pulse amplitude of GH is greatly diminished when compared to that of puberty.

Metabolic and associated factors

Various other factors influence GH secretion. Certain amino acids, including excitatory amino acids such as glutamate and aspartate, as well as others such as arginine, leucine and ornithine, increase GH synthesis. The effect of excitatory amino acids is primarily central, by influencing hypothalamic GHRH and somatostatin release. Free fatty acids also stimulate GH release. A decreased blood glucose concentration (hypoglycaemia) is a very potent stimulator of GH release, and as such is used clinically following the careful administration of insulin, to test for anterior pituitary function. Consequently, it is not surprising to find that fasting is a powerful stimulus for GH release. Furthermore, exercise is a potent stimulator of GH release, and it is quite likely that at least part of the effect is due to alterations in metabolic substrates such as glucose (Figure 3.10).

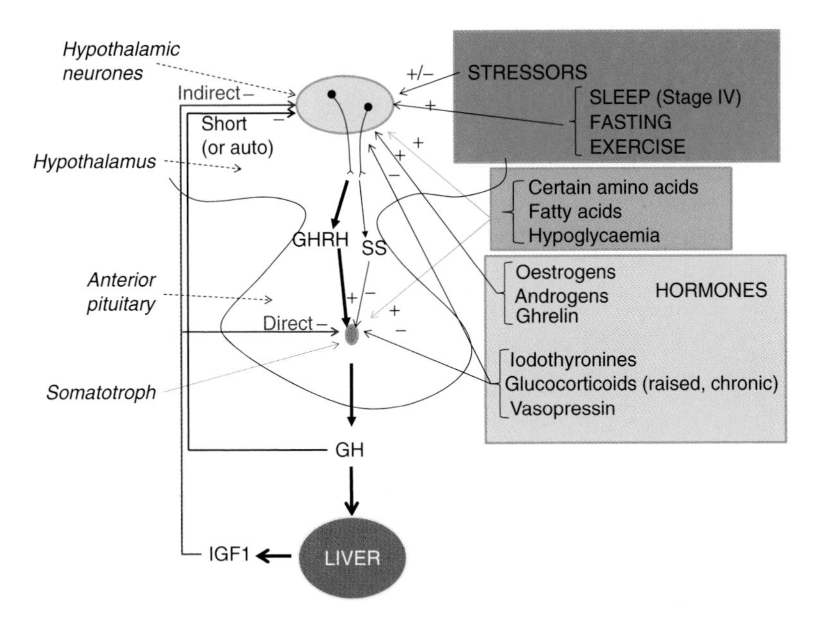

Figure 3.10 The control of growth hormone (GH) release, involving circulating influences on the somatotroph cells of the anterior pituitary by metabolic substrates and hormones as well as central effects on the hypothalamus. GHRH = growth hormone–releasing hormone and SS = somatostatin. Direct and indirect negative feedback effects by IGF1 and a short-loop negative feedback by GH are also shown. For further details, see section 'Metabolic and associated factors'.

Hormonal influences

Acute stressors such as physical exercise, trauma and hypoglycaemia increase glucocorticoid-stimulated GH secretion. In contrast, chronic raised levels of glucocorticoids, and longer term stressors such as emotional deprivation and depression, are associated with depressed GH secretion. Part of the glucocorticoid effect is probably central, involving an inhibition of GHRH pulses, but cortisol (the principal glucocorticoid in humans) also antagonises target tissue responsiveness to GH. Vasopressin also has an inhibitory effect on GH release and this could at least partly be associated with its release into the hypothalamo-adenohypophysial portal system (see Chapter 4).

Other hormones also influence GH release. Hyperthyroid patients have a decreased pulsatile release of GH and this situation is ameliorated once they have been made euthyroid, suggesting that raised iodothyronine levels have an inhibitory effect. Oestrogens and androgens are associated with stimulation of GH release, and this effect is particularly relevant at puberty.

Ghrelin is a 28–amino acid polypeptide initially synthesised as a much larger prohormone molecule which is produced mainly by gastric mucosal cells in the stomach and, to a lesser extent, in other tissues such as the intestinal tract, kidney and placenta. It is also produced within the hypothalamus, and small amounts appear to be produced by the anterior pituitary gland. Its name comes from an early observed effect of the molecule when first isolated, namely, the stimulation of GH release. It acts through GH secretagogue (GHS) receptors by a central effect on GHRH release, and also has a direct effect on the pituitary somatotrophs. Ghrelin also has other effects centrally such as in stimulating appetite, and is associated with obesity.

Feedback control

GH has a short, or auto, negative feedback effect on the hypothalamic release of GHRH, and it may even have an autocrine effect on its own production by the somatotroph cells. The other negative feedback loops influencing GH release involve IGF1 from the liver produced in response to GH stimulation. Thus IGF1 has a direct negative feedback effect on the pituitary somatotrophs producing GH as well as an indirect negative feedback by stimulating the hypothalamic production of the inhibitory somatostatin.

Interestingly, the somatotrophs in the anterior pituitary express interleukin 1 receptors suggesting that there is a potential feedback effect by this product of immune cells such as macrophages and monocytes.

Prolactin

Synthesis, storage and release

Prolactin has a close homology to GH. It is mainly synthesised in lactotroph cells of the adenohypophysis which comprise approximately 10% of all secretory cells in the anterior pituitary. One variant, called human placental lactogen (hPL), is produced by the placenta. The gene for prolactin is located on chromosome 6, and the expression product is a single-chain protein hormone comprising 199 amino acids in humans. The hormone is stored in intracellular granules from which it is secreted into the blood by exocytosis. As with other adenohypophysial hormones, the release of prolactin is pulsatile, the pulses being larger in women with regular menstrual cycles than in menopausal women and men. Circulating levels rise and fall modestly during each menstrual cycle after puberty, but rise markedly during pregnancy. There is also a circadian variation, with levels of pulsatile prolactin highest during sleep, and then decreasing during the rest of the day. Pulsatile levels of prolactin at night are associated with rapid eye movement (REM) sleep.

As with many other hormones including growth hormone, prolactin levels decrease with age in men and women.

Transport in the blood

Like GH, prolactin circulates in the blood in various sizes, of which the monomeric form is much more bioactive than the polymeric forms. In the circulation, the monomeric prolactin can be cleaved into smaller components which may have physiological effects of their own. Furthermore, there is a circulating binding protein for prolactin which corresponds to the extracellular part of the prolactin receptor.

Prolactin receptors and mechanisms of action

As with GH, the prolactin receptor is a transmembrane receptor with hydrophilic extracellular and intracellular domains in addition to the hydrophobic transmembrane portion. Prolactin receptors are found in many tissues including breast, pituitary, brain, liver, kidneys, adrenal cortex, prostate, ovaries and testes. The receptors dimerise when prolactin (or another ligand) binds to them, and the intracellular mechanism activated includes the phosphorylation of Janus kinase/signal transducer and activator of transcription (JAK/STAT) molecules. Subsequent increased transcription of new protein molecules results in the induction of prolactin's effects in its target cells.

Physiological effects of prolactin

Prolactin is quite ubiquitous in its actions, and while very often its effects may not be pronounced on their own, they seem to have important interactions with other hormones and with endocrine systems. This variety of different effects tallies with the equally ubiquitous distribution of receptors in many tissues of the body.

Lactation

For many years, prolactin was generally considered to have only one, essentially gender-specific, effect: the stimulation of milk production (lactogenesis) in the alveoli of the female breast, or mammary gland, during lactation. This is certainly the most important physiological effect of prolactin that we know of at present, and without any appropriate alternative source of nutrition for the baby it is essential for its survival in the first few months of life. This effect is inhibited during pregnancy by the very high circulating levels of oestrogens and particularly progesterone. At parturition, when the levels of gonadal steroids fall drastically, the inhibition is lifted and prolactin then exerts its lactogenic effect. However, if suckling by the new-born baby is not initiated soon after birth and repeated frequently

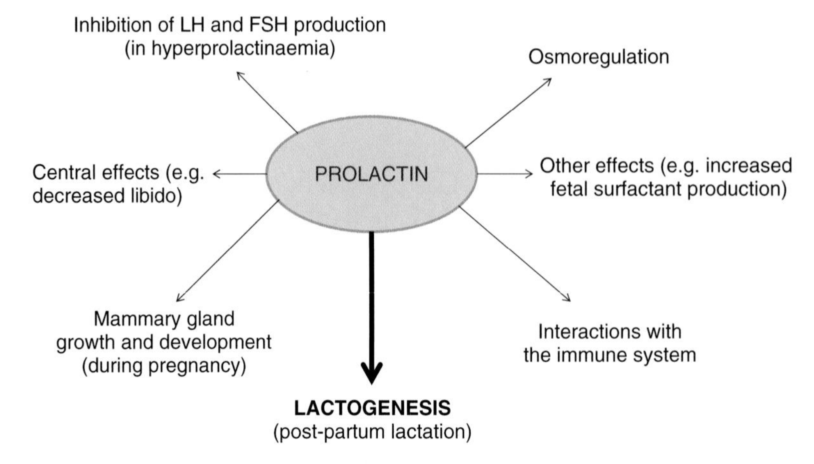

Figure 3.11 The general effects of prolactin, the principal physiological one being the stimulation of lactogenesis.

with short intervals, prolactin production is reduced and circulating levels return to normal (Figure 3.11).

Breast growth and development

Another effect of prolactin is the development and growth of the mammary gland during pregnancy, in combination with other hormones including oestrogens, progesterone, GH and locally produced IGF1. During pregnancy, the pituitary can more than double in size, and much of that growth is due to a proliferation of lactotrophs producing prolactin. Consequently, the plasma prolactin concentration increases steadily throughout pregnancy. Interestingly, the initial phase of mammary gland development during puberty is relatively independent of prolactin, but seems to require GH-mediated IGF1.

Immune system

Since prolactin is synthesised in males as well as females, and is also present in the general circulation in both genders, it has always been assumed that the hormone has non-gender-specific physiological effects. One such effect may be the interaction between prolactin and the immune system. Prolactin receptors are found in many tissues in the body, including lymphoid. Furthermore, prolactin is produced not only by the anterior pituitary lactotrophs but also by a variety of cells in the body including human lymphocytes and other cell types of the immune system. Indeed, there is evidence to suggest that prolactin stimulates the proliferation of various cell types, including T cells. Prolactin can also enhance cell survival, at least partly by inhibiting apoptosis (normal regulated cell death). Interestingly prolactin, and prolactin receptor, knockout mice do not exhibit major defects in their immune responses, suggesting that the prolactin-mediated effects are subtle. Indeed it has been proposed that one effect of prolactin is to counterbalance the inhibitory effect of glucocorticoids on the immune system and inflammation, for instance under stress conditions. Certainly, various stressors do stimulate prolactin, as well as corticotrophin, release from the anterior pituitary, and prolactin does appear to protect against glucocorticoid-induced lymphocyte cell death in *in vitro* studies.

Direct effects on the reproductive axis

Prolactin has a direct involvement with ovarian function in some mammals, such as certain rodents and ruminants, having a corpus luteum–maintaining, or luteotrophic, effect once ovulation has been induced. Indeed, prolactin receptors have been detected on ovarian granulosa cells in the pre-ovulatory stage. While prolactin may interfere with progesterone production during the late follicular phase, it may actually stimulate progesterone production during the luteal phase when it may also induce the synthesis of LH receptors. Prolactin may also interfere with the FSH-induced production of oestrogen from granulosa cells during the follicular phase. Prolactin does not appear to be luteotrophic in humans, however, and any influence on gonadal steroid synthesis is still unclear. For further information about the ovarian cycle and hormonal involvement, see Chapter 7.

In males, similarly, most data suggestive of prolactin influences on testicular function come from rodent studies. Prolactin appears to stimulate the synthesis of LH receptors on Leydig cells and may potentiate the effect of LH on testosterone production. Prolactin may also potentiate the effects of androgens on the prostate and seminal vesicles. Again there is little evidence for a physiological role in men.

Inhibitory (Hyperprolactinaemic) effects on the reproductive axis

In contrast to the direct stimulatory effects of prolactin on reproduction, there is clearly an inhibitory effect in humans, for instance when the hormone is produced in excess, and is present at high concentrations in the blood (hyperprolactinaemia). In this situation, prolactin impairs fertility in both sexes by an attenuation of LH pulsatile release from the pituitary, consequently associated with decreases in circulating gonadal steroid levels. Part of the effect may be a prolactin-induced decrease in the sensitivity of the pituitary gonadotrophs to hypothalamic GnRH. However, release of the hypothalamic pulse generator GnRH is also inhibited. In women this is associated with the loss of menstrual cycles (secondary amenorrhoea), as well as loss of libido. In men, hyperprolactinaemia is also associated with reduced libido, and decreased sperm numbers and motility (i.e. decreased fertility). Hyperprolactinaemia can be caused by pituitary tumours (adenomas), but it can also be a perfectly natural physiological occurrence, for instance in lactating mothers when circulating levels are high. The simultaneous inhibition of ovulation and menstrual cycles during lactation confers a natural contraception on the mother at a time when another pregnancy may be undesirable. It is important to appreciate that the contraceptive effect is dependent on the frequency and duration of lactation episodes. Should the frequency of breast feeding be reduced, then the contraceptive effect can be lost.

In men (and quite possibly women), it has been suggested that the physiological increased prolactin release at ejaculation (or climax) inhibits the dopamine-driven 'pleasure' stimulus which brought about the orgasm.

Osmoregulatory effects

Prolactin has osmoregulatory effects, at least in some vertebrates. For example, in some euryhaline teleost fish it is the hormone which is necessary to allow them to survive a change in their environment when they move successfully from salt water to fresh water. This effect is due to decreased water absorption and increased salt retention, through alterations in the permeabilities of gill, renal, intestinal and other tissue membranes. In mammals, a physiological role for prolactin in the regulation of salt and water balance is currently speculative. In rats, prolactin appears to have a diuretic effect which could be associated with a decreased proximal reabsorption of sodium ions. In the absence of circulating vasopressin, as in the Brattleboro rat with hereditary hypothalamic diabetes insipidus, prolactin appears to have an antidiuretic effect. One could speculate that this might be relevant during pregnancy when prolactin and placental lactogen levels are high, and when there is also an increase in vasopressinase enzyme which inactivates circulating vasopressin.

Control of prolactin release

Since the main physiological effect of prolactin is to stimulate lactogenesis in the mammary glands, it is not surprising to find that the chief stimulus for its release is stimulation of tactile receptors on the nipple of each breast by the new-born baby post-partum. The afferent nerve pathway from the tactile (stretch) receptors is multineuronal, ultimately reaching the hypothalamus (Figure 3.12).

Indeed, the hypothalamus exerts the principal controlling influence on prolactin release from the anterior pituitary lactotrophs. Of particular importance are tuberoinfundibular dopaminergic neurones from the arcuate nucleus which terminate on the walls of the primary capillary plexus located in the median eminence. The dopamine that is released here

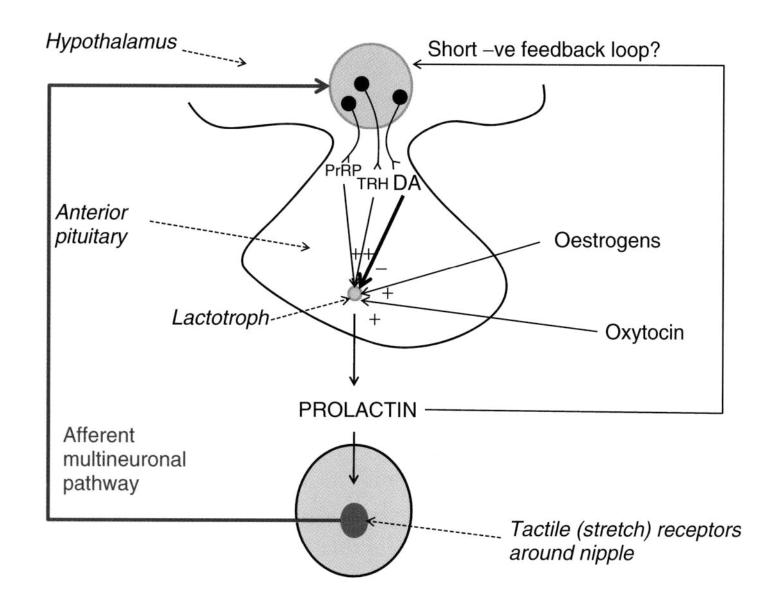

Figure 3.12 The main controlling influences on prolactin production, including the important afferent neural pathway stimulated by suckling which stimulates prolactin synthesis and release. PrRP = prolactin-releasing peptides, DA = dopamine, and TRH = thyrotrophin-releasing hormone.

enters the hypothalamo-adenohypophysial portal system through which it reaches its target cells, the lactotrophs. Dopamine inhibits prolactin release via D2 receptors, and this is the dominant effect exerted by the hypothalamus. As such, dopamine is a hypothalamic release-inhibiting hormone. Prolactin has a short (or auto) negative feedback on the dopaminergic neurones in the hypothalamus by stimulating tyrosine hydroxylase activity, which results in increased dopamine production. Thus the main drive for prolactin release has to be the central inhibition of the inhibitory dopamine pathway. However, the hypothalamus also has a lesser, but stimulatory, effect on prolactin release by means of TRH. As its name suggests, it is of greater importance with regard to the release of thyrotrophin from the anterior pituitary thyrotroph cells than on prolactin release (see Chapter 13) but is of use clinically. In addition, there are other hypothalamic substances (prolactine-releasing peptides, PrRP) that influence prolactin release, including the recently identified RFamide-related peptide-1 (RFRP-1), growth hormone and gonadotrophin-releasing hormones (GHRH and GnRH). Furthermore, the neurohypophysial hormones may also have an effect on prolactin release; certainly oxytocin appears to stimulate its secretion, and even vasopressin may exert some influence although its role in humans is unclear. The polypeptide copeptin which is co-released with vasopressin has been suggested to be a prolactin-releasing factor.

Various modulators also influence the release of prolactin at the level of the pituitary lactotroph, including other hormones. Oestrogens certainly stimulate prolactin synthesis and secretion, presumably explaining why women generally have higher levels of circulating prolactin than men, and why the frequency of secretory pulses is greater in ovulating women than in men and post-menopausal women. There is some evidence to suggest that both cortisol and the iodothyronines may have an inhibitory effect on prolactin release. Iodothyronines inhibit prolactin release indirectly, by exerting their normal indirect negative feedback effect on hypothalamic TRH release.

The glycoproteins

There are four main glycoprotein hormones all having a similar chemical structure, being derived initially from a common ancestral gene. They are the three anterior pituitary hormones luteinising hormone (LH), follicle-stimulating hormone (FSH) and thyrotrophin (TSH), as well as human chorionic gonadotrophin (hCG) which is usually produced by the placenta during pregnancy. Each hormone consists of two heterodimeric subunits, an α subunit which is common to all four hormones and consists of 92 amino acids in humans, and a β subunit which is specific for each hormone. The carbohydrate chains associated with both subunits are essential for physiological activity. The genes for these subunits are located on chromosomes 6, 11 and 19.

The gonadotrophins: Luteinising hormone (LH) and follicle-stimulating hormone (FSH)

Synthesis, storage, release and transport

The two gonadotrophins are very similar molecules synthesised in specialised cells called gonadotrophs which make up approximately 10–15% of the anterior pituitary. The β subunits for LH and FSH consist of 121 and 118 amino acids, respectively. Unusually, both LH and FSH are associated with the same population of adenohypophysial gonadotrophs indicating that their differential secretion must be independently regulated within the cells, the intracellular pathways probably being influenced by different, but specific, extracellular factors. There is the additional possibility that another population of cells specific for one or other of the two hormones may also exist, as in other species, although the evidence for this is lacking in humans. Within the cells are numerous granules which are either large (350–400 nm in diameter) or smaller (150–250 nm), found in vesicles within the cytoplasm.

The release of LH and FSH from the gonadotrophs following appropriate stimulation is by exocytosis, which occurs when vesicle proteins fuse with proteins in the cell membrane. Once in the general circulation, the gonadotrophins are transported in the plasma, with little or no binding to plasma proteins, to their target cells in the gonads.

As with other anterior pituitary hormones, gonadotrophin release is pulsatile, with a frequency of release being of the order of 1–2 pulses per 6 h. This pulsatile release is essential for their physiological activity, largely brought about by the similarly pulsatile release of GnRH from the hypothalamus. This is clinically relevant since an intravenous infusion of GnRH (or another GnRH receptor agonist) initially stimulates, but then inhibits, the release of pituitary gonadotrophins.

A recently identified gonadotrophin inhibitory hormone (GnIH) called RFamide-related peptide-3 (RFRP-3) has an inhibitory effect on GnRH-stimulated LH pulses. The neurones producing this hormone appear to be limited to the dorsomedial nucleus in the hypothalamus, at least in rodents. Fibre projections are mainly to GnRH neurones within the hypothalamus, indicating a central inhibitory role on GnRH release. However, there is some evidence to suggest that there are fibres terminating in the median eminence and it is possible that a direct effect is exerted on the gonadotrophs of the adenohypophysis.

Receptors and mechanism of action

The LH receptor is a large G protein–coupled receptor found in the gonadotrophin target cells, mainly in the testes and ovaries as well as uterus, where it also acts as the receptor for the structurally very similar glycoprotein placental hormone hCG. LH stimulates specific steroid hormone synthesis in its target cells.

The ligand binds to the highly glycosylated extracellular domain of the LH receptor. Binding to the receptor activates adenyl cyclase which stimulates production of the second messenger cyclic AMP (cAMP). This molecule in turn activates protein kinase A which then phosphorylates further intracellular proteins which bind to cyclic AMP response elements (CRE) on the DNA resulting in the synthesis of specific proteins which induce the physiological response to the initial LH (or hCG) stimulus. One essential protein which is activated following LH action is steroidogenic acute regulatory (StAR) protein which is involved in the translocation of the precursor steroid molecule cholesterol from outer to inner mitochondrial membrane, this being the rate-limiting step in steroid hormone synthesis.

The FSH receptor is another G protein–coupled transmembrane receptor found in the cell membranes of its target cells. Like LH, FSH binds to the glycosylated outer domain of the receptor, which then activates adenyl cyclase resulting in the synthesis of cAMP. Consequently, intermediary activated protein kinases and other intracellular molecules act as transcription factors which bind to specific cAMP response elements on DNA mediating the transcription of new proteins which induce the physiological actions associated with FSH.

The main LH and FSH target cells in the testes are the Leydig and Sertoli cells, respectively. In the ovaries they are the thecal and granulosa cells in the follicles, and the luteal cells in the corpus luteum. The actions these hormones induce in their target cells will be discussed more fully in Chapters 6 and 7 where male and female gonadal functions are considered.

Physiological effects of LH and FSH

LH and FSH are both involved in the regulation of steroid and gamete production in the testes and ovaries. Indeed they act in collaboration with each other in order to promote these gonadal effects, namely, the stimulation of steroidogenesis and gametogenesis.

In males

The adult testes consist mainly of Sertoli cells, which are linked together to form integrated structures called seminiferous tubules, and the Leydig cells which lie scattered in the interstitial spaces between the tubules. The Sertoli cells play an essential role in gametogenesis, which is the production of gametes called spermatozoa in males, a process specifically called spermatogenesis. The Leydig cells are the source of the main steroid hormone produced by the testes, the androgen testosterone. LH stimulates androgen production by the Leydig cells, while FSH stimulates the Sertoli cells which are intimately involved with the various stages of spermatogenesis. Indeed, FSH is required in order to begin the process of spermatogenesis, but once started the androgen testosterone is required as well. One protein product from the FSH-stimulated Sertoli cells is a hormone called inhibin.

In fetal males, the gonadotrophins are essential because they stimulate androgen production from the early Leydig cells, and also stimulate the Sertoli cells once they have been derived following activation of a specific sex-related gene on the Y chromosome (the *Sry* gene). The endocrine basis for sex determination is considered in Chapter 6.

In females

Each ovary consists of a stroma in which are found follicles containing ova, at various stages of development. The follicular cells consist of two types: the outer thecal cells, and the inner granulosa cells surrounding the ova. The thecal cells synthesise membrane receptors specific for LH (and hCG), and when stimulated by LH they produce androgens from the initial precursor cholesterol. In contrast, the granulosa cells synthesise FSH receptors and respond to this hormone by activating the enzyme aromatase which converts androgens reaching them from the outer thecal cells to oestrogens. The main oestrogen produced by the granulosa cells during the menstrual cycle (see Chapter 7) is 17β-oestradiol. The granulosa cells also synthesise inhibin following stimulation by FSH.

Control of release

Both LH and FSH are released in pulses and this form of release, controlled by the hypothalamic GnRH, is essential for their stimulatory effects on the gonads which are the main target tissues. However, overall control of, and differentiation between, LH and FSH production in the anterior pituitary are clearly related to the integration of signals from hypothalamic and peripheral factors. As mentioned earlier, the physiological importance of gonadotrophin inhibitory hormone (GnIH, also known as RFRP-3, remains to be clarified. It clearly inhibits the release of gonadotrophins from the anterior pituitary but it is likely that the main effect is by inhibiting the release of GnRH, maybe having a more specific effect on LH, rather than FSH, release.

In males, testosterone has direct and indirect negative feedback effects on the anterior pituitary gonadotrophs and the hypothalamus, respectively. It is likely that at the cellular level the main influence is exerted by the more potent androgen, dihydrotestosterone (DHT), produced from testosterone by the action of 5α-reductase in peripheral target cells. Inhibin from the Sertoli cells has a specific inhibitory effect on the production of FSH from the anterior pituitary, exerting direct and indirect negative feedback effects at adenohypophysis and hypothalamus respectively (Figure 3.13).

In women, oestrogens (and androgens) normally inhibit LH and FSH production via direct and indirect negative feedback loops to the anterior pituitary and hypothalamus (on GnRH production) respectively. There is evidence suggesting that RFRP-3 (GnIH) mediates the negative feedback influence of oestrogens, at least. In contrast, inhibin also has inhibitory effects, but specifically on FSH production (Figure 3.13). Activins stimulate gonadotrophs by autocrine and/or paracrine effects. During the luteal phase of each menstrual cycle, progesterone is produced by the corpus luteum (see Chapter 7) and this steroid also exerts a powerful negative feedback influence on LH and FSH production, acting by direct and indirect negative feedback loops at the anterior pituitary and hypothalamus, respectively. In addition to the normal negative feedback loops controlling LH and FSH release, however, the ovary differs from all other known endocrine systems by exerting an unusual positive feedback effect on both the anterior pituitary and hypothalamus. This is produced by 17β-oestradiol under very specific circumstances (once each menstrual cycle, at the end of the follicular phase). The consequence of the transient positive feedback loops is an LH (and an FSH) surge which is essential in order to promote the final process of ovulation.

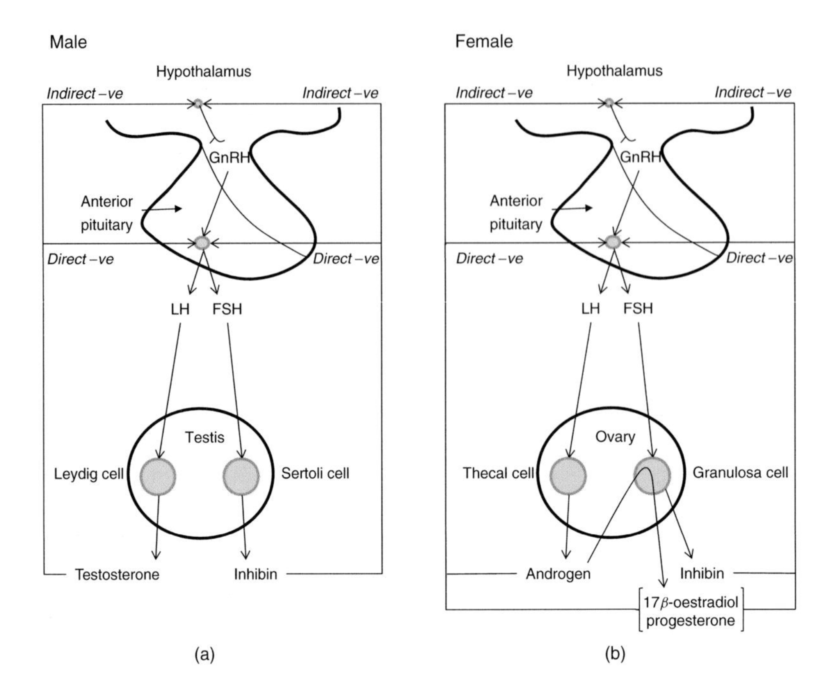

Figure 3.13 The negative feedback loops relating the hypothalamo-anterior pituitary axis with specific target cells in the gonads: (a) testes in males, and (b) ovaries in females.

This differential control over the hypothalamo-adenohypophysial system during each menstrual cycle is considered in more detail in Chapter 7.

Thyrotrophin

Synthesis, storage, release and transport

Thyrotrophin is synthesised in specific adenohypophysial cells called thyrotrophes, which make up approximately 5% of the cells of the anterior pituitary mainly in the anteromedial region. They tend to be smaller than other cell types, are irregularly shaped and have relatively small secretory granules. The common α-subunit is synthesised in thyrotroph, gonadotroph and placental cells, but its regulation is specific to each cell type. Thus the synthesis of the TSH α-subunit in thyrotrophs is specifically inhibited by triiodothyronine (T3; see Chapter 13) while the synthesis of the β-subunit (112 amino acids) is inhibited by the action of the intracellular iodothyronine receptor acting directly on the gene, which is located on chromosome 1. Glycosylation allows the molecule to fold and for the two subunits to link to each other. Hypothalamic TRH enhances the glycosylation process, while T3 tends to inhibit it. Fetal thyrotrophin is detectable by the 12th week of development.

The secretion of thyrotrophin is pulsatile as with the other anterior pituitary hormones, but because the pulses (frequency approximately every 2–3 hours interspersed with low basal secretion) are normally low in amplitude and the half-life of the molecule is relatively long (54 minutes), the diurnal variability of the hormone in the blood is somewhat reduced, with the peak levels appearing from around midnight to the early hours of the morning.

Release of thyrotrophin from the thyrotroph cells is by exocytosis. In the blood, this large glycosylated protein circulates mainly in the free unbound state, although there may be some binding to plasma proteins.

Thyrotrophin receptors and mechanism of action

Thyrotrophin (TSH) binds to its G protein–linked receptors on the plasma membranes of its target cells, which are mainly the follicular cells of the thyroid, but they are also found in the CNS. The mechanism of action of TSH is mediated by the intracellular second messenger cyclic AMP, formed by the action of the enzyme adenyl cyclase on adenosine triphosphate. The subsequent intracellular steps ultimately are associated with the synthesis and/or activation of intracellular proteins (e.g. thyroglobulin and hydrogen peroxidase) and the activation of the iodide pump which transports iodide from the plasma into the follicular cell against an electrochemical gradient.

There are other ligands for the thyrotrophin receptor which can mediate abnormal, uncontrolled stimulation through long-term binding: immunoglobulins such as the long-acting thyroid stimulator (LATS) which is an auto-antibody which targets the thyrotrophin receptor and produces excessive stimulation of the follicular cells. This massive stimulation of the follicular cells by the auto-antibodies is associated with the hyperthyroid condition called Graves' disease (see Chapter 13).

Physiological effects of thyrotrophin

Thyrotrophin (TSH) stimulates the general growth (activity) and maintenance of the follicular cells which form the follicles comprising the thyroid gland. It has various actions on the follicular cells, including the stimulation of the iodide pump, and the synthesis of the enzyme hydrogen peroxidase which is located in the cell (apical) membrane facing the central colloid. In addition, it stimulates the synthesis of intracellular thyroglobulin protein, the iodination of the thyroglobulin, the process of endocytosis by which mechanism iodinated thyroglobulin is brought back into the follicular cells from the colloid, and the migration of lysosomes towards the endocytosed colloid. The overall effect of TSH is to stimulate the synthesis, storage and release of the iodinated hormones, the iodothyronines, which are considered in detail in Chapter 13.

Control of release

The synthesis and release of thyrotrophin are mainly under the control of hormones from the hypothalamus. However, there are various other influences on the thyrotrophs including other hormones, a neural input and negative feedback loops.

Hypothalamic hormones

Thyrotrophin-releasing hormone (TRH): This tripeptide, sometimes called thyroliberin, is released from nerve terminals in the median eminence and is transported to its target cells in the anterior pituitary by the hypothalamo-adenohypophysial portal system. Its main target cells are the thyrotrophs but they also stimulate the lactotrophs. The TRH receptors belong to the family of G protein–coupled receptors (GPCRs) which span the membranes of their target cells. The second messenger system involves the membrane-bound enzyme phospholipase C which acts on membrane phospholipids to produce inositol triphosphate (IP3) and diacylglycerol (DAG) which, in turn, activate other components of the intracellular pathways. As with other hypothalamic releasing hormones, the stimulatory effect of TRH on thyrotrophin production is dependent on the pulsatility of its release.

Somatostatin: Somatostatin and its analogues have been successfully used in patients with thyrotrophin-producing pituitary tumours. The mechanism of action is unclear, but studies on *in vitro* cell preparations suggest that it reduces thyrotrophin production in response to TRH either by decreasing TRH receptor numbers on thyrotrophs or by decreasing the intracellular inositol triphosphate response to TRH. It also dampens the nocturnal increase in thyrotrophin release, partly through an effect on TRH neurones and partly by its influence on the thyrotrophs.

Other (peripheral) hormones

Various hormones appear to influence the hypothalamo-adenohypophysial -thyroid axis, in particular oestrogens such as 17β-oestradiol. This oestrogen appears to have a stimulatory effect on thyrotrophin production from *in vitro* anterior pituitary cell preparations but at least part of the stimulatory effect is indirect, via stimulation of TRH from the hypothalamus. It has also been suggested that part of the oestrogen effect on TRH release is due to an inhibition of the negative feedback exerted by iodothyronines. Glucocorticoids also influence thyroidal activity, particularly in fetal and early developmental stages of life, at least partly by a central action resulting in increased intracellular triiodothyronine (T3) formation in brain tissue. This effect could be relevant to the early maturation of the fetal brain.

Feedback control

The hypothalamo-pituitary-thyroid axis involves an important self-regulatory negative feedback system involving direct, indirect and short feedback loops. The thyroidal hormones, the iodothyronines thyroxine (T4) and triiodothyronine (T3), have a direct negative feedback effect on the anterior pituitary production of thyrotrophin (TSH) and an indirect negative feedback effect on the hypothalamic production of TRH. In addition, thyrotrophin exerts a short (auto) feedback effect on TRH production. Thyroxine can be deiodinated to the more active T3 molecule in many target cells including the adenohypophysial thyrotrophs.

The polypeptide corticotrophin (ACTH)

Synthesis, storage and release

Corticotrophin, also known as adrenocorticotrophic hormone (ACTH), is the only main adenohypophysial hormone which is small enough to be called a polypeptide as it is a single chain of only 39 amino acids. However it is initially synthesised as part of a much larger precursor protein molecule, 241 amino acids long, called pro-opiomelanocorticotrophin (POMC). This prohormone is synthesised mainly in the corticotroph cells of the anterior pituitary, but also in a few other cells, for example in certain hypothalamic neurones, in melanocytes of the intermediate lobe and in the placenta. As its name indicates, POMC is a precursor for opioids, melanocyte-stimulating hormone (MSH) and ACTH. The corticotrophs make up approximately 20% of the anterior pituitary, mainly in the central portion of the lobe. They are quite large and irregular in shape, with many granules in their cytoplasm. The POMC gene is located on chromosome 2. POMC has eight tissue-specific potential cleavage sites, some of which are acted upon in corticotrophs by a specific convertase enzyme. Cleavage occurs within the secretory granules following enzymatic action. Initially two molecules are produced by the initial cleavage: a precursor for ACTH, and β-lipotrophin (or β-lipotrophic hormone, β–LPH). The ACTH precursor is then further processed to produce ACTH itself and a pro-γ MSH which is subsequently cleaved to produce γMSH. The βLPH is also cleaved in tissues such as the intermediate lobe, certain hypothalamic neurones and placenta to produce γ-LPH and β-endorphin (see Figure 3.14). Beta endorphin is the precursor of an enkephalin, which is an endogenous molecule having morphine-like properties, and βMSH. Corticotrophin itself can be cleaved to form αMSH and corticotrophin-like intermediate peptide (CLIP). Peptide breakdown products of POMC such as ACTH and αMSH are called melanocortins. Their receptors are called melanocortin receptors, and they mediate various effects of the melanocortins and other ligands in target tissues such as the

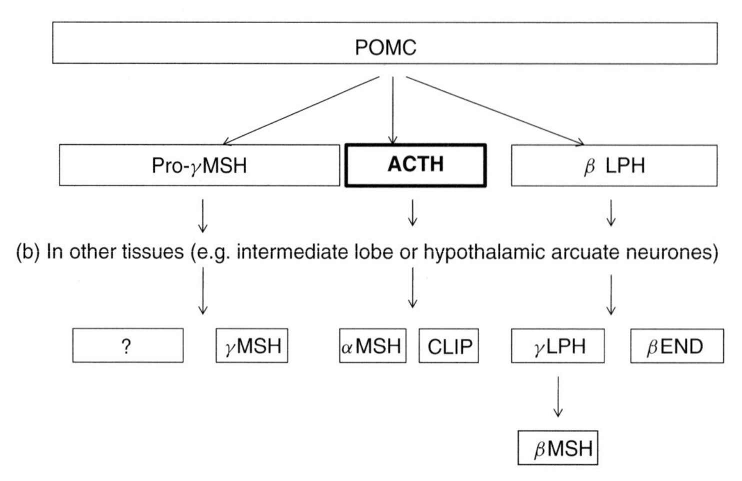

Figure 3.14 The cleavage products produced from the precursor pro-opiomelanocorticotrophin (POMC) in the corticotrophs of the anterior pituitary, and the further processing that takes place in other cells such as certain hypothalamic neurones and placenta (see text for details).

brain where they have important roles in the regulation of appetite and satiety (see Chapter 15).

The pigmentation associated with excessive production of the POMC precursor (as seen in Cushing's disease; see Chapter 5) has in the past been associated with the production of the various MSH molecule derivatives. However, it now seems to be caused by excessive circulating ACTH itself, binding to melanocortin receptors (MC1R) on the melanocytes.

When the corticotroph cells are appropriately stimulated, the secretory granules fuse with the cell membrane and the contents are released by exocytosis into the surrounding extracellular fluid and subsequently the general circulation. In the blood, corticotrophin does not bind to any plasma proteins, so it is transported to its target tissues as a free polypeptide. As with other adenohypophysial hormones, the release of ACTH is pulsatile. Furthermore, under normal circumstances, there is a clear circadian rhythm in its release, with the greater pulse peak concentrations being around the time of waking up in the morning (around 7:00–8:00 AM) and the lowest circulating levels around midnight. This circadian pattern of release is lost or greatly diminished when there is an excessive production of ACTH from an anterior pituitary adenoma. Clinically, this can be a useful indicator in determining the precise aetiology of Cushing's syndrome (see Chapter 10).

Corticotrophin receptors and mechanism of action

ACTH binds to its specific receptor which is a member of a group belonging to the rhodopsin family of seven-transmembrane, G protein–coupled receptors. That group comprises five different receptors for melanocortins (MCRs) such as ACTH and αMSH. Melanocortin receptor type 2 (MC2R) is selectively stimulated by ACTH. MC2R is found on the plasma membranes of cells in the inner two zones of the cortical regions of the adrenal glands, the zonae fasciculata and reticularis (see Chapter 10).

Once corticotrophin has bound to its receptors on the target cells, adenyl cyclase is stimulated and consequently cAMP is produced from ATP. This second messenger then induces intracellular pathways which mediate the actions of the hormone, these being essentially the synthesis of new proteins, such as the P450 enzymes necessary for the conversion of cholesterol to various steroids within adrenocortical cells, and other actions. In particular, ACTH stimulates the mitochondrial transfer of cholesterol, and its subsequent rate-limiting conversion to pregnenolone.

Physiological effects of corticotrophin

Corticotrophin action results in the increased synthesis of glucocorticoids, and to a lesser extent the adrenal androgens, from their precursor steroids, in the two inner zones of the adrenal cortex, the zonae fasciculata and reticularis (see Chapter 10). In humans, the principal glucocorticoid is cortisol. The hormone aldosterone, which is solely synthesised in the outer zona glomerulosa cells of the adrenal cortex, is only indirectly influenced by corticotrophin which enhances synthesis of its precursor. In addition to its effect on steroid synthesis, it also has a short negative feedback effect on the hypothalamic production of the relevant releasing hormones corticotrophin releasing hormone (CRH) and vasopressin (VP), and may even have an effect (albeit an ultra-short one) on its own production by acting on the corticotrophs in the adenohypophysis.

Control of corticotrophin release

The precise control of ACTH synthesis and release, as with other adenohypophysial hormones, is quite subtle. One major influence is exerted by the hypothalamus which releases at least two relevant neurosecretions into the primary capillary plexus in the median eminence. These are CRH and VP, both of which are expressed in, and released from, the same hypothalamic parvocellular neurones originating in the paraventricular nuclei. Available evidence indicates that the two hormones act synergistically, but there are specific situations (e.g. the influence of different stressors) when either one of these hypothalamic hormones has a dominant role. Stressors can be endogenous or exogenous, acute or chronic, and they can activate different pathways. Thus hypoglycaemia (an endogenous stressor) will involve

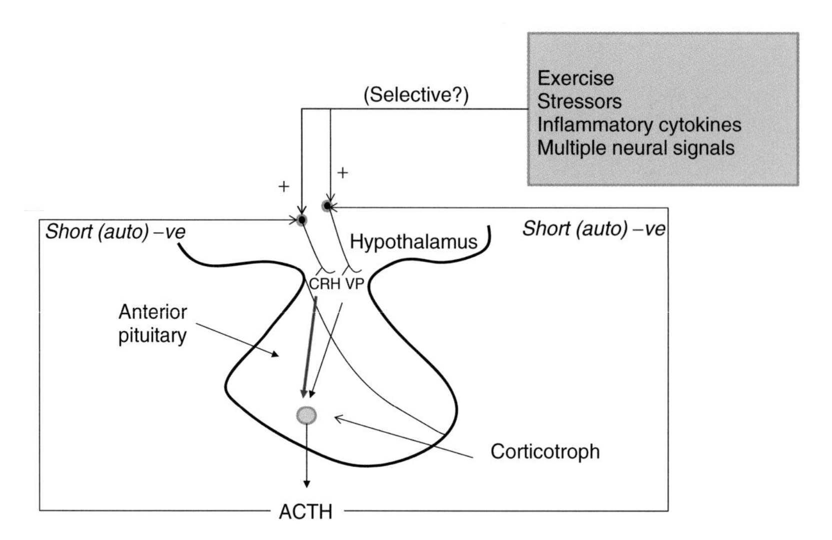

Figure 3.15 The hypothalamo-pituitary-adrenal (HPA) axis and its regulation.

a mass response from a variety of hormones, including corticotrophin, whose effects will indirectly via cortisol include a rise in blood glucose concentration. In contrast, an infection (an exogenous stressor) will involve not only the endocrine system, specifically the hypothalamo-pituitary-adrenal (HPA) axis, but also the immune system, the latter producing a variety of molecules such as certain cytokines which can modulate the overall response, at least partly by influencing the HPA axis. Exercise is a potent stimulus of corticotrophin release and highly trained athletes can have raised baseline circulating ACTH and cortisol levels. Other potent stress stimuli include haemorrhage, surgery and acute illness, the latter stimulus also being associated with the temporary loss of the normal circadian pattern of release. Psychological stressors, including emotional deprivation, also increase the circulating levels of ACTH (and cortisol).

Another important influence on ACTH production is provided by the negative feedback loops exerted by the hormonal response to its actions, namely, cortisol which is the principal glucocorticoid synthesised in the target tissue, the adrenal cortex. Cortisol exerts a direct negative effect on ACTH production by an action on the adenohypophysial corticotrophs as well as by indirect negative feedback on the hypothalamic neurones producing CRH. Furthermore, ACTH itself exerts a short (or auto) negative feedback effect on CRH release by an action on the hypothalamus (Figure 3.15).

The clinical state of psychological depression can also be associated with raised ACTH and cortisol levels in the circulation, as well as a loss of the normal circadian rhythm of release. One explanation for this is that there is a decreased sensitivity to the negative feedback effect exerted by cortisol at the hypothalamo-pituitary level. A major uncertainty remains about whether the overexpression of the HPA axis is causative or consequential.

Further Reading

Buckingham, J. (2010) Understanding the role of vasopressin in the hypothalamo-pituitary adrenocortical axis. In *Perspectives on Vasopressin*, ed. J. Laycock, pp. 230–56. Imperial Academic Press, London.

Dardenne, M., Smaniotto, S., de Mello-Coelho, V., Villa-Verde, D.M. & Savino, W. (2009) Growth hormone modulates migration of developing T cells. *Annals of the New York Academy of Science*, 1153, 1–5.

Krieksfeld, L. J., Gibson, E. M., Williams, W. P. II, Zhao, S., Mason, A. O., Bentley, G. E. & Tsutsui, K. (2010) The roles of RFamide-related peptide-3 in mammalian reproductive function and behaviour. *Journal of Neuroendocrinology*, 22, 692–700.

Yu-Lee, L.-Y. (2002) Prolactin modulation of immune and inflammatory responses. *Recent Progress in Hormone Research*, 57, 435–55.

CHAPTER 4

The Pituitary Gland (2): The Posterior Lobe (Neurohypophysis)

Embryological derivation and general structure

As mentioned in Chapter 3, the posterior lobe of the pituitary gland is derived from a downward growth of neural tissue from the developing brain. The embryological origin of the neurones is from neuroepithelial cells of the lining of the third ventricle which develop into nerve cells migrating down towards the optic chiasma and laterally to the paraventricular region close to the third ventricle, forming the supraoptic and paraventricular nuclei (SON and PVN), respectively. The posterior pituitary is comprised of the nerve axons descending through the pituitary stalk from their cell bodies in the hypothalamus, and numerous glial cells called pituicytes which make up the bulk of the neural lobe. This posterior, neural lobe of the pituitary gland is also called the neurohypophysis and is separated from the anterior lobe (adenohypophysis) by an intermediate lobe (the pars intermedia) which in humans is pretty well non-existent, except in pregnant women when there is a limited growth of this region.

The nerve axons penetrating the neural lobe are larger than those of normal neurones and are generally described as being magnocellular in order to distinguish them from normal (parvocellular) neurones. They are unmyelinated fibres, and have the distinguishing feature of swellings along the axons, particularly near the nerve terminals, called Herring bodies. The nerve terminals, and indeed many of the Herring bodies, are in close contact with the walls of a capillary network in the neural lobe. The capillaries are typically fenestrated (i.e. have 'windows') allowing for the passage of molecules released from the nerve endings, and probably from the Herring bodies, into the general circulation. Arterial blood reaches the posterior lobe capillary network from the inferior hypophysial artery which derives from branches of the posterior communicating and internal carotid arteries. Capillary blood then drains out via the cavernous sinus into the jugular veins.

Integrated Endocrinology, First Edition. John Laycock and Karim Meeran.

The magnocellular neurones have their cell bodies generally grouped together in two paired nuclei in the hypothalamus, the PVN and SON, which as their names suggest are close to the walls of the third ventricle and the optic chiasma, respectively. Each magnocellular neurone is associated with the synthesis of a polypeptide hormone which is released into the general circulation. That polypeptide is either vasopressin (VP) or oxytocin (OT), so the neurones are described as being either vasopressinergic or oxytocinergic. Most (approximately 80%) of the SON neurones projecting down to the neural lobe are vasopressinergic. The paraventricular nuclei, unlike the supraoptic nuclei, are actually composed of not only magnocellular neurones terminating in the neural lobe but also parvocellular neurones which project either to the median eminence (see Chapter 3) or to other parts of the brain. The neurosecretions from the parvocellular fibres terminating in the median eminence reach the cells of the anterior pituitary via the hypothalamo-adenohypophysial portal system described in Chapter 3. These neurosecretions act as hormones on their target cells in the adenohypophysis, the main one being vasopressin which acts as a corticotrophin-releasing factor on corticotroph cells. However, other parvocellular neurones of the PVN synthesise and release other molecules such as corticotrophin-releasing hormone (CRH), thyrotrophin-releasing hormone (TRH) and somatostatin, all acting on target cells in the adenohypophysis. In contrast, the parvocellular axons going to other parts of the central nervous system (CNS) release their neurosecretions across synapses to other neurones and hence they act as neurotransmitters or neuromodulators here (see Figure 4.1). The main central projections are to the limbic system, the brainstem and the spinal cord.

There are a few smaller groups of parvocellular vasopressinergic neurones originating elsewhere in the anterior hypothalamus, one particularly interesting area being the suprachiasmatic nucleus. This group of neurones, being associated with the circadian rhythms of various physiological systems in the body, is sometimes described as the region of the 'biological clock'. An important input to the suprachiasmatic nuclei is from the retinae of the eyes.

Synthesis, storage, release and transport of neurohypophysial hormones

The two hormones associated with the posterior pituitary, vasopressin and oxytocin, are both nonapeptides synthesised in the neuronal cell bodies of the hypothalamic SON and PVN. The genes for both hormones are located on chromosome 20 in humans, in a tandem arrangement but in reverse order regarding their transcription. The initial translation precursors are pre-prohormones which then enter the endoplasmic reticulum, losing the

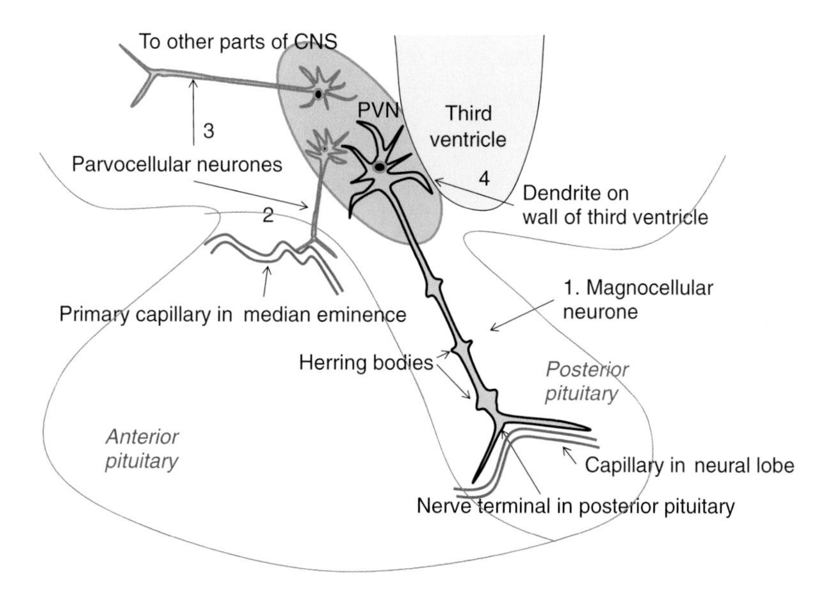

Figure 4.1 The hypothalamo-neurohypophysial system, illustrating magnocellular projections to the posterior pituitary (1), and parvocellular projections to the median eminence (2) and other areas of the CNS (3). PVN = paraventricular nuclei. Some PVN neuronal dendrites terminate on the walls of the third ventricle (4).

signal peptide in the process. The prohormones are then incorporated into vesicles in the Golgi complex where they are further processed. Provasopressin is processed into vasopressin, a 93–amino acid neurophysin II and a 39–amino acid glycopeptide called copeptin (see Figure 4.2). Pro-oxytocin is similarly processed within the vesicles to just oxytocin and a slightly different neurophysin I. Within the granules, the neurophysins remain bound to the nonapeptide hormones in the presence of the acidic pH of 5.5.

The vesicles are transported down the nerve axons by axoplasmic flow, involving the intracellular microtubular network and protein 'motor' molecules kinesin and dynein. This is a similar process in all neurones. They accumulate in the Herring bodies and the nerve terminals to provide stores of hormone ready for release. When the neurone is stimulated, an action potential is propagated down the axon causing an influx of calcium ions across the nerve terminals. The calcium ions, in conjunction with the microfilament network in the cytoplasm, bring about the movement of vesicles towards the cell membrane to which they fuse. The subsequent release of vesicle contents by exocytosis is accompanied by the separation of hormone from neurophysin in the presence of the more alkaline pH of the extracellular fluid and plasma. Accompanying the release of vasopressin and neurophysin II in equimolar amounts is the copeptin molecule.

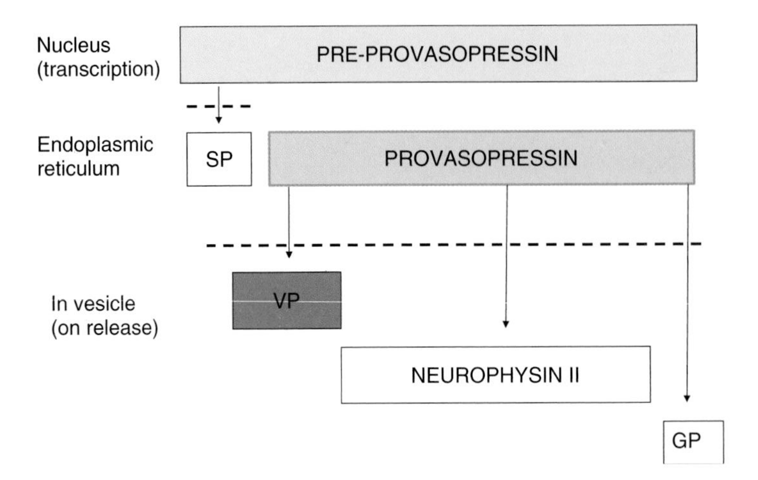

Figure 4.2 Diagram illustrating the various enzymatic cleavages of the initial pre-pro-vasopressin molecule. SP = signal peptide; VP = vasopressin; GP = glycopeptide (copeptin).

Currently, it is believed that the Herring bodies and even undilated parts of the nerve axon can also release the molecules by exocytosis. Furthermore, there is increasing evidence indicating that they can also be released at dendritic endings. This becomes of physiological importance because many of the PVN neurones have dendritic endings abutting the walls of the third ventricle. The potential release of the hormones into the cerebrospinal fluid of the third ventricle could explain some of their central effects, particularly in those parts of the brain devoid of vasopressinergic and oxytocinergic nerve endings but having abundant receptors (discussed in this section). The two nonapeptides each consist of a ring of six amino acids, with two cysteines linked together by a disulphide bridge, and a short chain of three amino acids. They only differ from each other by two amino acids, one in the ring and one in the chain (see Figure 4.3). The presence of the disulphide bridge in either molecule was originally believed to be necessary for the induction of the hormone's mechanisms of action. This is now known to be erroneous, since ligands have been synthesised which lack the bridge but can still bind to the receptors and exert physiological actions.

The firing rates of the magnocellular neurones are characteristic for either vasopressinergic or oxytocinergic fibres, with different patterns of phasic firing rates under basal or stimulated conditions. Although individual neurones fire independently of each other in bursts, the hormone levels in the blood nevertheless can increase or decrease overall because the overall firing rates increase or decrease with the changing stimulus. Thus,

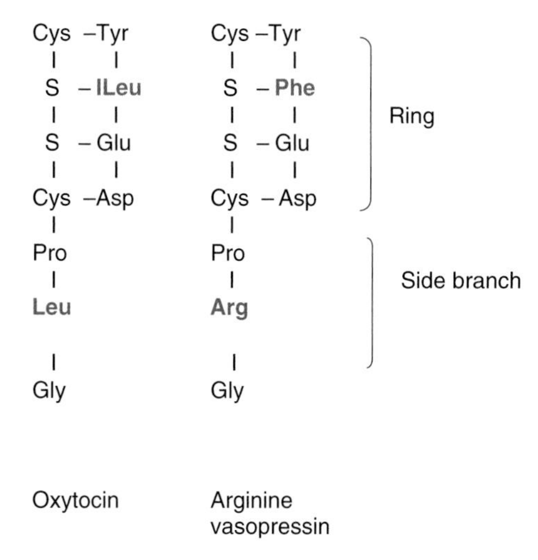

Figure 4.3 The amino acid structures of oxytocin and vasopressin, illustrating the ring and side branch common to both nonapeptides.

for example, an increased stimulation of vasopressinergic magnocellular neurones is associated with an overall increase in firing rate and a consequent increase in the quantity of vasopressin released into the general circulation in proportion to the stimulus. Furthermore, the hormones can be released not only from the magnocellular nerve terminals and Herring bodies but also from the dendrites into the surrounding extracellular space. Since the neurones also have receptors for these neuropeptides, it is becoming clear that some degree of local coordination of firing activity in adjacent fibres takes place.

Once in the circulation, oxytocin and vasopressin are transported unbound to their target tissues. They have relatively short half-lives, their $t^{1/2}$ being approximately 5 minutes. Most of the vasopressin in the circulation may be located within the platelets, which could have physiological relevance (see section 'Receptors and mechanisms of action').

In addition to the release of these neurosecretions into the blood, whether into the general circulation from the posterior pituitary or into the hypothalamo-hypophysial portal system, they can also be released at nerve endings across synapses within the CNS where they act as neurotransmitters. As mentioned in this section, there is also a third possible means by which these neurosecretions can be transported to their target cells, and this follows their release from dendrites which terminate on the walls of the third ventricle into the cerebrospinal fluid (CSF) which can then transport them to their central receptors.

Vasopressin (VP)

Receptors and mechanisms of action

Vasopressin receptors exist in at least three forms, known as V1a, V1b (also called V3) and V2. All these receptors span the target cell membranes with seven intramembranous domains, three extracellular and three intracellular loops (see Figure 4.4), and are members of a large family of G protein-coupled receptors (GPCRs). The V1a, V1b and V2 vasopressin receptor genes are located on chromosomes 12, 1 and X, respectively. The V1a, V1b and V2 vasopressin receptors consist of 418, 424 and 371 amino acids, respectively. There is a 45% sequence homology in the structures of these vasopressin receptors in humans. It is worth noting that vasopressin also has some small affinity for the oxytocin receptor (OTR) but oxytocin has little if any binding affinity for any of the vasopressin receptors.

When the V1b receptor was first identified as being different from the V1a receptor, the actual structures of the receptors were still unknown; in general, the receptors have many similarities. Differentiation between the two classes of V1 receptor was introduced as a consequence of the relative binding properties of different ligands being synthesised at the time. The nomenclature generally remains as it originally was, so they are still called V1a and V1b receptors, although we now know the structures and the genes involved, and can recognise them as truly different. Nowadays, there is an increasing move to call the V1b receptor the V3 receptor, but the original names remain very much in use.

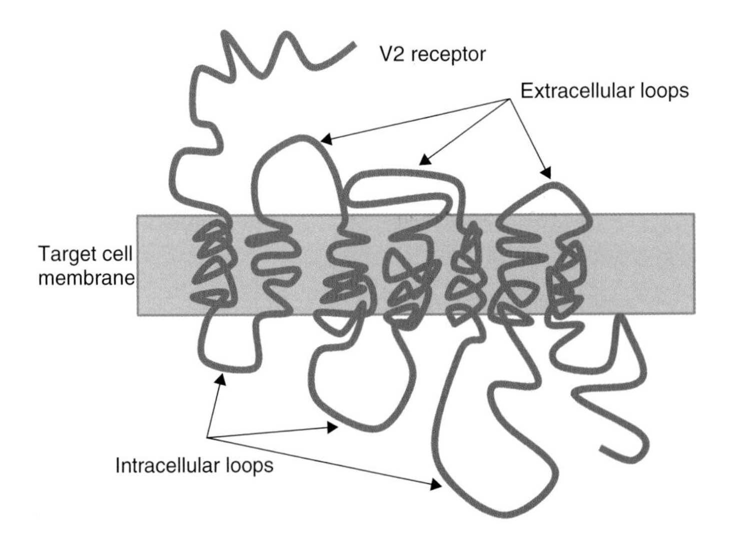

Figure 4.4 The general structure of the vasopressin V2 receptor illustrating the seven transmembrane domains and the three extracellular and three intracellular loops characteristic of this receptor family.

The V1a receptor is found in a variety of vasopressin's target tissues, including vascular smooth muscle, hepatocytes, cardiomyocytes, platelets, the adrenal cortex, the kidneys and the CNS (e.g. the hippocampus, septum and amygdala). In contrast, the V1b receptors are found on the corticotrophs of the adenohypophysis, as well as in the heart, lungs, thymus, mammary glands and the CNS (e.g. in the hippocampus, cortex, thalamus and cerebellum).

On the other hand, the V2 receptor is found in the principal cells of the renal cortical and medullary collecting ducts and the loops of Henle, and probably in vascular endothelial cells. They do not appear to be present in the CNS.

The V1 receptor and its signalling pathways

When vasopressin or another ligand binds to the V1a or the V1b receptor, the GPCR activates the membrane-bound enzyme phospholipase C which catalyses the hydrolysis of a membrane phospholipid called phosphatidylinositol 4,5 bisphosphate (PIP_2). From this precursor, two second messengers are produced: inositol triphosphate (IP_3) and diacylglycerol (DAG). IP_3 opens calcium channels in intracellular calcium stores such as the endoplasmic reticulum, allowing calcium ions to move into the cytoplasm where they can induce various intracellular effects. These include binding to the protein calmodulin, thus activating certain calcium- and calmodulin-dependent kinases which in turn phosphorylate other intracellular proteins (see Figure 1.13). The other second messenger, DAG, also phosphorylates intracellular proteins. One enzyme that is activated by DAG in the presence of calcium ions is protein kinase C (PKC) which then phosphorylates further intracellular proteins. Thus, each second messenger-activated pathway involves multiplying cascade effects within the target cell. Furthermore, the activation of cAMP response element binding protein (CREB) can be induced by calcium- and calmodulin-dependent kinase. CREB can then induce longer term effects by moving into the nucleus where it can influence gene expression and new protein synthesis.

The V2 receptor and its signalling pathway

The V2 receptor is also linked to a G protein, but this time the enzyme which is activated is an adenyl cyclase. This enzyme activates protein kinase A (PKA) which in turn phosphorylates other intracellular proteins including CREB (see Figure 1.14). A key molecule influenced by vasopressin through this V2 receptor–mediated pathway is a member of the aquaporin (AQP) family of molecules. These are proteins which can be inserted into cell membranes where they act as pores through which water can pass, up an osmotic gradient. The principal cells of the renal collecting ducts synthesise a vasopressin-dependent aquaporin (AQP2) which is essential for the

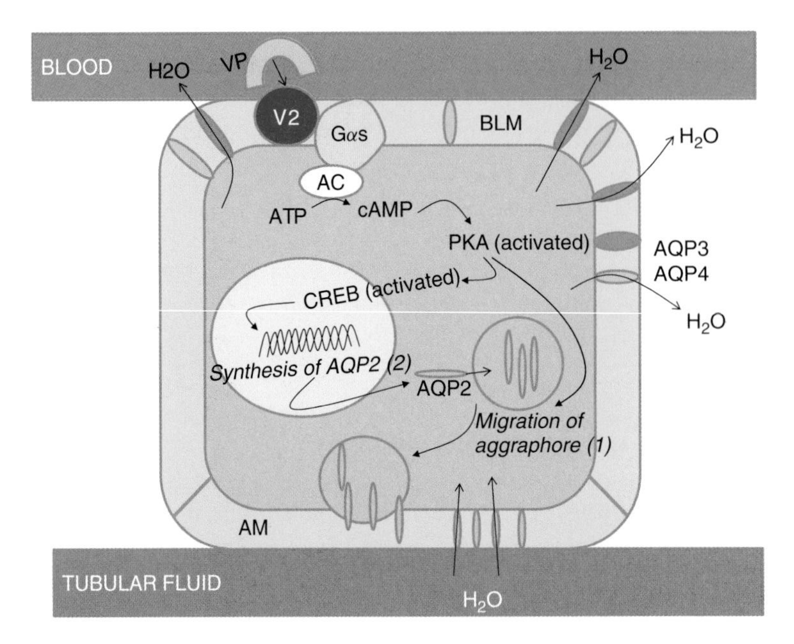

Figure 4.5 Diagram illustrating a collecting duct principal cell and the basic intracellular mechanism by which vasopressin (VP), having bound to its V2 receptor, activates the associated G protein which in turn activates adenyl cyclase (AC). This enzyme induces the formation of cAMP which then activates (phosphorylates) protein kinase A (PKA). AQP = aquaporin; BLM = basolateral membrane; AM = apical membrane and CREB = cAMP response element binding protein.

movement of water from tubular fluid into the cells. Vasopressin induces not only the synthesis of AQP2 but also its transport (within vesicles called aggraphores) to the apical membranes of the principal cells where they are inserted. Water moves out of the principal cells via different aquaporins in the basolateral membranes (AQP3 and AQP4), and one of these (AQP3) is also vasopressin sensitive (see Figure 4.5).

It is worth noting that the migration of aggraphores containing AQP2 to the apical membrane involves microtubules and microfilaments which are present in the region, and which appear to be influenced by vasopressin. As with all other receptors, a controlled trafficking system exists to remove the receptor from the cell membrane and either inactivate it or return it to the membrane for reuse. Furthermore, the fine control of AQP2 to and from the apical membranes is also regulated, at least partly by vasopressin although other factors also influence the process.

Physiological actions of vasopressin

The physiological actions of vasopressin can be related firstly to the sites from which this neurosecretion is released (see Figure 4.6). The magnocellular neurones which terminate within the neurohypophysis

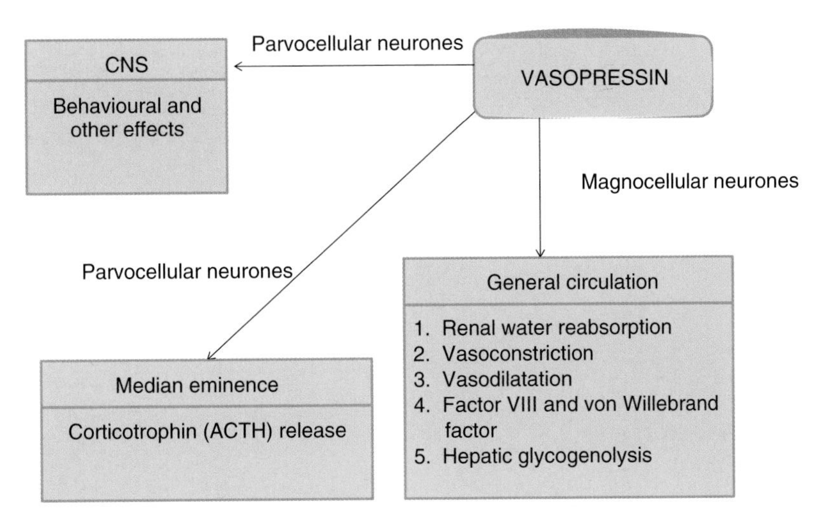

Figure 4.6 Some of the principal actions linked to vasopressin when released into the general circulation, into the primary capillaries of the median eminence, and either released centrally as a neurotransmitter or transported by the cerebrospinal fluid (see section 'Physiological actions of vasopressin' for details).

release vasopressin into the general circulation which transports it to its target cells in tissues throughout the body, for instance in the vasculature and the kidneys. The parvocellular neurones terminating on the walls of the primary capillary plexus in the median eminence release vasopressin into the hypothalamo-adenohypophysial portal system which transports it to its specific target cells, the corticotrophs, in the anterior pituitary. VP participates in the control of corticotrophinm (ACTH) release from the corticotrophs. In contrast, those relatively few parvocellular neurones which have extended axons reaching other parts of the CNS release vasopressin as a neurotransmitter, acting on other neurones across synapses and thus influencing specific brain pathways and activities.

The various actions of vasopressin can also be related to the different vasopressin receptors and their distributions in the various tissues and organs of the body (see Figure 4.7).

Peripheral actions

Antidiuretic effect: The peripheral actions of vasopressin are various, but the principal physiological effect without doubt is its stimulation of the water reabsorption process in the collecting duct, which acts as the final concentrating segment of the renal nephron. Without this regulatory system operating normally, that final concentrating ability is lost and the excretion of large volumes of dilute urine ensues, the clinical condition being called diabetes insipidus, meaning the excretion of large volumes of urine (diabetes) which is mainly water

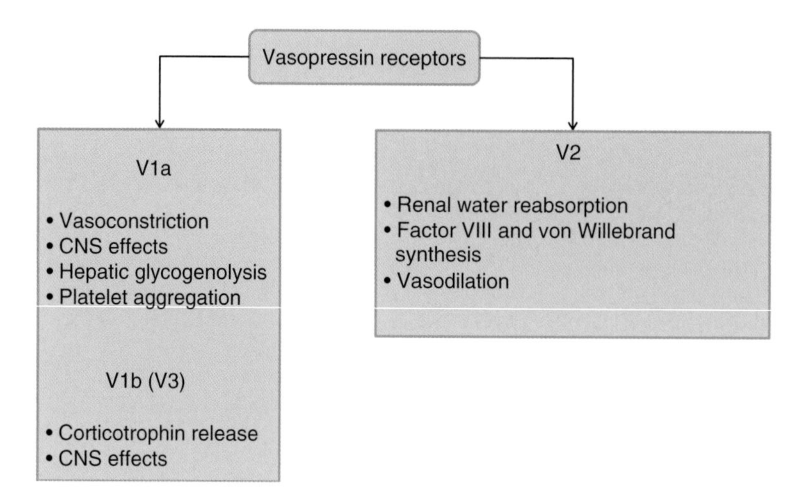

Figure 4.7 Diagram linking the principal actions of vasopressin to its different receptors.

and therefore tasteless (insipidus). Diabetes insipidus is considered in more detail in Chapter 5. Vasopressin binds to V2 receptors on the principal cells of the collecting ducts and initiates cAMP-mediated actions which, via PKA activation (see section 'Physiological actions of vasopressin'), results in the following (see also Figure 4.5):

1. the migration of aggraphores (vesicles) containing AQP2 proteins towards the apical membranes of the cells, where they are inserted. Providing there is a local osmotic gradient across the membrane, water molecules move into the cell from the tubular fluid via these AQP2 channels;
2. the longer term synthesis of new AQP2 molecules via the activation of CREB which enters the cell nucleus and promotes gene expression;
3. the degradation or recycling of the V2 receptor via β-arrestin activation (see Chapter 1);
4. the recycling of AQP2 back into the cell for subsequent degradation or recycling; and
5. the additional synthesis of another aquaporin, AQP3, which enables water to leave the cell via the basolateral membrane.

Other renal effects: Vasopressin has other renal effects, including the stimulation of sodium and chloride ion reabsorption from the ascending limb of the loop of Henle, and the stimulation of urea reabsorption via specific urea transporters in the thin descending limb of the loop of Henle and the collecting duct. These actions are important with regard to the maintenance of the renal countercurrent multiplier system which provides the high medullary interstitial concentration of solute necessary for the concentration of urine.

Vasoconstrictor activity: Vasopressin is one of the most powerful endogenous vasoconstrictors produced by the body. It induces this effect by binding to its V1a receptors in vascular smooth muscle, particularly in arterioles. By increasing vasoconstriction, the total peripheral resistance rises resulting in an increase in mean arterial blood pressure. The extent of the vasoconstrictor effect is tissue specific, being particularly effective in skin and splanchnic bed. Under normal hydrated conditions, however, vasopressin actually has little effect on arterial blood pressure. At least part of the reason is that there are equally powerful compensatory mechanisms which come into play to prevent the arterial blood pressure from rising. However, in situations such as dehydration, haemorrhage and other conditions of volume depletion, the pressor effect of vasopressin is of physiological relevance in maintaining, or raising, blood pressure to normal levels.

Vasodilator activity: One compensatory mechanism which prevents the usual vasoconstrictor activity of vasopressin from manifesting itself under normal circumstances is a kidney-independent vasodilation which decreases total peripheral resistance, resulting in a decrease in blood pressure. This can be observed when vasopressin is administered and V1 receptor activity is blocked with specific antagonists. Since V2 receptors are mainly associated with the kidneys, this suggests that vasopressin exerts its vasodilatory effect directly by an action on the vasculature. Indeed, it would seem that it is due to a V2 receptor–mediated effect on the vascular endothelial cells lining the blood vessels, involving the local release of the vasodilator molecule nitric oxide. This then acts on the adjacent vascular smooth muscle to cause relaxation and hence vasodilation.

Blood coagulation: Another endothelium-dependent V2-mediated vasopressin effect is the stimulation of molecules which participate in blood coagulation, namely, Factor VIII and the von Willebrand factor. This action is of clinical importance, and vasopressin is used in the treatment of certain forms of haemophilia.

Hepatic glycogenolysis: V1a receptors are also found on hepatocytes, and one action that vasopressin has on these cells is the stimulation of glycogenolysis. This is probably not of any great physiological significance, but since vasopressin can be considered to be a stress hormone (e.g. in response to haemorrhage or dehydration), then like all other stress hormones it contributes to the increase in the blood glucose concentration which is an essential element of the normal stress response.

Adenohypophysial action

The release of vasopressin from the parvocellular neurone terminals in the median eminence into the hypothalamo-adenohypophysial portal system is specifically directed at the corticotrophs which have V1b (V3) receptors. Some of these parvocellular neurones release vasopressin while others are believed to release both vasopressin and CRH. Yet others may release CRH only. The action of vasopressin on these cells is stimulatory regarding corticotrophin release, and generally it seems to work synergistically with CRH, greatly potentiating CRH activity. However, it is becoming clear that both peptides can function independently too, so that each hormone can have direct effects of its own. For instance, when corticotrophin is released and has its stimulatory effect on adrenocortical glucocorticoid (e.g. cortisol) production, then cortisol has an indirect negative feedback effect on CRH release. While glucocorticoids can probably also attenuate the release of vasopressin, this hormone may become the main driver for corticotrophin release in response to long-term stressors which are associated with raised circulating glucocorticoid levels and enhanced negative feedback on CRH release.

Central actions

Until recently it was generally assumed that any central effects of vasopressin were associated with those parvocellular vasopressinergic neurones which send their axons to specific regions of the brain such as the hippocampus. As indicated, this was rather perplexing since studies on vasopressin receptor distribution within the brain did not match the (rather sparse) VP-neuronal connections from the PVN. An exciting new development is the discovery that vasopressin is released from dendrites abutting the walls of the third ventricle, and that the concentration of vasopressin within the CSF is higher than in the plasma, the two concentrations seeming to change independently from each other. Thus the vasopressin would appear to be released as a hormone into the CSF which then transports it to various parts of the brain not necessarily innervated by vasopressinergic neurones. Vasopressin's effects on brain function are mediated by either V1a or V1b receptors. The main functions which are influenced by vasopressin are behavioural. There is intriguing evidence which suggests that vasopressin, and in particular V1 receptor distribution, correlates with the expression of certain social behaviours. For example, male prairie voles form a monogamous bond with females after mating, while montaine voles are promiscuous. The male prairie vole then becomes aggressive towards other males after mating. These differences in pair bonding and aggressive behaviours appear to correlate with the differing distributions of V1a receptors within the brain, and not with the distribution of vasopressinergic neurones. However, there are vasopressinergic projections to certain parts of the brain associated with aggressive behaviour, such as from the medial amygdala and the bed nucleus of the stria terminalis to the lateral septum, which also seem to be relevant and sex steroid dependent. Indeed, in rodents, at least some of the central vasopressin effects (e.g. on aggressive behaviour) are found in males and mothers with newly born litters. Male parental care in mice may also be influenced by vasopressin via V1a receptors. Central vasopressin may also play a role in social recognition, certainly in rodents. In humans, various vasopressin receptor gene polymorphisms have been associated with autism which can be described as a condition associated with decreased social functioning and communication, as well as restrictive practices, suggesting that this hormone has a central involvement in the manifestation of these social behaviours.

Other behaviours have also been linked to central vasopressinergic effects, such as learning and memory, as well as drinking behaviour. Whether they are related to hormonal actions of vasopressin transported to relevant parts of the brain by the CSF, or whether they may be linked to vasopressinergic neurone projections, is still a matter for conjecture. Furthermore, there are other central vasopressinergic effects which are likely to be of physiological significance, including an influence on cardiovascular regulation via projections to the brainstem.

Control of vasopressin secretion

As indicated, the most important physiological effect of vasopressin is to provide the ultimate regulation of fluid balance by its action on water reabsorption in the renal collecting ducts. Not surprisingly, the degree of hydration, as indicated by the osmolality of the plasma, is a key controlling factor for vasopressin production. Furthermore, the blood volume, which can also reflect the state of overall hydration, is involved in the control of vasopressin production, with changes in the blood volume being reflected by alterations in blood pressure. In addition, there are other regulatory factors which influence the secretion of vasopressin, mainly acting through neural pathways from other parts of the CNS (see Figure 4.8).

The osmoreceptor control system for vasopressin release is the main mechanism operating during normal physiological conditions. It is highly sensitive, and normal urinary concentrating ability occurs with circulating concentrations up to 12 $pmol.L^{-1}$ (maximum concentration achieved in the region of 1500 $mmol.kgH_2O^{-1}$ in humans). A much more powerful system for controlling vasopressin release is provided by the baroreceptors. However, while much greater plasma concentrations of vasopressin can be attained in response to changes in baroreceptor activity (as much as 500 $pmol.L^{-1}$), the system is much less sensitive and normally operates only when the blood pressure (or volume) has decreased markedly. This is clearly one mechanism for bringing about vasopressin release when the plasma osmolality may be normal but the circulating blood volume is reduced (e.g. during haemorrhage). The central mechanisms involved in

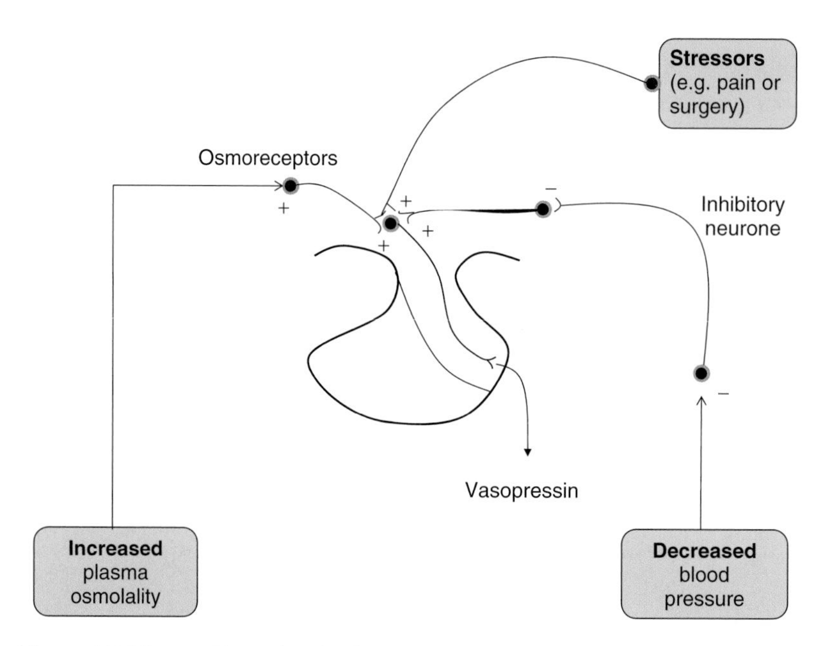

Figure 4.8 Diagram illustrating the three main controlling pathways for vasopressin release (see section 'Control of vasopressin secretion' for details).

regulating vasopressin release (e.g. in response to stressors) also operate without changes in plasma osmolality.

Osmoreceptors and vasopressin

An increase in plasma osmolality is associated with a rise in vasopressin release from the neurohypophysis into the general circulation. The change in plasma osmolality is detected by osmosensitive (stretch-responsive) cells that are in contact with the general circulation but also able to communicate with the vasopressinergic neurones in the hypothalamus. These osmoreceptors exist in certain structures called circumventricular organs which are found in the anterior hypothalamus close to the third ventricle, and which all-importantly have fenestrated capillaries, like the median eminence (see Chapter 2). These osmoreceptors therefore are specific neurones lying 'outside' the blood–brain barrier, hence in contact with diffusible molecules such as sodium ions in the blood. The neural structures are primarily the organum vasculosum of the lamina terminalis (OVLT) and the sub-fornical organ (SFO) from where axons contact the vasopressinergic neurones in the PVN and SON directly or indirectly. They may well also influence other parts of the CNS where they exert effects such as influencing drinking behaviour. The polypeptide angiotensin II, better known as being the product of an enzymatic cleavage of the circulating precursor angiotensin I in the blood (see Chapter 10), is also produced

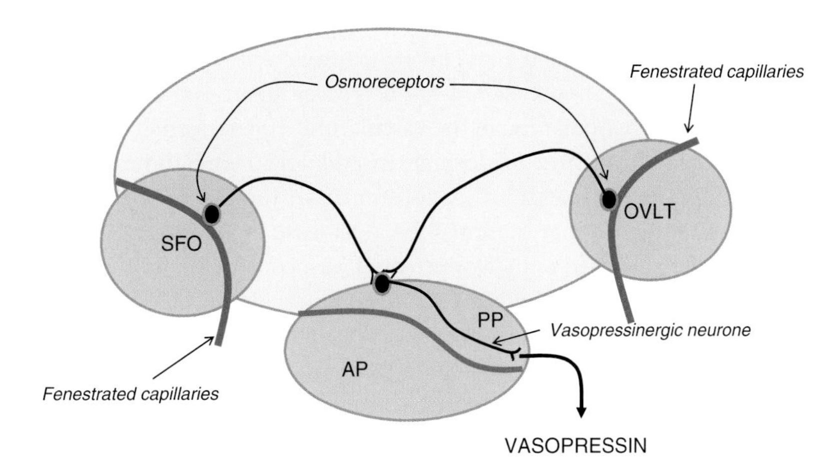

Figure 4.9 The osmoreceptor locations in the circumventricular organs and their relationship with the vasopressinergic neurones in the posterior pituitary. AP = anterior pituitary; OVLT = organum vasculosum of the lamina terminalis; PP = posterior pituitary and SFO = sub-fornical organ.

in the CNS and one specific site is the SFO. Angiotensin II from SFO neurones plays a role in regulating thirst and also has an influence on vasopressin release. Finally, it is worth noting that there is evidence to suggest that the firing rate of vasopressinergic neurones is also increased directly when the osmolality of the surrounding medium is increased, indicating that these neurones may be osmosensitive themselves (see Figure 4.9).

Baroreceptors and vasopressin

Stretch receptors which respond to changes in pressure (baroreceptors) are located in both the high-pressure (arterial) and low-pressure (venous) parts of the circulation. The high-pressure baroreceptors are stretch-sensitive nerve endings located in the walls of the carotid sinus and aortic arch, and their axons pass up the carotid sinus and vagus nerves to the brainstem from which other neurones innervate the cardiovascular centre and the hypothalamus, including the vasopressinergic neurones. In contrast, the low-pressure baroreceptors essentially respond to changes in venous blood volume, and therefore are usually called volume receptors. They are located in the right atrium of the heart, and certain of the large systemic veins such as the vena cava and hepatic veins. They too send their axons up to the brainstem and hypothalamus. Both groups of baroreceptors when stimulated influence not only the autonomic nervous system but also the release of vasopressin. Both the sympathetic nervous system and vasopressin release are inhibited in response to stimulated baroreceptors when the blood pressure (or volume) rises. The response is known as the baroreceptor reflex, and the decreased sympathetic activity occurs simultaneously with an increased parasympathetic activity. Consequently, the cardiovascular response includes a decreased heart rate and force of contraction, and a vasodilation of the vasculature which together normally restore the blood pressure to normal. In addition, stimulation of these receptors inhibits the release of vasopressin from the posterior pituitary, with the consequence that there will be a decrease in water reabsorption from the collecting ducts, and a decreased vasoconstrictor influence on the vasculature, all contributing to a restoration of the blood pressure and volume. Thus, stimulation of the baroreceptors is associated with an inhibition of vasopressin release, and a decreased stimulation, as seen in haemorrhage for instance, will result in an increased release of vasopressin.

Other regulatory factors and vasopressin

Vasopressin is also released by various stressors independent of any changes to plasma osmolality or blood volume. For example, pain and surgical procedures are both associated with very high circulating levels of the hormone. Furthermore, visceral information (e.g. food being eaten) is associated with increased vasopressin release before any food constituents have actually been absorbed, via vagal afferent projections to the brainstem and hence to the hypothalamus.

Oxytocin (OT)

Receptor and mechanism of action

Oxytocin simply has one receptor known as the OTR which is also a member of the family of GPCRs containing seven transmembrane domains. The gene locus for the oxytocin receptor is on chromosome 3. Transcription and subsequent translation produce the 389–amino acid sequence. The mechanism of action of oxytocin involves a G protein which activates the membrane phospholipid enzyme PLC which in turn activates the IP_3–DAG–calcium ion signalling pathways described elsewhere (see Chapter 1). One action which characterises oxytocin's contractile activity in smooth muscle target cells is the activation of myosin light chain kinase by the calcium–calmodulin complex.

Physiological actions of oxytocin

Like vasopressin, oxytocin has peripheral effects but more recently has also been shown to have potentially important central effects, in particular on social and maternal behaviours. Oxytocin is an interesting hormone because it is produced by the magnocellular neurones of males and females equally, but its main accepted physiological effects are seen in females, specifically during pregnancy at term, and during lactation. In males, various actions have been proposed, including the stimulation of ejaculation and the increased production of androgens by the testicular Leydig cells. And now, there is the growing appreciation that oxytocin has behavioural effects, and some of those are likely to be in males as well as females.

Uterine contraction

During the final stage of pregnancy, at the onset of labour, the uterus becomes sensitive to the contractile action of oxytocin, and this is associated with the occasion when oestrogens begin to dominate the local environment and there is an increased synthesis of oxytocin receptors. Until then, the progesterone-dominated uterus has been relatively insensitive to oxytocin and there is a 'hold' on oxytocin receptor synthesis. Oxytocin release is stimulated at this time by the increased stretch of the distended uterus, presumably once the fetus has reached a certain size (maybe determined by the ability of the placenta to provide sufficient nutrients). The stretch is detected by mechanoreceptors in the wall of the uterus and impulses pass up afferent fibres in the inferior hypogastric plexus, ultimately reaching the hypothalamus.

Oxytocin released from the maternal posterior pituitary acts on oxytocin receptors in the myometrial cells lining the uterine wall. The end result of its mechanism of action is the increase in calcium ion concentration within the smooth muscle cells which bind to the intracellular protein calmodulin. The increase in intracellular calcium ion concentration results from the IP3-mediated increased release of calcium from intracellular stores and probably also by an influx from the exterior via the opening of calcium channels which may be ligand or voltage gated. The calcium–calmodulin complex then activates the sliding actin–myosin filaments which bring about contraction of the myometrial smooth muscle (see Figure 4.10). The myometrial cells function to some extent as a syncytium, providing a spreading wave of contraction down the uterus from corpus to cervix. As with vasopressin, oxytocin release occurs in pulses due to specific synchronised bursts of neuronal firing probably controlled by the local release of hormone by dendrites and axonal swellings binding to receptors on the neurones. Furthermore, it is known that oxytocin is also synthesised in the uterus (and in certain other tissues such as ovary, testis, heart and pancreas) where it is likely to have similar localised effects.

The milk ejection reflex

Post-partum, suckling of the nipple by the baby stimulates large pulses of oxytocin release from the posterior pituitary into the general circulation. In an endocrine environment dominated by oestrogen but in the presence

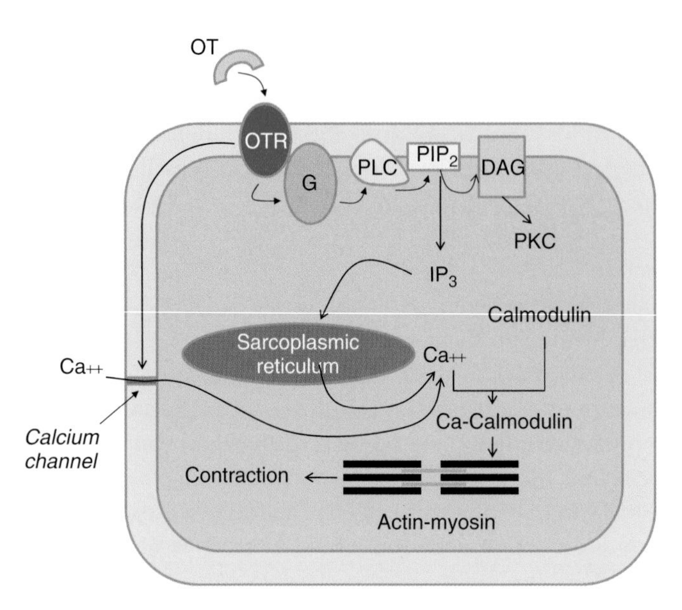

Figure 4.10 Diagram illustrating an oxytocin target cell (either a myoepithelial cell or a myometrial cell) where it initiates contraction. DAG = diacyglycerol; G = G protein; IP3 = inositol triphosphate; OTR = oxytocin receptor; PIP2 = phosphatidylinositol 4,5 bisphosphate; PKC = protein kinase C and PLC = phospholipase C.

of various other hormones including progesterone, iodothyronines and particularly prolactin, the alveoli and ducts of the mammary glands are full of newly synthesised milk. Surrounding the alveoli and the ducts are smooth muscle cells called myoepithelial cells which during lactation express many oxytocin receptors. The oxytocin stimulates the myoepithelial cells to contract in much the same way as the uterine myometrial cells described earlier, and the milk within the alveoli and ducts is ejected from the nipple into the mouth of the suckling baby. This neural-endocrine reflex is called the milk ejection reflex.

Central effects

Oxytocin receptors are found in various parts of the brain, and yet the oxytocinergic fibres extending to the brain are relatively few and restricted in their distribution, and they do not necessarily correspond to the sites where oxytocin receptors are present. This, and the fact that the basal oxytocin concentration in the CSF is generally higher than in the plasma, suggests that the release of this neuropeptide by the dendrites into the third ventricle is of some significance, as it is for vasopressin (see sections 'Synthesis, storage, release and transport of neurohypophysial hormones' and 'Vasopressin').

Oxytocin is associated with various behavioural effects in mothers, including the development of maternal bonding with her new-born infant, and an increased protectiveness towards her young. Females also develop oxytocin-dependent appeasing, less aggressive responses to external threats. Furthermore, oxytocin appears to attenuate endocrine and behavioural responses to various stressors, as indicated by a reduction in glucocorticoid production and less manifestation of behavioural responses such as fear (i.e. oxytocin appears to be anxiolytic). In humans, recent studies indicate that oxytocin may even play a positive role in aspects of behaviour, such as trust and generosity, in both sexes.

Control of release

The uterine contractions induced by oxytocin are believed to be initiated by the stretch of the uterus by the fetus having reached a certain critical size. Mechanoreceptors in the uterine wall are stimulated as it is stretched, and afferent nerve fibres activated. Neuronal connections in the hypothalamus result in the oxytocinergic neurones firing in response. The lactation process also involves an afferent neural pathway, this time originating from stretch-responsive mechanoreceptors around the nipple and terminating ultimately in the hypothalamus (Figure 4.11). In this case, the neural connections are such that not only is prolactin release induced, resulting

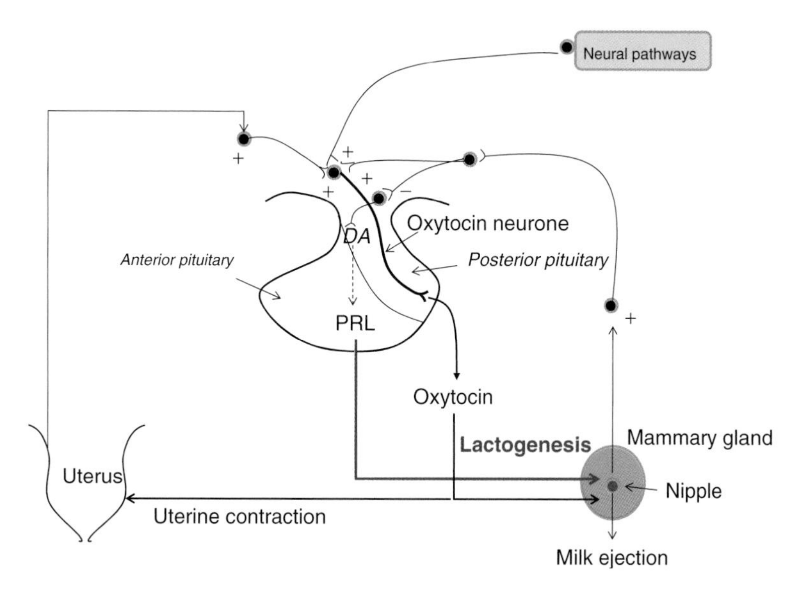

Figure 4.11 Diagram illustrating the control pathways which regulate the release of oxytocin. Also shown is the parallel release of prolactin (PRL) following stimulation of the same stretch receptors around the nipple of the breast (only the dopaminergic (DA) pathway, which is inhibited, is shown; for further details of this control system, see Chapter 3).

in new milk synthesis (see Chapter 3), but so too is oxytocin release, which stimulates the ejection of milk previously synthesised and stored as a consequence of the previous round of nipple stimulation. Thus when the nipple is stimulated, one meal for the hungry baby is released while the next is prepared!

The neural input which directs the central release of oxytocin is less clearly defined but will involve projections from other parts of the brain, including olfactory and visual pathways.

Further Reading

Laycock, J.F. (2010) *Prospectives on Vasopressin*. Imperial College Press, London.

CHAPTER 5

Diseases of the Pituitary Gland

The anterior pituitary (adenohypophysis)

Because the pituitary gland is surrounded by bone, a tumour growing within or near the gland can cause pressure on the cells causing them to fail. Any, or all, of the anterior pituitary cells can stop producing their hormones. Making the diagnosis of pituitary failure is difficult, because the normal levels of adenohypophysial hormones present in the circulation vary. Each of the hormones listed in Chapter 3 can be deficient, and a deficiency of any of them causes different symptoms, but none cause any signs. Pituitary failure is thus one of the most difficult diagnoses to make. The diagnostician can be helped if a number of these hormones are deficient. Once the pituitary gland is imaged with a magnetic resonance imaging (MRI) scan, a visible tumour can be classified as a microadenoma if the tumour is smaller than 10 mm, and a macroadenoma if it is 10 mm or larger. It is also important to realise that non-functioning microadenomas smaller than 6 mm are present in up to 10% of the normal population. These never cause harm because they are too small to cause hypopituitarism. It is also important to note therefore that if small microadenomas of the pituitary are found by chance (e.g. when a patient has a brain scan for something else), the patient can usually be reassured that no intervention is needed, but it is important to exclude them from the other diagnoses in this chapter. This can be difficult! For this reason, there are regional pituitary multidisciplinary teams (MDTs) which consist of several endocrinologists, a dedicated pituitary neurosurgeon, a pituitary radiologist, a radiotherapist and a pituitary histologist. They may be from several sites and meet regularly (often now using tele-radiology) to ensure that all treatment options including drugs and pituitary surgery are considered for all pituitary patients. The Hammersmith Multidisciplinary Endocrine Symposium is an annual national meeting to ensure best practice around the United Kingdom.

Integrated Endocrinology, First Edition. John Laycock and Karim Meeran.

Individual adenohypophysial hormone deficiencies

Corticotrophin (ACTH) deficiency

Corticotrophin is trophic for the zonae fasciculata and reticularis of the adrenal gland. Deficiency is thus associated with variable atrophy of the gland since this is the predominant part of the adrenal cortex. Lack of corticotrophin causes cortisol (i.e. glucocorticoid) deficiency. As there are other glucose-regulating hormones, an isolated adrenocorticotrophic hormone (ACTH) deficiency sometimes is associated only with tiredness, and manifests only with illness at times of 'stress' or infection. Unlike Addison's disease, where there is an absence of cortisol and aldosterone (causing hypotension and potentially an Addisonian crisis), cortisol deficiency alone does not usually cause hypotension (see Chapter 10).

Measuring corticotrophin levels in the blood can be difficult, because even in normal individuals who are well, concentrations can be so low as to be undetectable. Concentrations of both corticotrophin and cortisol are highest early in the morning, so the first test one should carry out is a blood sample at 9:00 AM for the measurement of cortisol and corticotrophin levels. Low levels are suspicious but not diagnostic without a stimulation test. To confirm a diagnosis of corticotrophin deficiency, the most popular test is the insulin (hypoglycaemic) stress test. In this test, fasting patients attend the clinic, and are then given a dose of insulin that will potentially cause a dangerous hypoglycaemia. The pituitary gland should be part of the counterregulatory response to hypoglycaemia, causing a large release of corticotrophin which in turn causes a significant release of cortisol. The standard response is for cortisol levels to peak at over 450–500 nM, depending on the assay used. If inadequate levels of cortisol are found, then the diagnosis of corticotrophin deficiency is confirmed.

Cortisol levels are appropriately low when a patient is exposed to excess exogenous synthetic steroids. Thus patients who are taking systemic prednisolone or dexamethasone will have low levels of corticotrophin and cortisol; this is not a true deficiency, however, but an appropriate suppression.

The insulin-induced hypoglycaemic test is the gold standard. Such an insulin-induced hypoglycaemia can be dangerous, so patients who have heart disease or epilepsy should not have this test. In such patients, a surrogate marker for corticotrophin deficiency in adrenal atrophy is used, namely, the lack of a cortisol response to a bolus injection of corticotrophin. This is commonly known as the short synacthen test, synacthen being a synthetic ACTH.

One other indirect test is the corticotrophin response to a metyrapone-induced cortisol deficiency. Metyrapone is an inhibitor of 11 hydroxylase, one of the rate-limiting enzymatic steps in adrenal cortisol synthesis.

The metyrapone test involves giving patients a large dose of metyrapone, and measuring the corticotrophin response the following morning. Reduced negative feedback should increase corticotrophin release, and this should be matched by a similar response in the concentration of the cortisol precursor 11 deoxycortisol.

Another indirect test of corticotrophin deficiency is the glucagon stimulation test. This can be used in individuals who cannot have insulin for safety reasons. If cortisol levels increase substantially with this test, then pituitary failure is excluded, but patients who fail to make adequate cortisol might simply have not responded well to glucagon. Glucagon stimulation is less reliable than an insulin stress test.

Once diagnosed, corticotrophin deficiency is treated with replacement hydrocortisone which can be given by mouth. The most common dose is 10 mg first thing in the morning, 5 mg at noon and 5 mg at 4:00 PM, mimicking the normal diurnal rhythm in cortisol secretion as outlined in Chapter 10.

Updated protocols for the dynamic pituitary function tests mentioned in this section can be found on the Imperial Centre for Endocrinology website: http://www.imperialendo.com.

Thyrotrophin (TSH) deficiency

This condition presents with some of the clinical features of hypothyroidism (secondary hypothyroidism; see Chapter 13). Patients have weight gain and tiredness, due to a fall in basal metabolic rate. Measuring the basal concentrations of TSH and thyroxine may miss the diagnosis, because the reference ranges for both are wide. Primary thyroid failure is much more common, and this diagnosis is made when the TSH is raised. A low level of TSH is much more difficult to be clear about, but the diagnosis of thyrotrophin deficiency should be suspected in any patient with a low concentration of free thyroxine (FT4) if the TSH remains in the reference range. Association with a deficiency of other pituitary hormones might be a clue as to the pituitary origin of the problem. Treatment is by replacement of thyroxine, administered by mouth at a dose of between 25 and 150 micrograms (mcg) daily. The dose is adjusted by monitoring the FT4 and the FT3 and occasionally using any symptoms the patient may have. Clearly one cannot use the TSH as a guide to therapy (unlike in primary thyroid disease, where the TSH is a very useful marker).

Growth hormone deficiency

Although GH is essential for normal growth and has a profound effect on final height, it also has many other metabolic effects. Thus GH deficiency in children causes short stature, but adults who develop GH deficiency have a change in metabolism. Some patients become tired and have other non-specific features that respond to GH replacement. Patients who might have GH deficiency have to undertake a questionnaire (the Adult GH Deficiency Assessment, or AGHDA, also at http://www.imperialendo.com) to determine whether they are likely to benefit from replacement or not.

GH deficiency needs to be proven by stimulation tests, and like with corticotrophin, the commonest gold standard test is the GH response to insulin-induced hypoglycaemia. GH deficiency is confirmed if the GH concentration following hypoglycaemia fails to reach 6 $mcg.L^{-1}$ (previously 20 $IU.L^{-1}$). Other less dangerous stimuli, such as exercise, a glucagon stimulation test or arginine, can be used. GH concentrations are usually close to zero in the well non-stressed individual, so a random low level is not helpful in making the diagnosis of GH deficiency. However, a random level of 6 $mcg.L^{-1}$ or more excludes GH deficiency.

As explained in Chapter 3, GH stimulates the production of IGF1 in the circulation, so patients with GH deficiency will also have a low concentration of circulating IGF1. IGF1 is bound to binding proteins, which can make the measured concentration of IGF1 difficult to interpret.

Children with GH deficiency require GH replacement to obtain a normal height. The dose of GH varies widely but can be between 0.1 and 0.9 mg daily. The dose is adjusted until the IGF1 is in the reference range, and the child is shown to be growing normally as evidenced by regular monitoring of growth rate.

It has been thought that adults do not require GH, but about 30% of adults who develop GH deficiency have metabolic changes and tiredness that warrant therapy. Again the dose varies between 0.1 and 0.6 mg daily, and the dose is increased until the plasma IGF1 level is normal. For adults, it is worth giving GH only if they actually feel a benefit, so this is assessed using the AGHDA questionnaire. Those adults who are confirmed to be GH deficient using the tests discussed in this section, and who score greater than 11 points on the AGHDA questionnaire, are more likely to stay on GH. Once GH therapy is commenced, those patients who perceive a benefit will find that the AGHDA score falls by 7 points.

Gonadotrophin (LH and FSH) deficiency

LH and FSH deficiencies manifest in females with amenorrhoea due to lack of ovarian secretion of oestrogen. In women older than 50, this may be confused with the menopause, and the diagnosis is usually not made in older women. In males, testosterone deficiency causes loss of libido and sometimes a reduction in the growth of facial hair. Unlike males with impotence (occurring in patients who have normal libido and vascular disease, and in patients who complain about the problem), males who have no libido rarely complain. Their partners or spouses might complain, but they themselves are not interested, so the diagnosis may be missed.

Treatment if necessary is by the administration of oestrogen in women and testosterone in men. The treatment for both conditions is identical to that for ovarian failure in women and testicular failure in men, so please see the relevant chapters (6 and 7) for options in therapy.

Hyperprolactinaemia

Prolactin is the only anterior pituitary hormone normally under tonic inhibition from the hypothalamus. Whereas a pituitary tumour causes deficiency of most anterior pituitary hormones, disconnection of the pituitary gland from the hypothalamus by the growing tumour causes a rise in circulating prolactin levels as the lactotrophs are no longer inhibited by dopamine. Prolactin causes galactorrhoea in the oestrogen-primed breast. Prolactin also suppresses the gonadotrophin-releasing hormone (GnRH) pulse generator, and thus causes amenorrhoea.

Prolactin levels due to disconnection hyperprolactinaemia rarely rise above 2000 $mU.L^{-1}$ (the reference range for normal prolactin levels being up to 600 $mU.L^{-1}$). Other causes of hyperprolactinaemia include the administration of dopamine antagonist drugs for other conditions such as travel sickness or psychiatric diseases such as schizophrenia.

Prolactinomas, another major cause of hyperprolactinaemia, are further discussed in section 'Prolactinomas', but it is important to note that distinguishing disconnection hyperprolactinaemia from a prolactinoma is not as easy as one would expect.

Treatment of the hyperprolactinaemia can be with dopamine agonists as with prolactinomas, but the aim of therapy with non-functioning tumours is simply to suppress the prolactin, as non-functioning tumours do not shrink with dopamine agonists (unlike prolactinomas). Either bromocriptine (1–15 mg three times daily) or cabergoline (0.5–1.5 mg 1–3 times weekly) can be used. If the high prolactin level has been suppressing the production of gonadotrophins, then there will be a recovery of the gonadotrophin axis provided that the gonadotrophs themselves are not affected by the tumour. If they are, then treatment with oestrogen or testosterone as discussed in section 'Gonadotrophin (LH and FSH) deficiency' will suffice as there is no point in trying to normalise the prolactin in that situation.

Anterior pituitary hypofunction (panhypopituitarism)

Large non-functioning pituitary tumours can produce anterior pituitary hypofunction by compressing the pituitary cells against the bone of the *sella turcica*. Several of the hormones discussed here can be affected, and it is not uncommon for patients to be deficient in all of these except prolactin. In patients who are deficient in several hormones, the most important one to replace is hydrocortisone. Once that has been initiated, further tests should be performed to check that the other hormones remain low. If they do, then thyroxine should be replaced next, and then either oestrogen in females or testosterone in males. Prolactin is never replaced. Once adequate replacement of hydrocortisone, thyroxine and sex steroids have been made, the patient should be re-investigated for GH deficiency, and before GH replacement is considered, an AGHDA questionnaire needs to be filled in. Patients who score more than 11 points on the questionnaire warrant a trial of GH replacement. Overall, treatment is by hormone replacement with hydrocortisone, thyroxine and oestrogen or testosterone as appropriate. The doses used are the same as for the individual hormone failures mentioned here.

Tumour size matters

Patients with confirmed hypopituitarism should have a pituitary MRI to determine whether the cause is a pituitary tumour. Almost all such tumours are benign, but can cause trouble because of the limited space within the pituitary fossa. If a pituitary adenoma is found, clearly it can be removed, although many pituitary adenomas never cause a clinical problem. As mentioned, pituitary tumours are known as macroadenomas if they are 10 mm or larger in diameter, and microadenomas if they are smaller than 10 mm. Figures 5.1 and 5.2 show a pituitary macroadenoma on MRI that is just pressing on the optic chiasm.

Figures 5.3 and 5.4 show visual fields: in Figure 5.3, the optic chiasm has not yet been compressed and the fields are normal. In Figure 5.4, there is the beginning of a bitemporal hemianopia (see Chapter 3).

Patients with hypopituitarism always have lifelong follow-up checks for monitoring purposes. In addition to monitoring of the replacement hormones, surveillance for recurrence or growth of a pituitary tumour must continue.

Monitoring of hydrocortisone therapy can be empirical as stated, with 10 mg on awakening, 5 mg at noon and 5 mg at 4:00 PM. This regimen will suit most patients but there are some who are either fast or slow metabolisers of oral hydrocortisone, because of differential first-pass hepatic metabolism. Some of these patients need higher doses of hydrocortisone, and to prove it, a hydrocortisone day curve may be necessary. This involves measuring plasma concentrations of cortisol throughout the day while patients are on a replacement dose of hydrocortisone.

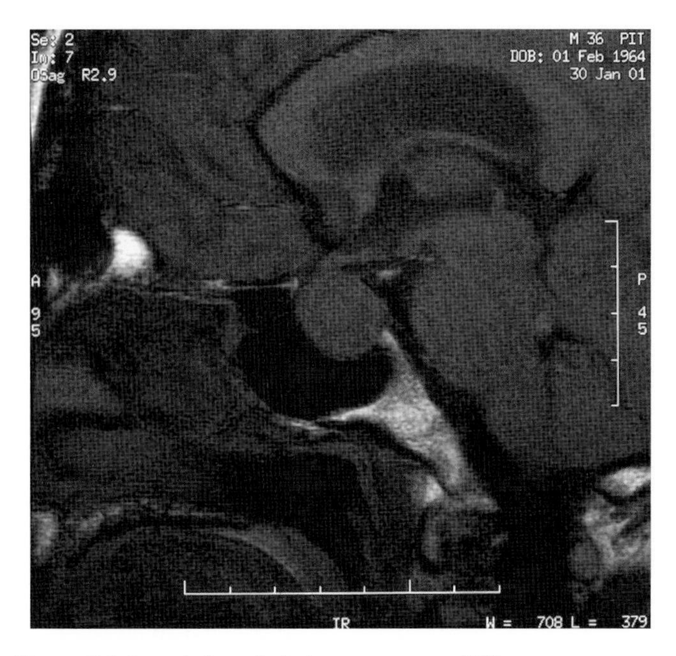

Figure 5.1 Lateral view of pituitary tumour on MRI.

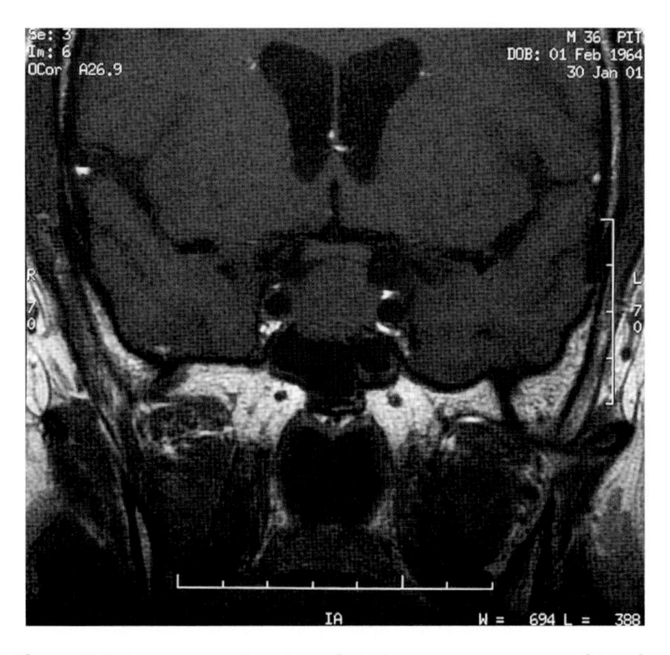

Figure 5.2 Anteroposterior view of pituitary tumour just touching the optic chiasm.

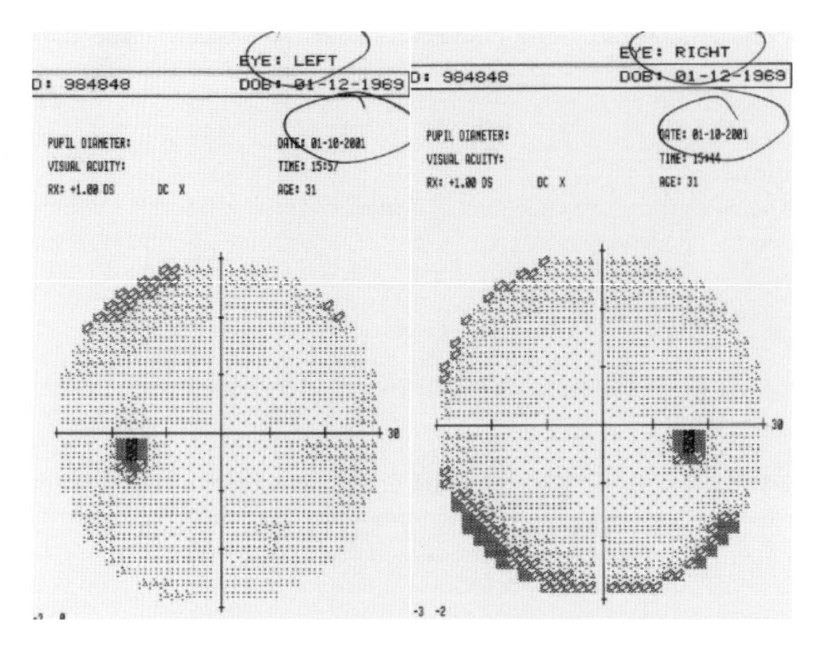

Figure 5.3 This patient has normal visual fields despite the large tumour that is just touching the optic chiasm.

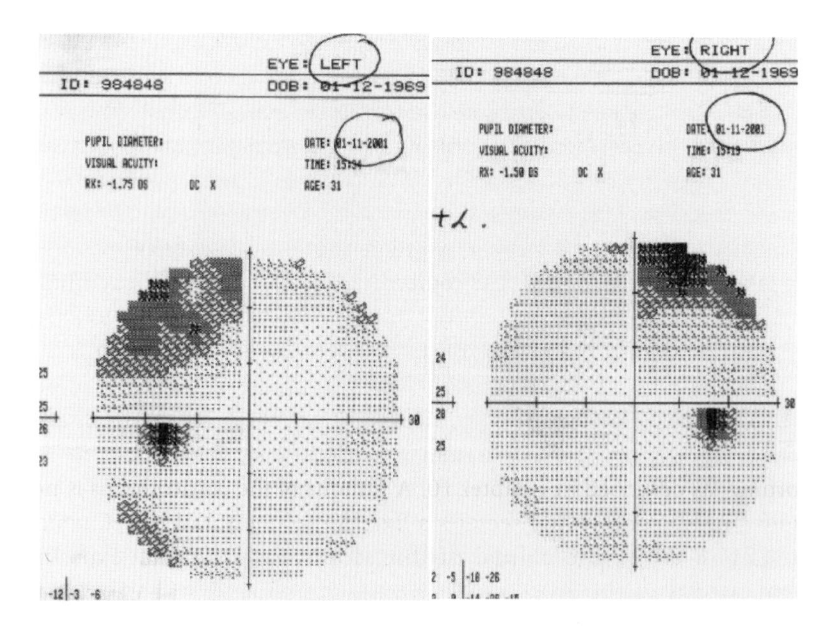

Figure 5.4 This is the abnormal visual field in the patient with a bitemporal hemianopia.

Non-functioning pituitary adenomas

So called non-functioning pituitary tumours are usually made up of gonadotrophs that fail to secrete their gonadotrophins. This should be confirmed histologically and markers of proliferation checked to have some idea of the risk of recurrence. This is performed by the histologist in the MDT.

The posterior pituitary is rarely affected by anterior pituitary tumours. However, transphenoidal surgery can result in damage to the posterior pituitary, and diabetes insipidus may result (see section 'Lack of Vasopressin: Diabetes Insipidus (DI)').

Anterior pituitary hyperfunction (hyperpituitarism)

Any of the cells of the pituitary can form a tumour and some secrete hormones in excess. Most tumours of the anterior pituitary are 'non-functioning' and do not secrete hormones. Commonly they stain for gonadotrophins, but very few actually secrete them. Thus most tumours that come from the gonadotrophin line in fact cause hypopituitarism as described above.

Cushing's disease

Clinical features

Cushing's disease is specific to a pituitary tumour producing uncontrolled, excess quantities of ACTH. Patients become gradually unwell, as the cortisol level slowly rises. Initially, the main effect is on the diurnal rhythm of cortisol secretion which becomes less pronounced. The total exposure of the body's cells to cortisol rises. This causes a gradual increase in appetite, and patients put on weight. Fat synthesis is promoted over protein synthesis, and patients develop muscle weakness, with a proximal myopathy becoming apparent. In addition, skin heals poorly, and wounds can take a long time to heal. The combination of reduced protein synthesis with rapidly increasing girth causes stretch marks in the skin which may be red (see Figure 5.5). Other features are listed in Table 5.1.

This condition is specifically caused by a tumour of the corticotrophs of the pituitary gland. Normal corticotrophs secrete ACTH in a pulsatile manner and with a diurnal rhythm so that they are most active in the morning, as described in Chapter 10. A tumour of the corticotrophs is not under such regulation, but secretes a fixed large amount of ACTH, which results in a loss of the diurnal rhythm. The consequent total exposure of the patient to glucocorticoids is thus increased, although the plasma level of cortisol during the day may be within the normal reference range. Thus a random cortisol is unhelpful in making the diagnosis. The time to measure the circulating cortisol concentration when Cushing's

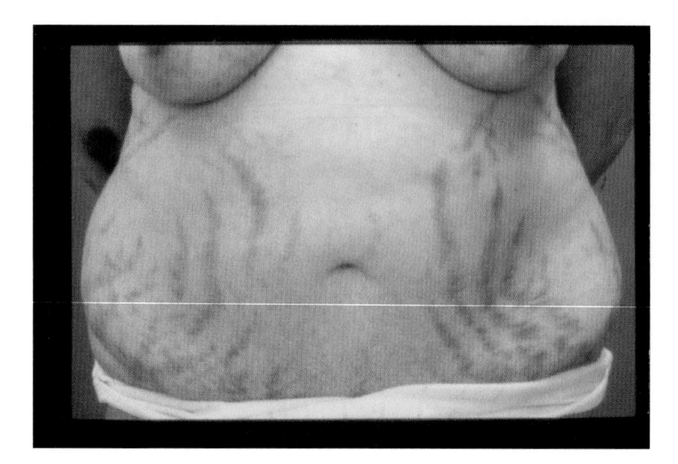

Figure 5.5 Slide of abdomen.

Table 5.1 Clinical features of Cushing's syndrome.

Increase fat deposition, in particular on face (moon-like facial appearance)
Interscapular fat pad (buffalo hump)
Thin skin with poor wound healing
Red striae, as in Figure 5.5
Proximal myopathy (difficulty climbing stairs)
Osteoporosis
Hypertension

disease is suspected is to take a sample when you expect the level of cortisol to be low. This is best done in a sleeping patient, at midnight. In practice, it can be difficult to ensure the patient is asleep, and for this reason there are a number of other confirmatory diagnostic tests that are used. The most popular of these is the dexamethasone suppression test, where oral dexamethasone (a synthetic glucocorticoid) is given in an attempt to suppress the hypothalamo-pituitary-adrenal (HPA) axis. Failure of suppression suggests Cushing's disease.

Cushing's syndrome

Exposure to excess cortisol, or to excess glucocorticoids from conditions other than a pituitary tumour, causes a syndrome that is similar to Cushing's disease, but where the cause is uncertain, the term Cushing's *syndrome* rather than Cushing's *disease* is used. Cushing's syndrome can be caused by Cushing's disease, but there may be other causes of the increased exposure to glucocorticoids. For example, it can be caused by the patient taking oral glucocorticoids as a treatment for another (e.g. inflammatory) condition. When patients present with clinical features of Cushing's syndrome, therefore, it is important to ensure they are not taking steroids before embarking on any other investigations.

Cushing's syndrome can also be caused by an adrenal adenoma making cortisol autonomously (see Chapter 10), or rarely by a malignant tumour (e.g. in the lung) making 'ectopic' ACTH. Patients with an adrenal adenoma will have an undetectable ACTH, so that is easy to exclude. However, distinguishing ectopic ACTH from pituitary-dependent Cushing's disease has always been more difficult to ascertain. In the past, the high-dose dexamethasone suppression test was used to make this distinction. This is because a pituitary corticotroph adenoma remarkably retains some feedback control, so that when very high-dose dexamethasone is administered there would be some suppression in cortisol production that does not occur with malignant ectopic ACTH. The high-dose dexamethasone suppression test is no longer required or used because it has too high a false-positive rate and also because petrosal sinus sampling is now relatively easily performed.

Once Cushing's disease or syndrome is confirmed, localisation is then attempted. To exclude an ectopic source of ACTH, one needs to compare the level of ACTH in the petrosal sinuses with that of the peripheral circulation (see Figure 5.6). This can not only confirm that the pituitary is indeed the source of the problem, but also suggest on which side of the pituitary the tumour is. Unlike other pituitary tumours, Cushing's tumours are often so small at presentation that they cannot be seen on an MRI scan.

Blood is taken from the veins draining the pituitary and a dose of corticotrophin-releasing hormone (CRH) is administered. A corticotroph adenoma exhibits an exuberant response to CRH, and the resultant release of ACTH can be measured in the catheter. Prolactin levels can also be measured in the samples to ensure the correct positioning of the catheters. The graph in Figure 5.7 is from a patient with a left-sided Cushing's adenoma.

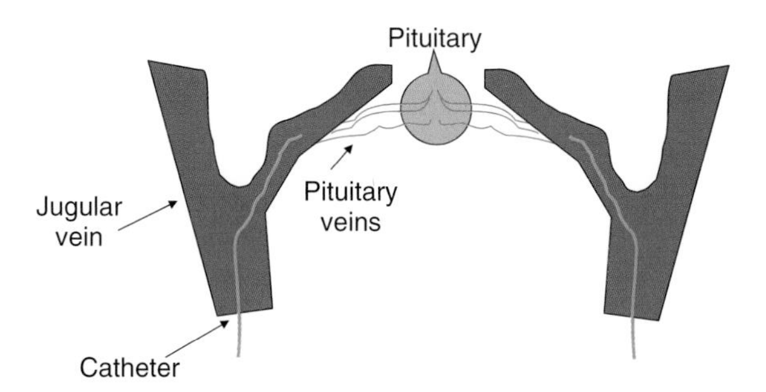

Figure 5.6 Inferior petrosal sinus sampling (IPSS) with corticotrophin-releasing hormone (CRH) stimulation.

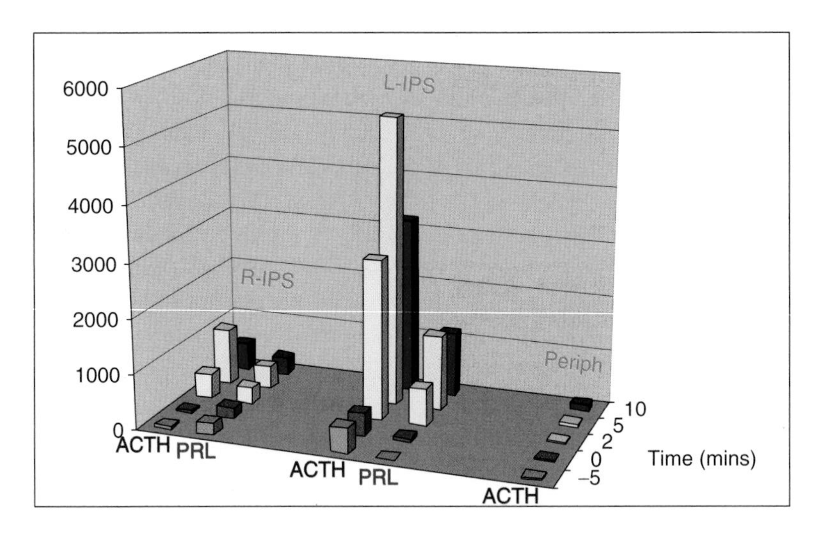

Figure 5.7 IPSS results. Note: ACTH = adrenocorticotropic hormone; PRL = prolactin; R-IPS = right inferior petrosal sinus; L-IPS = left inferior petrosal sinus and Periph = peripheral.

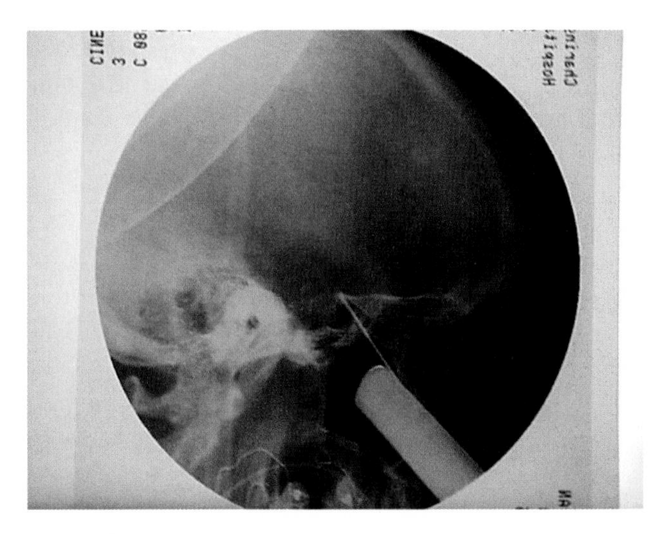

Figure 5.8 This is an X ray performed during transphenoidal surgery to check that the drill is pointing towards the pituitary gland.

Treatment of Cushing's disease

This should be managed by the MDT. Options include surgery to the pituitary gland itself to try to remove the source of ACTH (see Figure 5.8). However, if the tumour cannot be easily seen, other options are available. Radiotherapy can be effective.

Bilateral adrenalectomy can be performed: this will control the excess cortisol levels, but will leave a small chance for the tumour in the pituitary gland to grow. Such a growing Cushing's adenoma after adrenal surgery was first described by Nelson, and now has the name Nelson's syndrome. The suggestion has been that corticotroph adenomas do retain some negative feedback from cortisol (as evidenced by the effects of the high-dose dexamethasone suppression test) so that when the adrenals are removed, the fall in cortisol takes the brake off the normal suppression of the pituitary tumour, which then grows fast. More recently, the risk of such rapid growth has actually been shown to be low.

Drugs are not very useful due to their side effects, although they can be effective in the short term. The most commonly used are metyrapone (an 11-hydroxylase inhibitor) and ketoconazole (an antifungal that happens to inhibit a number of enzymes in the adrenal cortex). Because Cushing's syndrome can result in poor wound healing, patients with severe Cushing's syndrome may be offered drugs for a few weeks before pituitary or adrenal surgery, to improve the outcome of surgery. This was particularly important in the days of the open adrenalectomy, where a fairly large scar needed to heal. Pre-operative steroid suppression may no longer be required where laparoscopic or retroperitoneal surgery for adrenalectomy is being performed, although it is essential to check that the surgeon is planning to use that route. Further details of adrenalectomy are given in Chapter 10.

Prolactinomas

Tumours of the lactotroph lineage secrete large amounts of prolactin. Prolactin turns off the GnRH pulse generator, so sex steroid deficiency is an important (but often missed) clinical feature of a high prolactin production. Prolactin will also cause galactorrhoea of the oestrogen-primed breast, so women, and some men with breast hypertrophy (gynaecomastia), complain of milk production. The lack of sex steroids in females causes amenorrhoea, and the combination of amenorrhoea with galactorrhoea in women who are not pregnant makes it imperative to measure the prolactin. Prolactinomas grow over time, but if the above clues are picked up early, an MRI scan is likely to show a small microprolactinoma.

In males the diagnosis can be more difficult to make because the clues of galactorrhoea and amenorrhoea do not occur. Instead males develop testosterone deficiency, and hence loss of libido. The lack of testosterone causes a lack of sexual drive which is unlike the impotence which is seen in patients with diabetes, for example. In the latter case, diabetic patients have a normal sex drive (libido) because of sufficient circulating testosterone, but seek medical attention for their erectile dysfunction (impotence) usually due to associated cardiovascular problems. In the male with a prolactinoma, the patient does not perceive any problem at all, and occasionally a spouse might complain about 'lack of interest' and encourage referral, or eventually the patient presents with a large prolactinoma and features of hypopituitarism, as described in earlier section. Some audits and textbooks suggest therefore that female patients have microprolactinomas, and male patients have macroprolactinomas. This is probably true only because we fail to diagnose the males when their tumours are small.

The combination of a raised circulating prolactin level with a visible pituitary tumour on the MRI scan suggests that there is a prolactinoma, but it is essential that diagnoses of disconnection hyperprolactinaemia or other causes of hyperprolactinaemia are excluded. The size of the tumour and the relative height of the prolactin profile might be helpful. For example, a prolactin level of over 6000 $mU.L^{-1}$ can only be due to a prolactinoma. Levels between 1000 and 6000 $mU.L^{-1}$ are more difficult to interpret, and can be due either to a small microprolactinoma, or to a large tumour causing disconnection from the hypothalamus. Patients should be reviewed by the MDT, and clues such as the rate of tumour growth or the rate of response to a dopamine agonist challenge can be helpful with regard to eventual treatment.

Treatment of prolactinomas

A trial of dopamine agonists should be used. Cabergoline or bromocriptine usage has the most long-term data and result in both normalisation of the prolactin level and shrinkage of the tumour. This is because the lactotrophs have dopamine receptors. If the tumour does not shrink, it is important to reconsider the diagnosis, as it is possible that the patient always had disconnection hyperprolactinaemia, and this differential diagnosis can be difficult to confirm. For patients who cannot tolerate dopamine agonists (e.g. those with schizophrenia who are made much worse by dopamine agonists), surgery of the pituitary gland and/or radiotherapy as described in section 'Treatment of Cushing's disease' for Cushing's disease are options. The choice for all these treatments should be discussed by an MDT as described in section 'The anterior pituitary (adenohypophysis)'.

Acromegaly

Tumours of somatotrophs secrete excess GH, and in children this causes gigantism, because they simply grow faster than normal to reach heights of up to 2 m or more. However, in adults, in whom the epiphyses of the long bones have fused, linear growth is unlikely but other tissues undergo hypertrophy where possible, and a very stereotypic appearance occurs. More importantly, the metabolic effects of excess GH cause increased risk of diabetes, hypertension and vascular disease resulting in a raised morbidity.

Figures 5.9 through 5.12 show a patient with acromegaly. Figure 5.9 shows a view from the front, and Figure 5.10 provides a lateral view,

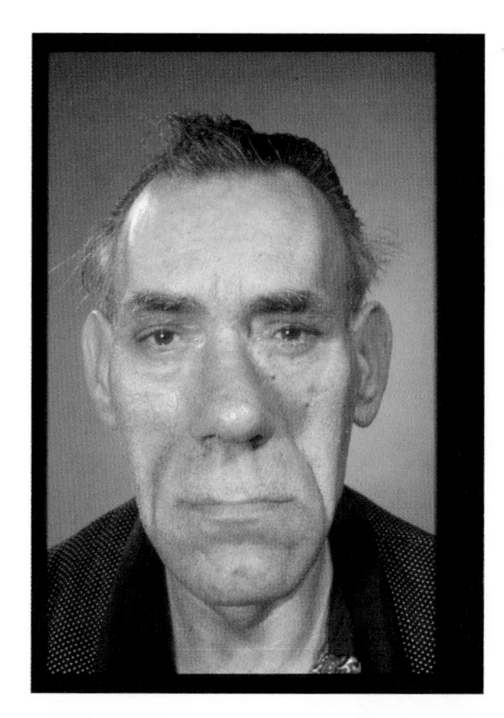

Figure 5.9 Face of acromegaly patient.

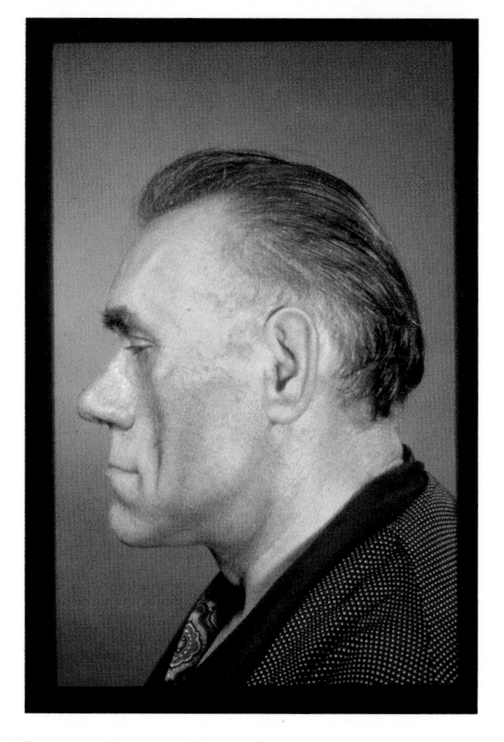

Figure 5.10 Lateral face of acromegaly patient.

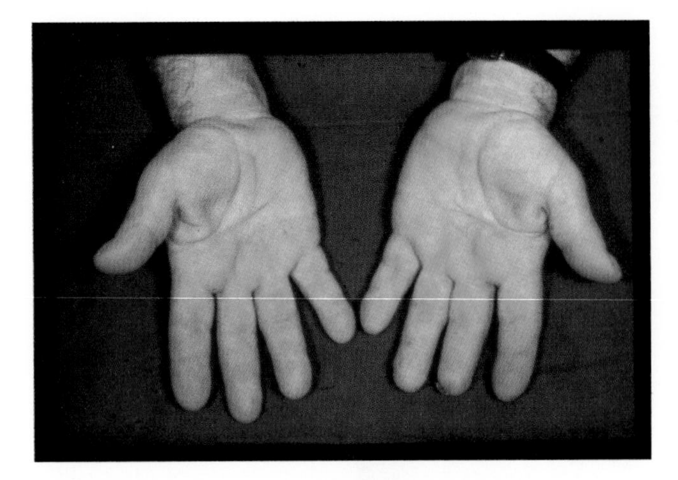

Figure 5.11 Hands of acromegaly patient.

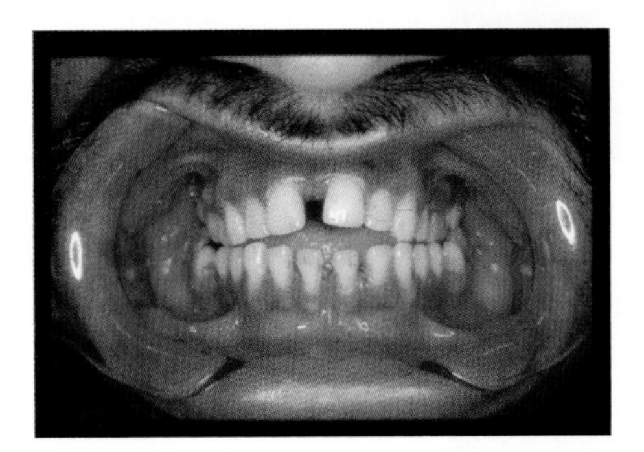

Figure 5.12 Teeth of acromegaly patient.

showing clearly the coarse features, a prognathic jaw and frontal bossing. The patient's hands are shown in Figure 5.11, while Figure 5.12 shows a patient with acromegaly who has had increased growth of the jaw, and hence has now got widely spaced teeth.

Because the secretion of GH is pulsatile in normal individuals, a random blood GH concentration is likely to be unhelpful unless it is zero, in which case it excludes acromegaly. A high level could be due either to acromegaly, or to the chance of hitting the top of a pulse of normal GH secretion. If the diagnosis is suspected, suppression of GH secretion using a glucose tolerance test should be performed. GH levels fall to zero during glucose tolerance testing in normal individuals, but this does not happen in patients with acromegaly. A somatotroph adenoma paradoxically will have the opposite response to the raised glucose level, and may result in an increase in GH levels which will confirm the diagnosis. The final diagnostic test is to measure the level of IGF1, which is used as a marker of long-term GH exposure. This confirmatory test is essential because starvation can give a false positive response (i.e. a failure of the glucose tolerance test to suppress GH). In addition, the levels of GH and IGF1 need to be measured to monitor response to treatment.

As with other pituitary conditions, patients with acromegaly should be discussed at the MDT, and can then be offered surgery, radiotherapy or drugs. A few patients respond to dopamine agonists, but more patients respond to somatostatin analogues such as lanreotide and octreotide. However, unlike with prolactinomas, shrinkage of somatotrophinomas does not occur, making surgery a more viable early option for such patients. Following transphenoidal surgery (by which the tumour is removed by suction through a hole made in the body of the sphenoid bone approached up through the nasal sinus), all the patient's pituitary axes need to be reassessed. If the GH levels are low enough, no further therapy will be warranted. Some patients progress to requiring radiotherapy and drugs. All patients with acromegaly require annual pituitary review, to ensure that they are on optimum pituitary hormone replacement therapy if needed, as described in section 'The anterior pituitary (adenohypophysis)' regarding pituitary failure.

Thyrotrophinomas (TSHomas)

These are extremely rare tumours, but present with features of hyperthyroidism and failure to suppress the blood TSH level. Because the features of hyperthyroidism may be subtle, the diagnosis may be difficult to make, and patients may be thought to have primary thyroid disease. Such tumours fail to respond to TRH, so a TRH test gives a flat response; this is an essential test to perform in order to confirm the diagnosis and it helps to distinguish primary thyroid disease (and the also rare condition of thyroid hormone resistance) from a TSHoma.

Once confirmed, patients should be discussed at a pituitary MDT, and all the treatment options available can be used as discussed throughout this chapter.

The posterior pituitary gland (neurohypophysis)

Excess vasopressin: The syndrome of inappropriate antidiuretic hormone hypersecretion (SIADH)

Very rarely, posterior pituitary hyperfunction presents as the syndrome of inappropriate antidiuretic hormone hypersecretion (SIADH), defined as a plasma vasopressin concentration which is inappropriate for the given plasma osmolality. Most commonly this is not truly a pituitary problem, but is caused by an ectopic source of vasopressin (often a lung cancer) which secretes vasopressin (ADH) autonomously. The increased renal water reabsorption results in a reduction in the volume, and an increase in the concentration, of urine. The consequence of the increased water reabsorption is a decrease in plasma sodium concentration (hyponatraemia). Very occasionally patients can have transient pituitary-driven SIADH that lasts about a week, starting around 10 days after pituitary surgery. A well-described, but difficult to explain, pattern of events is for a transient diabetes insipidus to develop immediately following pituitary surgery for anterior pituitary disease, this being managed with careful fluid balance. Approximately 10 days later, some patients have an episode of SIADH that lasts about a week, and should be managed with fluid restriction. Patients who are unaware of this sometimes feel thirsty and drink too much water, resulting in transient hyponatraemia. An adequate fluid restriction is always sufficient to control this, and as it is transient any other therapy is likely to cause an increased morbidity. For example, the use of ADH antagonists is contra-indicated, as this will result in diabetes insipidus as soon as the patient normalises, and will confound any further diagnostic tests. There are extremely rare cases of pituitary SIADH following the fracture of a skull.

Lack of vasopressin: Diabetes insipidus (DI)

Failure of the posterior pituitary to produce vasopressin results in a central form of diabetes insipidus (named because of the excretion of a large volume of 'insipid', i.e. tasteless, urine). This presents as a polyuria and polydipsia (consequential increased drinking) and it is important to exclude the much more common diabetes mellitus as a possible cause. Diabetes insipidus can be due to post-pituitary surgery, where sometimes the posterior lobe also has to be removed or is damaged. Metastases from other parts of the body can sometimes present with pituitary failure, including posterior pituitary failure. Benign anterior lobe disease usually does not affect the posterior pituitary. Fractures of the skull have also been known to cause diabetes insipidus.

Another form of DI is associated not with the lack of vasopressin but with end-organ (i.e. kidney) failure to respond to it. This is called nephrogenic DI. It is quite rare but can be familial (e.g. receptor defects), or be caused by certain drugs used to treat other conditions (e.g. lithium).

The diagnosis of diabetes insipidus should be confirmed with a water deprivation test, which usually consists of a period of 2–3 hours with complete water restriction and regular (hourly) measurements of urine volume and osmolality. The body weight is usually monitored hourly too, since a patient with DI will continue to lose large volumes of water and runs the risk of serious dehydration. Careful continuous monitoring is therefore required in a patient suspected to have DI, and the test is halted immediately if the patient is losing weight. Once the diagnosis is confirmed, treatment is with a vasopressin analogue such as des-amino D-arginine vasopressin (DDAVP). As the name describes, this molecule has had the arginine in the vasopressin molecule modified to give it a longer half-life, so that patients do not need administration of the drug every 5 minutes, as they would do with native vasopressin. By using a D-amino acid (rather than L-arginine as occurs in nature), the molecule is relatively resistant to the usual degradative pathways. Patients usually take DDAVP intranasally twice daily, and by giving vast doses, a tablet of DDAVP is just about viable. Most of an oral dose of DDAVP is digested, but enough survives to be absorbed and be therapeutically useful. The oral dose is over 10 times that of the intranasal dose. As a trial, 0.5–1.0 mcg of DDAVP can be given subcutaneously or intravenously. This dose and route are sometimes used in patients immediately post-operatively. Following transphenoidal pituitary surgery, some patients have transient diabetes insipidus, and one or two doses may be used in the days following surgery if the patient starts to become hypernatraemic, until the pituitary recovers. Patients who have permanent diabetes insipidus are usually offered intranasal therapy at a dose of about 10 mcg per nostril once or twice daily, titrated to urine output and plasma sodium level. A tablet of 100 mcg is also available, and patients need between one and three of these up to four times daily because most of the drug is destroyed in the digestive system.

Oxytocin deficiency or excess

Oxytocin deficiency does not cause a clinical problem. Theoretically this will become an issue only in patients who are undergoing parturition and breast feeding, but since pregnancy does not occur spontaneously with pituitary failure, those who lack oxytocin may never know because they will not have become pregnant in the first place. In addition, oxytocin analogues are now regularly used in normal obstetric practice (because this has reduced maternal morbidity and mortality substantially by preventing postpartum haemorrhage), so that one would never know if a patient is oxytocin deficient.

Likewise, the condition that might be associated with an excess of oxytocin, for whatever reason (e.g. an ectopic tumour), is unlikely to present with any symptoms, and therefore it is clinically unknown.

Appendix: Clinical Cases

Clinical Scenario 5.1

Mrs Jones complains that she has been putting on weight for the last year. She has also noticed that an injury she had to her shin took a long time to heal. She has found it increasingly difficult to climb stairs due to a combination of muscle weakness and weight increase. She has recently found it increasingly difficult to sleep. She goes to see her GP, Dr Moffat, for advice.

He notices that her face is round and that her proximal muscles are quite weak. Between her scapulae he notices a pad of fat, and that she has indeed put on weight around her torso. He also notices red "stretch marks" over the abdomen. Her blood pressure is raised at 190/110 and she has some glycosuria when he tests her urine with a dipstick.

Questions

Q1. What is the likely diagnosis?
Q2. What investigations should be performed?
Q3. What treatment is required?

Answers to Clinical Scenario 5.1

A1. This sounds very much like Cushing's syndrome, and this patient has many of the features given in Table 5.1. The glycosuria suggests diabetes mellitus, so the patient should also have a fasting blood glucose measurement (see Chapter 15) to clarify whether or not she has diabetes, and, depending on the result of that, a glucose tolerance test. Electrolytes (sodium and potassium) should also be checked, as the patient is at risk of hypokalaemia which should be replaced.

A2. To confirm the diagnosis of Cushing's syndrome, initially a 24-hour urine collection can be used to determine if the overall cortisol secretion rate is increased. In addition, the diurnal cortisol level should be checked by measuring the cortisol during the day (particularly at 9:00 AM) and, most importantly, when the patient is asleep at midnight, when the cortisol level should be very low (<50 nM). It is important not to tell patients that they will have a blood sample taken at night, as they might not sleep if they suspect this. The ideal is to admit the patient to hospital, and to check the midnight cortisol from an intravenous line. Alternatively a sample can be taken immediately after waking the patient, and before the HPA axis will have had time to increase cortisol secretion (i.e. in less than 10 minutes).

The next test is a low-dose dexamethasone suppression test, followed by a high-dose dexamethasone suppression test.

The results for this patient were:

24-hour urine free cortisol level: 800 nmol/24 hours (normal should be less than 270 nmol/24 hours).

The midnight sleeping cortisol was 450 nM (normal if truly asleep (which she was in hospital) should be <50 nM).

The 9 AM cortisol was 650 nM and after 48 hours of low-dose dexamethasone (where she was given 0.5 mg of dexamethasone every 6 hours) the 9 AM cortisol was 500 nM. She was then given a high-dose dexamethasone suppression test (2.0 mg every 6 hours for 2 days) and the cortisol at the end of that test had fallen to 200 nM. Although there was a significant suppression of the cortisol, it was still detectable.

These results suggest that the patient has pituitary-dependent Cushing's disease. The results given here of Mrs Jones should be compared with those of Mr Leroy in Chapter 10, and also with her friend Mrs. Peters in 'Clinical Scenario 5.2'.

To confirm a pituitary source, Mrs Jones underwent inferior petrosal sinus sampling (IPSS), which confirmed a pituitary source. She also underwent a pituitary MRI which also showed a small pituitary lesion (see Figure 5.1).

A3. Treatment options include transphenoidal pituitary surgery, where the pituitary tumour is removed. If there is to be a delay to pituitary surgery, the patient and her endocrinologist might elect to start a short course of metyrapone and ketoconazole to reduce cortisol levels before the operation. Usually pituitary surgery causes a very small scar indeed, so that drugs are often not used before surgery.

Another option is to leave the small pituitary lesion alone, but to remove both adrenal glands, which will stop the production of cortisol. However, the patient will then require hydrocortisone and fludrocortisone for life, and the primary pituitary tumour might grow over time (Nelson's syndrome). With regular modern MRI scanning, this can be managed in a straightforward manner. Most patients will not have a problem with Nelson's syndrome.

Clinical Scenario 5.2

Mrs Peters is a close friend of Mrs Jones, and has noticed the dramatic improvement in the clinical state of Mrs Jones after her pituitary operation. She too has noticed an increase in weight over the previous few years, and also thinks that it is becoming increasingly difficult to climb stairs.

She went to see Dr Moffat, who by this point was becoming quite an expert at suspecting Cushing's syndrome. She didn't quite have any stretch marks and her face was not round but Mrs Peters already had high blood pressure, and already had had diabetes for several years so Dr Moffat referred her to the local endocrine unit, suspecting that Mrs Peters also had Cushing's syndrome.

Questions

Q1. What is the likely diagnosis?
Q2. What investigations should be performed?
Q3. What treatment is required?

Answers to Clinical Scenario 5.2

A1. Although this patient also needs investigations for Cushing's syndrome, it is important to recognise that simple obesity can look very similar.

A2. This patient's results revealed a 24-hour urine free cortisol level of 250 nmol/24 hours. The midnight sleeping blood cortisol measurement was 45 nM, which is normal and excludes Cushing's syndrome. A 9:00 AM cortisol was also checked and the result came back at 650 nM. This confirmed that she had a normal diurnal rhythm, with a high cortisol in the morning and a low one at midnight. At this point the patient was told that she did not have Cushing's syndrome and that she thus did not require an operation.

She insisted on getting a second opinion, and a low-dose dexamethasone suppression test was carried out at another hospital. This revealed a baseline 9:00 AM cortisol of 590 nM which fell to 35 nmol/l at 9:00 AM after 48 hours of low-dose dexamethasone, confirming that she did not have Cushing's disease. She was still insistent on having more tests, and went to a private hospital and paid for a pituitary MRI scan. Unfortunately this revealed an incidental 3 mm pituitary microadenoma, which was of no consequence, but made her more worried.

A3. What treatment is required? Mrs Peters needs to be reassured that she does not have Cushing's syndrome. She also needs to be reassured that the incidental lesion on her pituitary MRI is a normal finding in about 10% of the general population, and that surgery is not required. She is likely to go on to have annual MRIs, although these are being carried out only to reassure the patient, and to prove that the incidental pituitary lesion is not growing.

She needs referral to a dietician and an improvement in the control of her diabetes. A weight-reducing diet, if she sticks to it, should be successful at enabling her to lose weight, and more importantly to lose what she thinks is a Cushingoid appearance.

It is crucial that she does *not* have pituitary surgery, however tempting it is to the patient. Pituitary surgery might result in hypopituitarism, and she will go on to need a number of replacement hormones as discussed at the start of this chapter.

It is also important to recognise and address Mrs Peter's concerns, as she is at risk otherwise of going from hospital to hospital until she finds someone who will remove her pituitary. She needs help with weight reduction, which is becoming more and more difficult. The mechanism of control of feeding and appetite is covered more fully in Chapter 16.

Further Reading

Assié, G., Bahurel, H., Coste, J., Silvera, S., Kujas, M., Dugué, M.-A., Karray, F., Dousset, B., Bertherat, J., Legmann, P. & Bertagna, X. (2007) Corticotroph tumor progression after adrenalectomy in Cushing's disease: a reappraisal of Nelson's syndrome. *Journal of Clinical Endocrinology & Metabolism*, 92 (1), 172–9.

Martin, N.M., Dhillo, W.S., Banerjee, A., Abdulali, A., Jayasena, C.N., Donaldson, M., Todd, J.F. & Meeran, K. (2006) Comparison of the dexamethasone-suppressed corticotropin-releasing hormone test and low-dose dexamethasone suppression test in the diagnosis of Cushing's syndrome. *The Journal of Clinical Endocrinology and Metabolism*, 91 (7), 2582–6.

Meeran, K., Hattersley, A., Mould, G. & Bloom, S.R. (1993) Venipuncture causes rapid rise in plasma ACTH. *The British Journal of Clinical Practice*, 47 (5), 246–7.

CHAPTER 6

The Gonads (1): Testes

Introduction

An essential feature of all life forms is the ability to provide for the continuation of the species. In vertebrates, this process of reproduction involves the fusion of two special cells called gametes, one provided by the male and the other by the female of the species, each containing only half the normal number of chromosomes (haploid) that are present in all other eukaryotic (diploid) cells. Specifically, it is the genetic material present in these two haploid cells which combines to generate a new diploid cell by the process of meiosis. The result will be the creation of a new cell containing a unique assortment of DNA; in humans the normal chromosome number is 46, arranged in 23 pairs with 22 being somatic (autosomal) and the other being the sex chromosomes. Subsequent mitotic divisions will ultimately result in the creation of a new individual.

The male and female haploid gametes are the spermatozoa and the ova respectively, and the fusion process normally between one spermatozoon and one ovum is called fertilisation. Aspects of this process, and the growth and development of the subsequent embryonic and fetal stages which take place during pregnancy, are considered in Chapter 9.

Spermatozoa are produced by the male sex organs (gonads) called the testes, while ova are produced by the female gonads called the ovaries. The production of spermatozoa and ova is under the control of gonadal steroid hormones. Thus the gonads in both genders have two related functions: gametogenesis and endocrine steroidogenesis. In this chapter and Chapter 7, the testes and ovaries respectively will be considered in some detail.

The embryonic development of the testes

Initially in the embryo, primordial germ cells migrate from the yolk sac to the genital ridge of mesoderm on the dorsal wall of the developing abdominal cavity. Subsequent proliferation of the cells of the genital ridge results in the formation of undifferentiated gonads, alongside the

Integrated Endocrinology, First Edition. John Laycock and Karim Meeran.

developing mesonephros which gives rise to the developing urinary system. The undifferentiated (indifferent) gonads consist of mesenchymal cells which will form the matrix of the gonadal tissue, and primordial germ cells which migrate into it. At this stage primitive male and female reproductive tracts are both present in the fetus, called Wolffian and Mullerian ducts respectively. At around the sixth week, the paired indifferent gonads in a male fetus develop into testes, and this is all brought about by the presence and expression of a special gene on the Y chromosome, called the *S*ex-determining *R*egion on the *Y* chromosome (the *SRY* gene). If this gene is present, and if it is activated, then the cell becomes a Sertoli cell. Sry protein is produced, and this acts as a transcription factor inducing the synthesis of a protein called Mullerian Inhibitory Hormone (MIH). Precisely what activates the *SRY* gene at this particular stage of development is unknown, but certainly production of MIH immediately blocks any further development of the primitive Mullerian ducts which disintegrate, to be absorbed into the surrounding tissue. Meanwhile, other cells in the early testes develop into Leydig cells and these begin to secrete androgens including testosterone, which stimulate the further development of the Wolffian ducts into the male reproductive system. This consists of long, coiled seminiferous tubules lined by the Sertoli cells, and these come together to form the rete testis from which the vasa efferentia arise. These then merge into another long, coiled section called the epididymis which finally becomes the smooth muscle-lined vas deferens. The vas deferens from each of the two testes finally joins the urethra close to the point where it leaves the bladder (Figure 6.1).

From the above, it is apparent that the initial expression of the *SRY* gene on the Y chromosome results in the development of the testes from the indifferent gonads, activation of which leads not only to regression of the female Mullerian ducts but also to the production of androgens which will initiate the formation of the male phenotype. Once the indifferent gonads have been converted to testes, the *SRY* gene is no longer expressed. Very occasionally, both ovarian and testicular tissues are found to be present in certain individuals. These true hermaphrodites generally arise because they have a mix of both XY and XX cells. It is also possible (but very rare) to have both XX sex chromosomes and yet to develop testes and subsequently male characteristics. In this case, it is generally because some of the Y chromosome containing the *SRY* gene has been translocated from the Y chromosome to the X chromosome during meiosis in the father. Likewise, it is possible for this *SRY*-containing region on the Y chromosome to be deleted or mutated so that it is not expressed in genetic (i.e. XY) males. In this case, the indifferent gonads become ovaries and the phenotype develops along the female line.

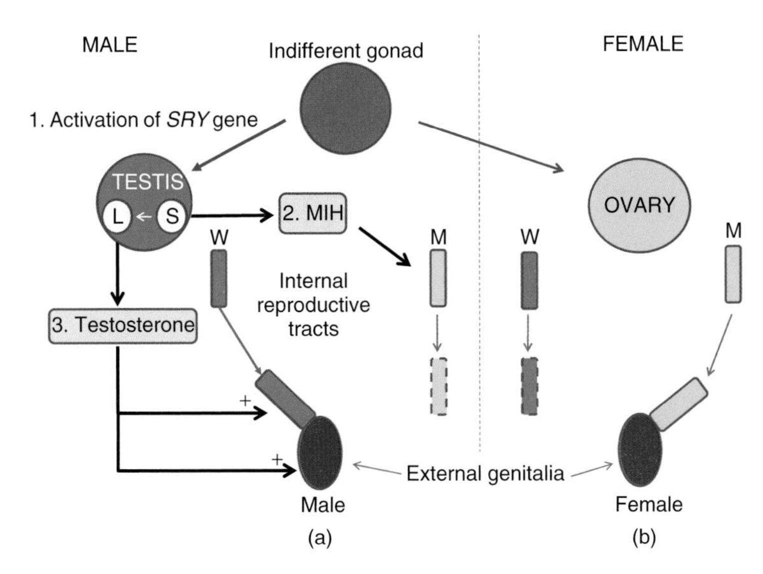

Figure 6.1 Diagram illustrating (a) the activating role of the *SRY* gene in a male (1) in Sertoli cells in the conversion of the indifferent gonads to testes (L = Leydig cells, S = Sertoli cells). (2) The Sertoli cells produce Mullerian Inhibitory Hormone (MIH) which inhibits further development of the Mullerian internal female reproductive tract (M = Mullerian), while the Leydig cells produce testosterone (3) which stimulates the development of the male internal reproductive tracts (Wolffian = W) and the external genitalia to the male form. (b) In the female, there is no Y chromosome hence no *SRY* gene, and development of the Mullerian internal reproductive tract and the external genitalia develop along the female line unimpeded.

The genetic basis of spermatozoon production

The production of haploid gametes in both sexes initially involves numerous mitotic divisions of diploid germ cells each containing a pair of sex-determining chromosomes. Normally, male cells have one X and one Y chromosome while in females there are two X chromosomes. In males, the germ cells remain quiescent until puberty when they are activated; at this stage they are called A1 spermatogonia. These cells then undergo a number of mitotic divisions, each time producing clones of identical cells. The number of cell divisions is species dependent. The final one of these mitotic divisions produces B spermatogonia, and these become primary spermatocytes which are still diploid (44XY). Not all germ cells enter each cyclic development phase; some germ cells return to the quiescent state until they too are stimulated to develop.

Primary spermatocytes are formed in the basal fluid space surrounding the seminiferous tubules. They are individual cells with their own nuclei, but they remain linked together by cytoplasmic bridges, and indeed will

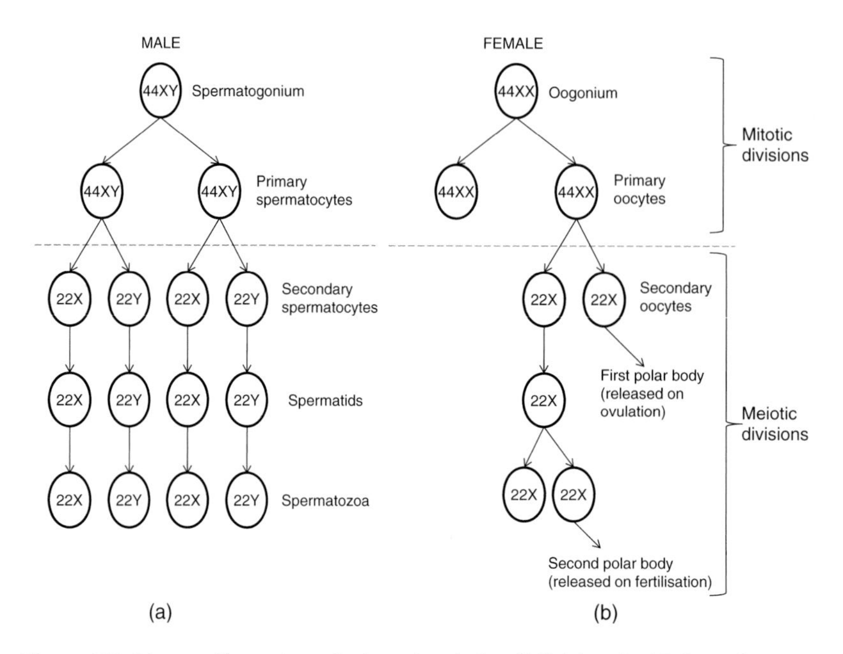

Figure 6.2 Diagram illustrating mitotic and meiotic cell divisions in (a) the male germ cell line and (b) the female germ cell line. For further details see this chapter for male cell divisions, and Chapter 7 for female cell divisions.

remain so throughout all subsequent meiotic divisions until they reach the final spermatozoon stage. They then undergo meiotic division to produce haploid secondary spermatocytes containing either 22X or 22Y chromosomes. At this stage, they penetrate between the Sertoli cells forming the seminiferous tubules by pushing through the zonal junctions linking each Sertoli cell to its neighbours. The secondary spermatocytes become attached to the Sertoli cells by gap junctions and specialised connections called ectoplasmic specialisations. The Sertoli cells play an important role in the final development process, and envelop the later stages of spermatocytes. Here, under the influence of Sertoli cell secretions, they become spermatids which develop rudimentary tails. The spermatids then continue to develop into spermatozoa which finally enter the seminiferous fluid, which is continually being secreted by the Sertoli cells (Figure 6.2).

The length of each cycle, from activated spermatogonium to spermatozoon, is species dependent, taking 64 days in humans. In some animals such as rats, different sections of the seminiferous tubule are found to have cells at a specific stage of development so that one section will have nothing but spermatozoa while the adjacent sections will have spermatids, for example. Therefore, in addition to the cyclic spermatogenic cell development there is also a spermatogenic wave as seen along the length of the seminiferous tubules, at least in some species. In man, however, any single cross-section of a seminiferous tubule is likely to show various stages of development, in wedges.

The endocrine basis for phenotype differentiation

A crucial point to appreciate is that once the *SRY* gene has been expressed and the gonads have become testes, subsequent 'maleness' is due to the hormones produced by the developing testes: MIH and the androgens. As mentioned earlier, MIH from the Sertoli cells inhibits further development of the Mullerian ducts, which subsequently regress, while androgens from the interstitial Leydig cells stimulate the further development of the Wolffian ducts and also the development of the external genitalia along the male line (penis and scrotum).

On occasion, a genetic male develops perfectly normally, with testicular formation and production of MIH and androgens, but there is a genetic defect which decreases tissue-specific androgen sensitivity. Thus the female Mullerian ducts regress quite normally, but the androgen-sensitive tissues such as the Wolffian ducts and external genitalia fail to respond to the rising androgen production. In this case, both reproductive tracts fail to develop normally (remnants may be present to differing degrees) and the external genitalia remain female. This condition, which used to be known as testicular feminising syndrome, is now called complete androgen insensitivity syndrome (CAIS). The degree to which the tissues are insensitive to androgens can of course vary, and a partial insensitivity to androgen syndrome can also develop, where the overall picture is less clear-cut. The opposite situation is also known to arise on occasion; in this case, a genetic female develops in the presence of large amounts of circulating androgens produced, for instance, by the fetal (or maternal) adrenals. In this case, the Mullerian and Wolffian duct systems can both develop to a varying extent, and the external genitalia are stimulated to become male. This condition is called congenital adrenal hyperplasia (see Chapter 9).

The anatomy and structure of the testes

In the developing human male fetus, the two testes initially lie within the abdominal cavity, but they then migrate downwards and out of the body to end inside the scrotal sac. A fibrous structure called the gubernaculum attaches the testes to the posterior abdominal wall. As the fetus elongates between weeks 10 and 15, the gubernaculum is pulled downwards together with the attached testes until they pass through the inguinal canal to reside finally in the scrotum. This developmental movement does not depend on androgens but does appear to require the presence of MIH which is believed to act on the gubernaculum. On the other hand, the passage of the testes through the inguinal canal is stimulated by androgens. Not all mammals have descended testes, but in man the surrounding temperature is usually at least 4–7°C lower than the inner core temperature and this seems to be optimal for spermatogenesis, although androgen production is unaffected. If the testes fail to descend, they remain 'hidden' within the body (cryptorchidism) and this is associated with infertility.

Arterial blood is provided by the testicular arteries which elongate as the testes descend down through the abdominal cavity. The surface of each testis receives its blood supply via the superficial and external pudendal branches of the femoral artery as well as from the perineal branch of the internal pudendal artery and the cremasteric component from the inferior epigastric artery. The blood leaving each testis passes through the corresponding veins: the femoral vein, the internal pudendal vein and the epigastric vein. Within each testis, the arterial blood passes through a coiled internal spermatic arterial system and leaves via the closely associated venous pampiniform plexus which drains blood into the venous system. The close proximity of arterial and venous systems within the testis provides a heat exchange system which is beneficial in keeping the temperature of the incoming blood lower and the departing blood higher than would be the case otherwise. Furthermore, the scrotal epithelium has a high density of sweat glands which also function to cool the testes when stimulated. The lymphatic vessels terminate in the inguinal lymph glands. The nerve supply to the testes is provided by inguinal branches of the lumbar nerve, perineal branches of the internal pudendal nerve and the pudendal branch of the posterior femoral cutaneous nerve. The spermatic cord is the name given to the bundle of arteries, veins and reproductive tract (vas deferens) from each testis passing through the inguinal canal.

The basic structure of the testis is the seminiferous tubule, which consists essentially of Sertoli cells. They continuously secrete seminiferous fluid which enters the lumen of the tubule and flows down, towards the rete testis and onwards to the vas deferens. Each seminiferous tubule is surrounded by a peritubular layer comprising a basement membrane lining a layer of myoid cells and fibrocytes. This does not represent much of a barrier to the movement of fluid from interstitium to seminiferous tubule lumen, even though it does separate the interstitial fluid, which equates with blood and lymph, from the basal fluid immediately surrounding the Sertoli cells (see Figure 6.3). What does provide a major barrier to the ready diffusion and movement of water-soluble molecules from basal to luminal fluids is the presence of multiple tight and gap junctions which completely encircle the Sertoli cells, linking them all together. Interestingly, this physical barrier develops only at puberty, when spermatogenesis begins. It probably functions to prevent the movement of spermatozoa into the blood where they could induce an immune response, and also to prevent the

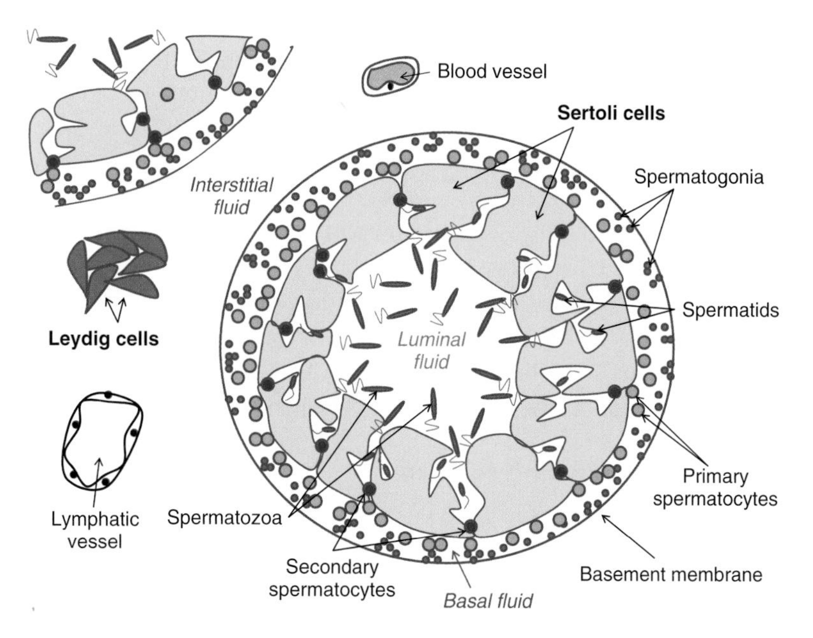

Figure 6.3 Diagram illustrating a section through a seminiferous tubule and the various stages of spermatogenesis beginning with the spermatogonia in the outer basal fluid, then the primary and secondary spermatocytes, the spermatids and the spermatozoa. The interstitial Leydig cells are the source of testosterone. For further details, see section 'The anatomy and structure of the testes'.

non-selective passage of molecules into the seminiferous tubules. Thus the later stages of spermatogenesis take place in a specialised environment (the luminal, or seminiferous, fluid) regulated by the selective transport and secretory properties of the Sertoli cells. The Sertoli cells synthesise receptors for follicle-stimulating hormone (FSH), and consequently respond to circulating levels of this hormone from the anterior pituitary. The Sertoli cells, in addition to producing many other substances including nutrients and other components present in the seminiferous fluid, are also endocrine cells, for example producing the polypeptide hormone inhibin. They also contain the enzymes necessary for converting testosterone from the Leydig cells to either the more potent androgen dihydrotestosterone or to the oestrogen 17β-oestradiol. The endocrine Leydig cells are located within the interstitial fluid compartment in between the seminiferous tubules.

The testicular hormones and their receptors

There are various hormones synthesised by testicular cells but the ones that are most closely associated with the testes, and in essence define their endocrine function, are the androgens. Other hormones such as the activins and inhibins are associated particularly with the control of the hypothalamo-adenohypophysial-gonadal axis. They will be described briefly, in this chapter and also in Chapter 8 regarding the ovaries.

The androgens

The androgens are a family of hormones which can be considered to represent the male and 'maleness' but they are nevertheless also synthesised in the ovaries of women. Furthermore, androgens are synthesised to a limited extent in the adrenal glands (see Chapter 10) and while these adrenal androgens have little importance in the male, their production here is relevant in females, particularly if over-produced.

Synthesis, storage and release

The androgens are steroids, the initial precursor being cholesterol. In the testes, the cells that synthesise androgens are the Leydig cells, located in small groups within the interstitial fluid present between the seminiferous tubules. Cholesterol can be synthesised from acetate in the Leydig cells, or it can reach the cells from the blood. The pathway followed from cholesterol resulting in the production of androgens is shown in Figure 6.4. There are two particularly important features to appreciate from this pathway: firstly, progesterone, a sex hormone in its own right and of importance in the female, is synthesised from cholesterol early on in the pathway and acts as a precursor to all the other steroids; and, secondly, the androgens themselves act as precursors in the synthesis of oestrogens which are sex hormones of particular importance in females (see Chapter 7). The first

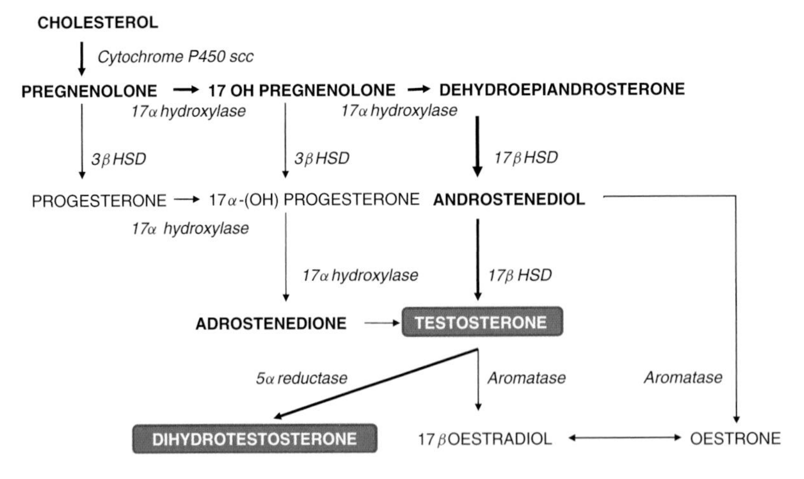

Figure 6.4 Diagram illustrating the synthesis pathway for the gonadal steroid hormones. HSD = hydroxysteroid dehydrogenase.

androgens to be synthesised are the weak dehydroepiandrosterone (DHEA) and androstenedione, and these are both precursors for the main testicular androgen, testosterone. Much of this testosterone enters the blood, and in an adult man 4–10 mg is secreted each day. It also reaches the lymph, and it is via this route that it reaches the male accessory sex glands such as the seminal vesicles and prostate. However, some testosterone also reaches the Sertoli cells of the seminiferous tubules. Since testosterone and the other sex hormones are all steroids and therefore lipid soluble (lipophilic), they can presumably cross cell membranes with ease. Testosterone can readily penetrate into the Sertoli cells forming the seminiferous tubules, and can also enter the luminal (seminiferous) fluid.

While testosterone is a potent androgen in its own right, it can also be converted by the enzyme 5α-reductase to an even more powerful androgen called dihydrotestosterone (DHT). The enzyme is present in many androgen-responsive tissues, so it is essentially a peripheral conversion that takes place where an enhanced effect is appropriate. Key target tissues include the prostate, the skin and sebaceous glands. As indicated the testicular Sertoli cells are also key target sites where this conversion can take place.

In addition, testosterone can be converted to oestrogen. Some tissues contain the aromatase enzyme which converts it to the potent 17β-oestradiol. Although oestrogens are particularly important in females, they have effects in males also. For instance, it appears that the sexual differentiation of the brain and the development of some male behaviours (e.g. mounting behaviour in rats) are actually associated with oestrogens locally synthesised from androgens acting within a specific window of time in the developing brain. Another effect, also identified quite recently in rats, indicates that the conversion of testosterone to 17β-oestradiol by Sertoli cells and its subsequent passage into the seminiferous fluid is important because it stimulates the reabsorption of water from the epididymis, resulting in the concentration of spermatozoa in this segment of the male reproductive tract. Furthermore, oestrogenic activity is linked to physiological effects such as bone resorption and the fusion of the epiphyses, as well as increased plasma lipids and adiposity, in males as well as females.

Steroid hormones being lipid soluble are capable of moving out of their cells of production as soon as they are synthesised. This means that the initiation of their synthesis occurs only when there is a stimulus for activation of the relevant enzymes involved in the process. The consequence is that there is no cellular storage form which might be readily available for release.

The synthesis of androgens is already apparent by the 10th week in the male fetus. Levels of production are always greater in males than in females (normally) and they are relatively high in the early stages of development *in utero* peaking at 2 $ng.ml^{-1}$ around 13–15 weeks followed by a decrease and plateau by 6 months, with a second peak occurring just after birth (levels as high as 3 $ng.ml^{-1}$) before levels finally decline after a few months to remain low until puberty. At this stage androgen levels rise, reaching concentrations as high as 9 $ng.ml^{-1}$, and remain at that level throughout adult life, with some decrease occurring with age, from the 60s onwards.

A feature of androgen levels in the blood is that they are very variable, as are those of many hormones. First, there is a circadian rhythm such that levels tend to be higher at night and in the morning than in the afternoon and evening. Second, the testicular androgens are released in pulses, determined by the controlling influence of the hypothalamo-adenohypophysial axis. A single random blood sample therefore may not give a correct indication of the true androgen production status.

Transport in the blood and seminiferous fluid

Once released from the Leydig cells, testosterone can enter the bloodstream readily, where it binds to a globulin protein called sex hormone–binding globulin (SHBG). As its name implies, this plasma protein also binds other 'sex hormones' such as the oestrogens and progestogens in the circulation. Not only does the binding of these hormones protect them from immediate inactivation, but also it provides an important storage form in the blood. There is a dynamic equilibrium between the various hormones and the binding protein, and this is important in appreciating that if one steroid hormone increases in the circulation, it can displace another from the common plasma protein. All the sex hormones are also bound to some extent to the circulating plasma albumin.

Testosterone can also readily pass through the Sertoli cell barrier to enter the seminiferous fluid. Here too it binds to a binding protein, but this one is slightly different in that it is more selective, and is called androgen-binding protein (ABG).

Androgen receptors

Steroids being lipid soluble, they can presumably readily traverse cell membranes and enter the nucleus which is where the androgen receptors are located. These receptors act as transcription factors when activated by the binding ligand (e.g. the androgen). The androgen receptor complex binds to a specific region of the target cell DNA ultimately inducing new protein synthesis. Within the target cell, in the presence of 5α-reductase, testosterone can be converted to the twice more potent dihydrotestosterone molecule. Both testosterone and DHT bind to the androgen receptor, the DHT molecule having the greater affinity for the receptor. Within some target cells, the testosterone can be aromatised to 17β-oestradiol by aromatase.

Actions of androgens

The androgens have many actions, particularly in males. As mentioned, they are essential in stimulating the fetal development of the internal male reproductive tract and the external genitalia, thereby establishing the male phenotype. They will also have central effects at this stage regarding the formation and density of synapses in specific parts of the brain, contributing to the development of male behaviours, for instance.

At puberty, androgens are released in much greater quantities, and contribute to the important developmental changes that arise at this time of adolescence (see Chapter 8). It would appear that some of testosterone's effects on bone are actually mediated by the conversion product 17β-oestradiol.

In the adult male, from puberty onwards, the androgens have various important effects.

Spermatogenesis

Androgens are necessary for the continued maintenance of the testes and for the process of spermatogenesis. Androgen production is under the control of the hypothalamo-adenohypophysial axis (see section 'Control of testicular function'), so it follows that spermatogenesis is also influenced by the relevant hormones from the hypothalamus and anterior pituitary. Indeed, the synthesis of androgens by the interstitial Leydig cells is stimulated by the anterior pituitary hormone, luteinising hormone (LH). However, the initiation of spermatogenesis seems to be crucially dependent on the actions of follicle-stimulating hormone (FSH) from the anterior pituitary; without it, even in the presence of androgens, the spermatogenic process does not start. The FSH binds to its own receptors in the Sertoli cells and induces various activities such as the synthesis of relevant proteins including the androgen receptor, and the mobilisation of energy substrates, which are all necessary for the initiation of the spermatogenic process. Another Sertoli cell polypeptide produced following FSH stimulation is the hormone inhibin (see section 'The activins and inhibins').

Maintenance of the male accessory sex glands

The spermatozoa released on ejaculation are in a seminal fluid medium called semen which is produced not only by the seminiferous tubules but also, and to a much larger extent, by various glands which open into the male reproductive tract. These accessory sex glands include the seminal vesicles, the prostate, the ampulla of the vasa deferentia and the bulbourethral glands, all of which provide constituents in the seminal fluid.

Once the seminiferous fluid reaches the epididymis, the spermatozoa become concentrated 100-fold as water is reabsorbed, in response to the oestrogens produced by the aromatisation of testosterone by aromatase enzymes in the Sertoli cells. The densely packed spermatozoa ($50 \times 10^8 . ml^{-1}$) are then held in the muscular reservoir provided by the vas deferens. Each vas deferens passes up and round the back of the bladder where it joins the duct from a seminal vesicle to form the ejaculatory duct which passes through the prostate, to enter the urethra (prostatic urethra section). Further fluid volume and various constituents such as nutrients are provided by the secretions of the seminal vesicles (60% of total volume), the prostate (25–30%) and the bulbourethral (also known as Cowper's) glands so that the final ejaculate contains spermatozoa at a lower concentration (see Figure 6.5). While some animals produce varying, sometimes considerable, volumes of ejaculate (e.g. the boar can produce volumes up to 500 ml), the human male can manage only 3–5 ml. The concentration of spermatozoa in the ejaculate can also vary considerably, with the ram producing an ejaculate containing up to $5000 \times 10^6 . ml^{-1}$, the boar up to $300 \times 10^6 . ml^{-1}$ and the human in the range of $50–150 \times 10^6 . ml^{-1}$.

The constituents produced by the various accessory glands also vary, and to some extent vary from species to species. For instance, in man, nutrients such as fructose and sorbitol come mainly from the seminal vesicles while inositol and glycerophosphorylcholine are mainly produced

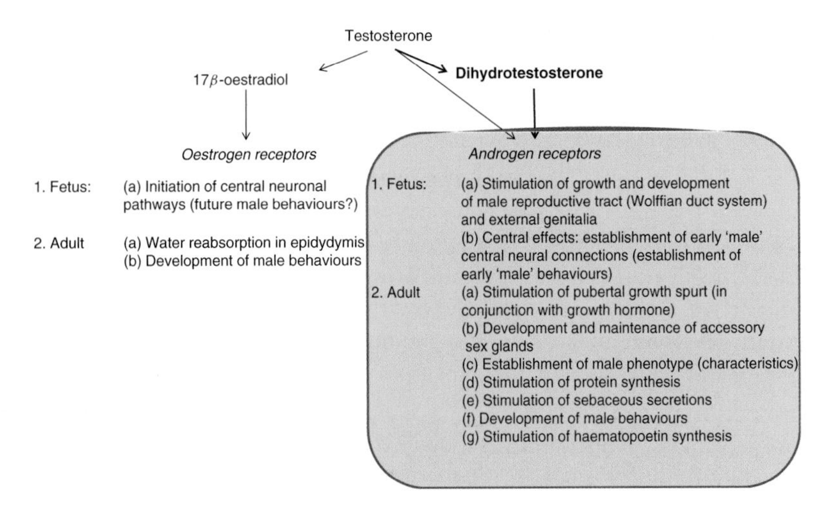

Figure 6.5 Diagram indicating the main physiological effects, in fetus and adult, of testosterone and its metabolic derivatives, the more potent androgen dihydrotestosterone and the oestrogen 17β-oestradiol.

by the epididymis. The pH of the semen is alkaline and this neutralises the more acid pH of the vaginal secretions following ejaculation. Furthermore, coagulants and anticoagulant components are present in the semen, allowing the semen to clot once deposited in the vagina and later to liquefy permitting spermatozoa to pass through the cervix, maybe at a steadier and more regular rate.

Secondary sex characteristics

The male phenotype is the result of androgenic actions on the organs and tissues of the body. Thus, shoulder breadth, narrow waist, (normally) lesser fat distribution, more muscular build, facial and body hair and receding scalp hair with age are all general masculine features associated with testosterone and DHT. Increased sebaceous secretion is also associated with androgens; acne, a common problem in juveniles, is due to the increased levels of testosterone (and other androgens) during adolescence.

Protein metabolism

Androgens are anabolic steroids. They stimulate protein synthesis by a genomic action. Consequently, males are generally larger and more muscular than females at birth as well as during subsequent growth and development, with larger internal organs such as heart and liver. Indeed, as indicated in Chapter 8, androgens in conjunction with growth hormone are responsible for the pubertal growth spurt in males (and to a lesser extent in females in whom oestrogens play the dominant role).

Haematopoeisis

Androgens stimulate the renal production of erythropoietin, a hormone which stimulates the proliferation of erythrocyte progenitor cells resulting in increased numbers of circulating erythrocytes, and consequently accompanying haemoglobin levels.

Central effects

It is clear that androgens have effects on the brain, with certain male behaviours such as competitiveness and aggression being associated with them. They are also particularly important in directing central nervous system development, *in utero* and in the neonate, in what is classically described as 'male type' behaviour. Interestingly, in rodents at least, some of these developmental effects in males may actually be due to oestrogens synthesised in neurones by aromatase enzymes converting androgens to oestrogens intracellularly. Another central effect must be the indirect negative feedback exerted by testosterone, and locally produced DHT, on the hypothalamic neurones which synthesise gonadotrophin-releasing hormone (GnRH) as described in the section 'Control of testicular function'.

The activins and inhibins

These glycoproteins have effects on the hypothalamo-adenohypophysial-gonadal axis in both genders so they are considered here as well as in Chapter 7.

Synthesis, storage, secretion and transport in the blood

Activins and inhibins belong to a large group of heterodimeric glycoproteins which also include various growth factors such as tumour growth factor-beta (TGF-β). They have molecular weights of the order of 30K, comprising two chains of amino acids linked by a single disulphide bond. Two forms of inhibin exist, identified as Inhibins A and B, and they are closely related to the other hormones called activins, of which various dimeric forms exist. Activins and inhibins are synthesised in males and females. In males inhibin is synthesised by Sertoli cells when stimulated by adenohypophysial FSH. In the circulation, inhibins probably circulate in the free unbound state. In contrast, activins are synthesised by the gonadotrophs in the anterior pituitary where they have an autocrine effect, stimulating FSH production and sensitising the cells to Gonadotrophin Releasing Hormone (GnRH). Activins are also produced by the Sertoli cells and germ cells, where they again have an autocrine or paracrine stimulatory effect.

Follistatins are glycoproteins which are also produced by gonadotrophs, as well as by folliculostellate cells, in the anterior pituitary. They bind to activin molecules, blocking activin activity.

Receptors and mechanism of action

Because inhibins tend to interfere with the actions of activins, it may be that they act via activin receptors. However, there is increasing evidence for specific inhibin-binding proteins in cell membranes, betaglycan and inhibin-binding protein/p120 (IBP/p120), both of which may be associated with activin receptors and hence mediate their activity. Little is known about the mechanism of action of inhibins.

Actions

In males, the main known action of inhibin is to inhibit the production of GnRH and FSH from the hypothalamus and andenohypophysis, respectively. They therefore form indirect and direct negative feedback loops on GnRH and FSH production. At least part of their action is to selectively block the effect of activins which stimulate FSH production by the gonadotrophs (also see Chapter 8 for actions in females).

Control of release

The inhibins are closely linked to the hypothalamo-adenohypophysial axis, and while other factors are likely to influence their production, such as gonadal steroids, the main controlling influence is FSH. Control of activin production is still unclear.

Control of testicular function

Both testicular functions of spermatogenesis and steroidogenesis are predominantly under the control of the hypothalamo-adenohypophysial axis. The predominant endocrine influence on the Leydig cells and their production of androgens is LH from the anterior pituitary gland (see Chapter 3). This glycoprotein hormone from the gonadotroph cells binds to its receptors on the Leydig cell membranes and exerts its effects on enzyme synthesis by means of its intracellular second messenger, cAMP. LH itself is under the direct control of the polypeptide GnRH from specific neurones in the arcuate and preoptic nuclei of the hypothalamus. The testes indirectly influence their own production of androgens by the negative feedback loops exerted by testosterone (and DHT) on the anterior pituitary gonadotrophs (direct negative feedback) and the hypothalamic neurones (indirect negative feedback). All-important is the pulsatile nature of GnRH release, with rapid or slower pulse frequencies preferentially stimulating LH and FSH pulsatile secretions, respectively.

Control of Sertoli cell function is also exerted by the anterior pituitary which produces FSH. This hormone, also a glycoprotein hormone (see Chapter 3), acts on its receptors on the Sertoli cells and exerts control over Sertoli cell activities by means of the same intracellular second messenger cAMP. One of the functions of the Sertoli cells under FSH control is to synthesise the protein hormone inhibin which exerts a negative feedback on the hypothalamo-adenohypophysial axis, having a direct negative inhibitory feedback on FSH production by the adenohypophysial gonadotrophs, and an indirect negative feedback on the hypothalamic GnRH neurones (see Figure 6.6). The adjacent germ cells may also have an influence on inhibin production by the Sertoli cells.

Coitus and penile erection

The process of successful intercourse, culminating in ejaculation, depends on the ability of the male to produce and maintain an erection of his penis. An inability to produce an erect penis results in impotence. The vital state of penile erection for intercourse is mainly due to central control over the arterial inflow of blood into the organ. Thus psychogenic factors (e.g. visual, tactile and olfactory) are particularly important in determining the successful development of penile erection, from the flaccid through the tumescent stage to a full erection. However, nocturnal erections (or at

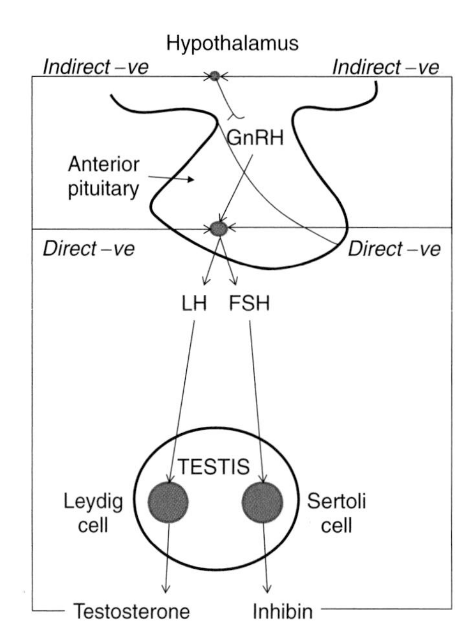

Figure 6.6 Diagram illustrating the control of testicular endocrine function, showing the hypothalamic gonadotrophin-releasing hormone (GnRH), the anterior pituitary luteinising- and follicle-stimulating hormones (LH and FSH, respectively) as well as the direct and indirect negative feedback loops involving testosterone (and its peripherally converted derivative dihydrotestosterone) and inhibin. For further details, see section 'Control of testicular function'.

least tumescence), which occur during rapid eye movement (dream state) sleep, are associated with large pulses of testosterone being released into the circulation following activation of the hypothalamo-adenohypophysial system. Exactly what the relationship between the raised testosterone levels and nocturnal penile erection is remains unclear. Indeed, it is possible that the influence of androgens on penile erection may actually be due to its central effects.

The most direct stimulation of the penis is a reflex involving sensory (afferent) fibres activated via tactile receptors along the length of the organ, which passes to the spinal cord in the internal pudendal nerves. In addition there are descending pathways from the brain where all stimuli are integrated, involving the limbic system. These descending pathways linking the brain to the penis are of tremendous importance, as indicated by many of the causes of impotence (e.g. stressors) in males. Efferent pathways include (i) parasympathetic fibres in the pelvic nerve, (ii) sympathetic fibres in the hypogastric nerve and (iii) efferent somatic fibres in the pudendal nerve.

The actual process of getting an erection involves the vasculature of the penis. Arterial blood enters the penis mainly from the internal pudendal artery into the cavernosal arteries. Some blood passes along the length of the penis along the dorsal artery which provides blood to the glans and the skin of the penis, and has no role in the development of the erection. The arterial blood then enters two interconnected sinusoidal spaces called the corpora cavernosa which run the length of the penis and are surrounded by a tough fibrous capsule which provides rigidity to the entire structure. Venous blood then passes out of the penis, ultimately into the pudendal vein. The urethra, surrounded by a venous sinusoidal space called the corpus spongiosum, passes along the lower part of the penis to its orifice at the tip of the glans (see Figure 6.7).

Under resting conditions the penis is flaccid, and blood flow entering the organ is restricted by the tonic inhibition exerted by the sympathetic innervation to the arterial smooth muscle. Indeed, the administration of β-adrenergic blockers results in the development of a state of tumescence. When the penis is stimulated, either by the reflex or from the brain as a result of psychogenic stimuli, then increased parasympathetic activity via the pelvic nerve induces a direct dilation of the arterial walls. This is a rare

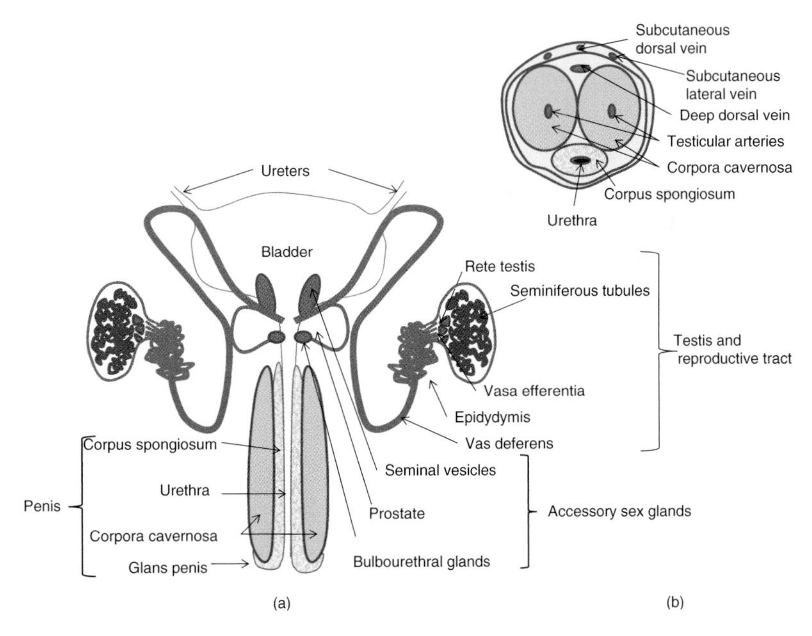

Figure 6.7 Diagram illustrating (a) a longitudinal section through the penis, testes, bladder and related glands and tracts, and (b) a cross-section through the penis illustrating the arrangement of tissues around the blood vessels and urethra. See section 'Coitus and penile erection' for further details.

example of parasympathetic control over a special part of the vasculature. Consequently there is an increase in blood flow into the corpora cavernosa which expand until they cannot any further, because of the fibrous sheath surrounding them. The pressure exerted by the expanded corpora cavernosa collapses the vein resulting in a decreased venous outflow. The result of this clever piece of 'bioengineering', driven by an increased inflow and a consequent decreased outflow of blood, is that the penis becomes engorged with blood and becomes erect. Without some form of protection, the increase in pressure within the penile vasculature would also cause the collapse of the central urethra, through which the semen would normally be ejaculated. Such a collapse is prevented because of the nature of the corpus spongiosum surrounding the urethra. As its name implies, this sinusoidal cavity has a structure that gives it a 'spongy' appearance, and it does not collapse as completely as the corpora cavernosa, this allowing the urethra to remain patent (Figure 6.7). Any androgenic effects on penile erection appear to be related to the spontaneous pulsatile release of testosterone driven by the hypothalamo-adenohypophysial axis (e.g. nocturnal erections) and do not appear to be of physiological relevance, at least regarding the act of copulation.

Stimulation of the penile arterial vessels by the activated parasympathetic nerve fibres is induced by the release of the neurotransmitter acetylcholine and vasoactive intestinal peptide (VIP), both of which are associated with the synthesis of nitric oxide. Nitric oxide actually exerts its effect by stimulating the synthesis of cyclic guanosine monophosphate (cGMP), the actual vasodilator in vascular smooth muscle. cGMP is inactivated by a specific intracellular enzyme, phosphodiesterase (type 5). The drug sildenafil (Viagra), which inhibits this enzyme, blocks the degradation of cGMP thus prolonging and enhancing the vasodilatory effect of nitric oxide. Maintenance of a full erection with this drug can be beneficial in patients suffering from erectile dysfunction due to specific causes.

The purpose of penile erection is to provide the means to deliver the spermatozoa present in semen into the vagina of the receptive female. While the penile erection is mainly under parasympathetic control, the semen is ejaculated following sympathetic stimulation of the vas deferens' smooth muscle which contracts, as do the accessory sex glands including the prostate. Sympathetic tone is then also restored to the penis, which becomes flaccid once more.

Diseases of the testis

Klinefelter's syndrome

The commonest cause of primary testicular failure is a chromosomal abnormality with an extra chromosome (47 XXY) called Klinefelter's syndrome, where primary testicular failure occurs as an adult. This has a prevalence of about 1 in 600 to 1 in 1000 males. It was first described in patients with small testes, azospermia, gynaecomastia and elevated circulating gonadotrophin levels, particularly FSH. It is likely that many cases remain undiagnosed and untreated throughout life. Patients will only present to medical attention if they try to father a child, or complain of the gynaecomastia, or if it is discovered during screening. In one study, it was found in 12% of patients presenting with azospermia.

Patients with Klinefelter's syndrome may present to medical attention at around the age of puberty, and tend to have a normal childhood. The early stages of puberty are normal. It appears that the growth velocity can be slightly higher than normal before puberty, and the final height may therefore be greater than average, with a mean adult height at the 80th centile for normal men. They have relatively long arm and leg span compared to their overall height, and are usually tall and thin. Examination of the external genitalia reveals small, firm testes. The testicular volume is measured using an orchidometer, which basically enables the comparison of the testes to standard ovoid spheres. The normal adult testicular volume is 15–25 ml, but patients with Klinefelter's syndrome often have volumes of between 1 and 4 ml.

Gonadotrophin levels start to rise at puberty, whereas testosterone levels, which are normally exuberant through puberty, seem to reach a plateau at about 10 nM. The reference range for testosterone is 10–26 nM, usually peaking in early puberty. After the age of about 30 years, testosterone levels gradually fall, and loss of libido occurs. The high levels of circulating gonadotrophins (LH) increase aromatisation of testosterone to oestradiol, which increases marginally. However, the sensitivity of oestrogen receptors for oestradiol is about 1000 times higher than the testosterone receptor is for testosterone. The reference range for oestradiol in females is thus measured in pM, with levels of between 500 and 1000 pM being normal for adult females. This corresponds to levels of between 0.5 and 1.0 nM. Thus less than 1% aromatisation of testosterone in Klinefelter males (with 10 nM testosterone), will result in levels of 100 pM oestradiol, which will cause significant gynaecomastia. The breast is extremely sensitive to these concentrations of oestradiol.

Patients who want treatment for their gynaecomastia will need plastic surgery or liposuction, as well as testosterone replacement therapy. Once present, gynaecomastia will not respond to testosterone replacement.

The IQ of boys with Klinefelter's syndrome is slightly lower than normal. In one study Klinefelter boys had a mean IQ of 92 compared to a control population of 103, with 100 being the average for the population as a whole.

Occasionally, pregnancy in the partner can be achieved from testicular biopsy followed by intra-cytoplasmic sperm injection. Exceptionally, spontaneous paternity has been described from a Klinefelter father, and when sperm are present in the ejaculate, the majority are normal.

Klinefelter patients with chromosome mosaics (47,XXY/46,XY) may show very few clinical symptoms. Testicular histopathology in adult men with Klinefelter's syndrome classically shows germ cell aplasia, total tubular atrophy or hyalinising fibrosis, and relative hyperplasia of Leydig cells. However, in some adult Klinefelter patients, foci of spermatogenesis up to the stage of mature testicular sperm can be detected.

Primary testicular failure

Primary testicular failure may be due to autoimmune damage to the testes, although investigations are often negative. Autoimmune testicular failure occurs more commonly in individuals who have other autoimmune endocrine diseases, such as hypothyroidism, Addison's disease and pernicious anaemia. Patients who have primary testicular failure with other autoimmune diseases may have polyglandular autoimmune syndrome type 2. As with Klinefelter's syndrome, blood testosterone levels fall, and gonadotrophin levels rise significantly.

Patients may suffer from loss of libido, tiredness and anaemia, and they may notice that they need to shave less frequently.

Testicular cancer

Patients may notice a lump in a testis. If confirmed, the testis will need to be removed, and if both testes are removed (in cases of bilateral carcinoma) then testosterone replacement will need to be instigated. Testicular tumours occur most commonly between the ages of 20 and 40. Most are malignant and most arise from germ cells. Teratomas are said to occur with a peak age incidence of 20–30 years, while seminomas occur in slightly older individuals with a peak incidence of 30–40 years.

Epididymitis and epididymo-orchitis

Epididymitis relates to the swelling or inflammation of the epididymis. This occurs most commonly between the ages of 19 and 35. Unprotected sex or having multiple sex partners increases the risk of infectious epididymitis.

The epididymis is a long, narrow, tightly coiled tube attached to the upper part of each testicle and located at the posterior aspect of the testis, that accommodates the storage of sperm as they maturate and are transported from the efferent ducts to the vas deferens. Patients can develop fever, chills, tenderness or a heavy sensation and pain in the scrotum. The scrotum may become red, swollen and warm. The inflammation may be a result of bacterial (especially gonorrhoea) or chlamydial infection.

Epididymitis must be differentiated from testicular torsion which also causes acute pain.

Testicular torsion

If the testis rotates about its axis and the spermatic cord twists, the blood supply can become compromised. This results in abdominal pain, because the nerve and blood supply to the testis originates in the abdomen. The testicle may infarct, and once that happens, it needs to be removed. However, if it is discovered in time, this requires emergency surgery to protect the vascular supply to the testicle. It presents most commonly in adolescents, but can present at any age.

Varicocoele

A varicocoele is a collection of varicose veins in the scrotum. It is common, affecting one in fifteen young adult males. It is more common on the left because the left testicular vein enters the left renal vein, where the pressure is slightly higher than on the right. It feels like a 'bag of worms', and if painful, should be treated surgically.

Hydrocoele

A hydrocoele is the result of excessive fluid in the tunica vaginalis (the serous space surrounding the testis). The testis cannot be palpated as it lies within the fluid collection. It usually presents as a soft, non-tender and cystic swelling in the scrotum which transilluminates if one can get above the lesion.

Primary hydrocoeles occur in the absence of disease in the testis. They tend to be large and tense, and are more common in young boys. Secondary hydrocoeles represent a reaction to some form of testicular pathology. They tend to be small and lax. They are more common in adults and occur because of testicular tumours or infection.

Androgen insensitivity syndrome (AIS, or testicular feminisation)

This is a condition caused by a defective testosterone receptor. This is coded for on the X-chromosome, so the condition is X-linked. The androgen receptor in the pituitary is also insensitive, so the pituitary gland is not suppressed by testosterone. Thus the pituitary gland makes large amounts of gonadotrophins, unchecked by testosterone. Circulating levels of LH, FSH and testosterone are therefore all very high. Some of the testosterone is converted to oestradiol, and the patient is thus exposed to functionally active oestrogen which is regulated by the pituitary.

Because all of the patient's testosterone receptors do not work, there are no external male features evident. The patient is thus a phenotypic normal female with testes that synthesise testosterone (which has no biological activity in the patient) and which can be aromatised to oestradiol, which takes the patient through a normal apparent female puberty, with normal breast development. The testes may be palpated in the groin. Pubic and axillary hair (growth of which are stimulated by testosterone) are usually absent.

In complete AIS, the patient externally appears to be a normal female, apart from palpable testes in the groin. If not noticed, patients may not present till they become aware of primary amenorrhoea. They have normal external female genitalia, but a short blind-ending vagina, without a uterus.

Patients may go through a normal (female) puberty and the testes should ideally not be removed until puberty is complete. Once puberty is complete, some authors suggest gonadectomy to reduce the risk of testicular malignancy, assuming that there is a risk of this occurring in an undescended testis. If gonadectomy is performed, patients will require support, with hormone replacement therapy, this being an oestrogen-containing preparation. A progestogen is not required, as there is no uterus present.

Testosterone replacement

Esters of testosterone such as Sustanon (containing testosterone propionate, testosterone phenylpropionate, testosterone isocaproate and testosterone decanoate) have been around for many years. This is one of the most cost-effective treatments. Injections are given monthly, and peak and trough levels of testosterone should be monitored. Other options include oral testosterone undecanoate (restandol, 40 mg tds), buccal testosterone, testosterone gels that need to be applied daily to an area of skin, and Depot injections every 3 months that are significantly more expensive than the monthly injection. Testosterone implants (600 mg every 6 months) are also very cost effective, but the cost of the procedure needs to be considered.

For all these methods of replacement, monitoring of the plasma testosterone level is most effective as it is almost impossible to suppress the gonadotrophins without administering a large excess of testosterone.

Further Reading

Fafioffe, A., Ethier, J.-F., Fontaine, J., JeanPierre, E., Taragnat, C. & Dupont, J. (2004). Activin and inhibin receptor gene expression in the ewe pituitary throughout the oestrous cycle. *Journal of Endocrinology*, 182, 55–68.

Gillies, G.E. & McArthur, S. (2010) Estrogen actions in the brain and the basis for differential action in men and women: a case for sex-specific medicines. *Pharmacological Reviews*, 62, 155–98.

CHAPTER 7

The Gonads (2): Ovaries

Introduction

As with the testes in the male, the female ovaries are essential for the procreation of the species. Similarly, the ovaries have two main functions: gametogenesis and steroidogenesis. The mature haploid gametes are called ova, while the main ovarian steroids are hormones called progestogens and oestrogens, the latter being derived from androgenic precursors. The production of ova is cyclical and is regulated by the gonadal steroids. In females the production of ova, which may be fertilised by male spermatozoa usually following intercourse, is only part of the procreation process. Indeed, should an ovum be fertilised, then implantation of the developing blastocyst into the uterine wall requires the simultaneous cyclic development of the uterus, specifically the endometrial lining.

These regular ovarian and endometrial events take place in all mammals. They are called oestrous cycles in some animals such as rodents, in which each cycle is 4 days long, or menstrual cycles in primates such as humans in whom each cycle lasts approximately one month. Not surprisingly, the two component cycles, ovarian and endometrial, are closely linked by an endocrine regulation involving the hypothalamus, the anterior pituitary and the ovaries (the hypothalamo-adenohypophysial-ovarian axis).

If the ovum is fertilised, then the process of implantation is followed by the early development of an embryo (up to the eighth week in humans) by which time clear features become apparent. Subsequent development is then related to the fetus which continues to grow in association with the placenta, which itself develops partly from the fetus and partly from the mother. The whole development phase takes place inside the uterus (womb) of the mother in all mammals, right up to birth, and is called pregnancy. The endocrine control of the events of pregnancy, the birth of the baby and the subsequent process of lactation is considered briefly in Chapter 9.

Integrated Endocrinology, First Edition. John Laycock and Karim Meeran.

Embryonic development of the ovaries

As indicated in Chapter 6, initially undifferentiated gonads are formed and then, in an XY male the *SRY* gene is expressed in Sertoli cells of the developing testes. In the absence of *SRY* gene expression (i.e. in an XX female), the gonads become ovaries and the internal reproductive tracts (called Mullerian ducts) develop to form the Fallopian tubes, uterus and upper part of the cervix. However, while in the male fetus the primitive medullary sex cords (specifically the developing rete testis) establish contact with the germ cell cords, in the female the medullary cords degenerate so no direct contact is made between the developing germ cells and the Mullerian duct. The consequence of this is that later in life, when the ova mature, there is no tubular transport system available along which they can enter the Fallopian tubes. Indeed, the tubular system in the female becomes the follicles surrounding the ova. Thus they have to be shed from the follicles (and ovaries) into the peritoneal cavity, to be caught by the fimbriae of the Fallopian tube openings (see section 'The ovarian cycle'). Meanwhile, again in the absence of testosterone from the developing testes in a male, the external genitalia automatically develop along the female line, with the formation of the outer section of the vagina, the labia and the clitoris. It would seem that *in utero*, unlike the male, the female fetus develops without the need for ovarian hormones, and the female phenotype develops spontaneously, by default.

The genetic basis of ovum production

In the male, spermatogenesis is initiated by androgens produced from the Leydig cells at puberty and throughout the rest of a male's life. Spermatozoa are produced continually in their millions (see Chapter 6) the process being dependent upon some of the spermatogonia returning to the quiescent state until reactivated at a later stage. In the female fetus, the initial mitotic divisions in the germ cells (oogonia) result in the formation of approximately 6 million primary oocytes (diploid number of chromosomes, i.e. 44+XX) which then enter the first of the meiotic divisions until the prophase stage, when the process is arrested. There are two events that can then take place: firstly, the number of oocytes diminishes steadily as they become absorbed into the surrounding stromal tissue by a process called atresia; and, secondly, from puberty ova (within their follicles) enter cyclic developmental processes called ovarian cycles. Regarding the total number of oocytes, at the time of birth there are probably only about 2 million, while at the start of the regular menstrual cycles in a woman (at menarche, the time when the first ovulation takes

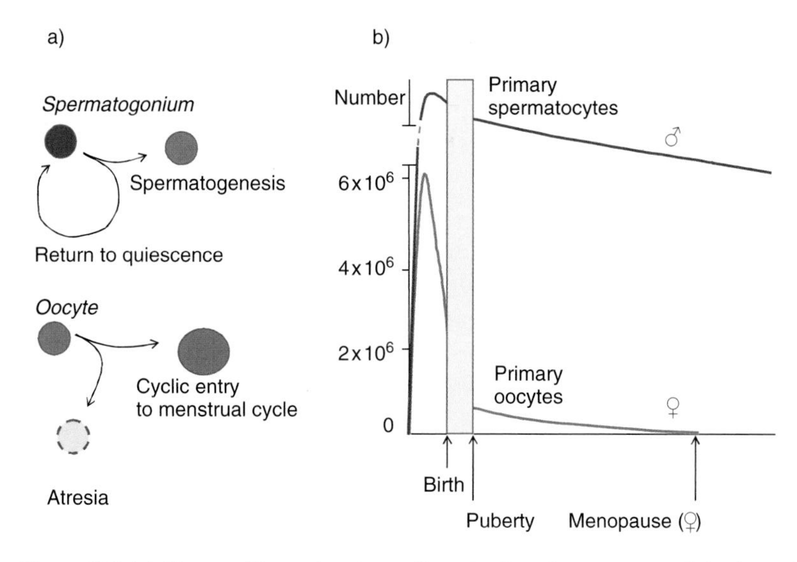

Figure 7.1 (a) Diagram illustrating the cyclic activation of spermatogensis in the male (upper) and female (lower); (b) diagram illustrating the productive capacities of men and women during the life cycle, indicating the continuous production of spermatozoa in men and the cyclic production of ripe eggs during the reproductive lives of women.

place), there are probably fewer than 250 000. During each menstrual cycle, the number of developing follicles entering further development is around 15–20 from each ovary, the number being particularly influenced by the amount of follicle-stimulating hormone (FSH) in the circulation. Unlike men, the reproductive life of a woman is dictated by the diminishing number of follicles, so that by approximately the age of 50, women have no oocytes left, reproductive capacity is lost and they are consequently infertile. This time of life is known as the menopause (Figure 7.1) and is associated with the lack of oestrogens. The period of time over which the menstrual cycles become at first irregular and then finally cease, together with the accompanying diminishing physiological and psychological effects of oestrogens, is called the climacteric.

The final maturation of the ovum during each menstrual cycle is associated with the restart of the meiotic process that was arrested at prophase, resulting in the formation of two cells. While both cells are haploid, containing 22XX chromosomes, one of these, called the secondary oocyte, contains most of the cytoplasm. The other is called the first polar body and is essentially a bag containing the other chromosomes, which then dies. The secondary oocyte then enters the second meiotic division which again becomes arrested, this time at the metaphase stage. The secondary oocyte is then released at ovulation. Final completion of the second meiotic division, with the formation of a second polar body which is also discarded, only takes place if, and when, the ovum is fertilised.

The anatomy and structure of the ovaries and related structures

The ovaries are oval structures (dimensions approximately 4 × 2 × 1 cm) located on either side of the uterus (the womb). They lie within the abdominal cavity in the pelvis, each attached to the posterior wall by the suspensory ligaments and to the uterine broad ligament by the ovarian mesentery, a wide fold of the peritoneum extending from the uterus to the wall of the pelvis on either side known as the mesovarium. The area of each ovary which is attached to the mesovarium, and through which the blood vessels, lymphatics and nerves enter, is called the hilum. The arterial blood supply to the ovaries is derived from branches of the ovarian and uterine arteries which penetrate the stroma of the central medulla as small spiral arteries, finally reaching the outer cortex. Capillary blood collects in a large venous plexus called the pampiniform plexus, before passing out of the ovary via the ovarian vein which forms at the hilum. The nerve supply to the ovaries is autonomic, mainly sympathetic fibres, arising from a number of sources including the intermesenteric and hypogastric nerves.

As indicated, each ovary consists of an outer cortex and an inner medullary stroma comprised mainly of connective tissue. Aligned along the outer cortical edge of the ovary is the layer of primordial germ cells, frozen at the prophase divisional stage. Within this layer, also in the cortical region, are found the primary oocytes, each surrounded by a ring of granulosa cells and an outer basement membrane, the *membrana propria*, together forming primordial follicles (Figure 7.2). Until puberty, the primordial follicles remain at the arrested prophase stage or undergo atresia. From puberty onwards, however, a few of the primordial follicles enter a process of development each day. As these follicles begin to grow by more than 10-fold, each primary oocyte doubles in size and secretes a layer of glycoproteins which forms the surrounding *zona pellucida*. The granulosa cells, now forming an avascular layer several cells deep, are linked to each other by gap junctions, and to the central oocyte via cytoplasmic processes which penetrate the *zona pellucida* and form gap junctions with the oocyte surface. Meanwhile, the developing follicles accumulate a layer of cells from the stroma called thecal cells at which stage they are called primary (or preantral) follicles. The spindle-shaped thecal cells proliferate to produce two distinct layers, the theca interna around the granulosa cells, and an outer fibrous theca externa. The granulosa cells of the primary follicles then

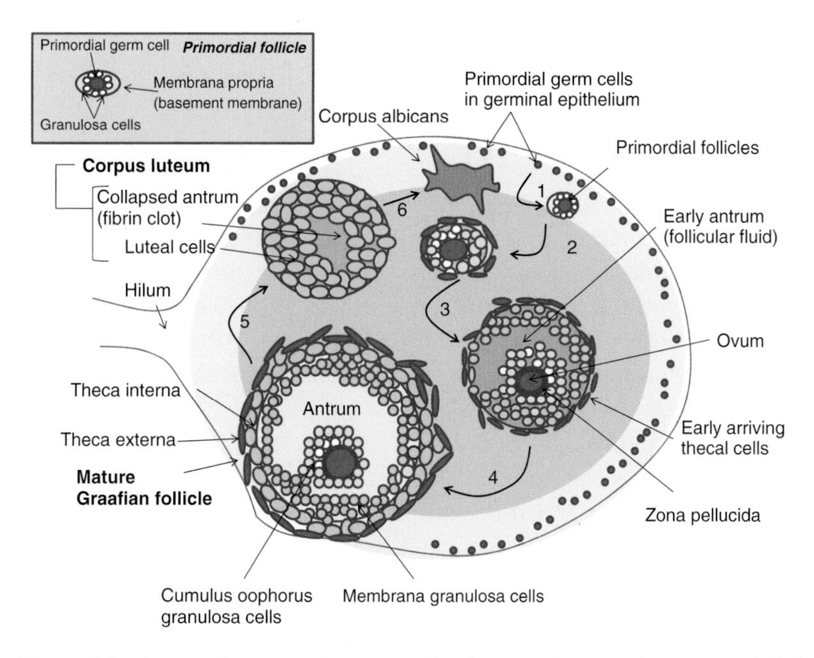

Figure 7.2 Diagram illustrating the various developmental stages of an ovarian follicle. (1) Formation of primordial follicles from primordial germ cells (see also inset). (2) Development of the preantral follicle with its early accumulation of outer thecal cells. (3) Formation of a growing antral follicle with the follicular fluid collecting in the centre. (4) Formation of the single large Graafian follicle. (5) After ovulation, the formation of the corpus luteum. (6) By the end of the cycle, the corpus luteum is inactive and becomes the corpus albicans which will become absorbed into the stroma.

begin to produce a viscous fluid containing mucopolysaccharides which accumulates in the centre forming an antrum. This stage of development is known as the antral stage. The fluid secretion separates an inner ring of granulosa cells surrounding the oocyte, called the *cumulus oophorus*, from the outer layer of granulosa cells, contact between the two layers being maintained by a narrow 'stalk' of cells (see Figure 7.2). Further development of one (usually) of the antral follicles, the largest, results in that one becoming the pre-ovulatory, or Graafian (after de Graaf, a Dutch anatomist), follicle. All the others that entered that particular cycle stop developing any further and undergo atresia.

While it is still unclear what signal induces the continuous trickle of primordial follicles to enter each menstrual cycle, subsequent development of the preantral and antral follicles, and survival of the Graffian follicle, all occur due to endocrine signals produced not only by the developing follicles themselves but also by hormones from the hypothalamo-adenohypophysial axis.

The menstrual cycle

The cyclic production of follicles, with their ripening oocytes, which occurs regularly throughout a woman's reproductive life from puberty to the menopause, is called the menstrual cycle. Each cycle lasts approximately one month hence its name (from the Latin for month, *mensis*). The average length is 28 days, although it can vary from 22 to 40 days. The beginning of each cycle (day 1) actually marks the start of the final process of the previous cycle. The menstrual cycle in fact can be considered as two interrelated cycles: (i) ovarian, and (ii) endometrial. These two cycles are related by the production and activity of specific hormones.

The ovarian cycle

The ovarian cycle is comprised of three parts: (i) an initial follicular phase, (ii) the release of a ripened oocyte (or ovum) and (iii) the subsequent luteal phase. Day 1 of each menstrual cycle is taken as the first day of menstruation, when the endometrial lining of the previous cycle is shed. The loss of blood and necrosed tissue which pass out of the body through the vagina is an easily identifiable marker indicating that menstruation has begun. Soon after the beginning of each ovarian cycle which is the start of the follicular phase, some of the preantral follicles start to grow in size, and develop into antral follicles under the influence of increasing levels of Follicle Stimulating Hormone (FSH) and luteinising hormone (LH) from the anterior pituitary. The multiplying thecal cells synthesise LH receptors and in response to LH synthesise steroid hormones called androgens in a series of conversions starting from the precursor molecule cholesterol. Cholesterol itself is either synthesised from acetate within the cells, or reaches the cells from the circulation. The two main androgens synthesised by the thecal cells are testosterone and mainly its precursor, androstenedione. Meanwhile, granulosa cells synthesise FSH receptors, and thus can respond to the increasing FSH level in the circulation by stimulating the synthesis of an aromatase enzyme. The androgens produced by the thecal cells reach the nearby granulosa cells and are converted to hormones called oestrogens by aromatisation. The principal oestrogen produced by the ovaries is 17β-oestradiol, with small amounts of a precursor oestrogen called oestrone also being released. As the antral follicles grow, the granulosa cells begin to synthesise oestrogen receptors (ERs) and, in the presence of the increasing amounts of oestrogens being produced locally, respond with further growth and development (Figure 7.3).

From days 1 to 6, the oestrogen levels begin to rise in the follicular fluid which is being secreted by the granulosa cells, and also in the general circulation. As the oestrogen levels in the circulation increase, they gradually exert an increasing inhibitory effect on the production of FSH and LH. As FSH levels in particular begin to fall, those antral follicles

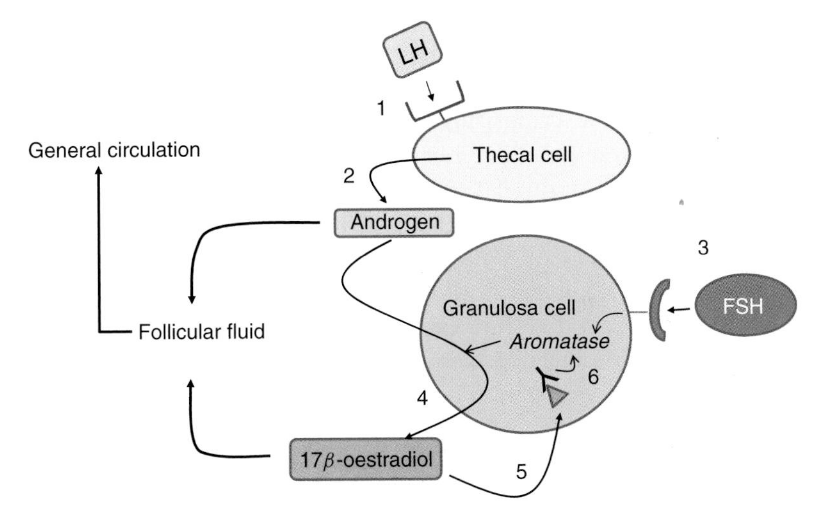

Figure 7.3 Diagram illustrating the collaboration between thecal and granulosa cells in the synthesis of 17β-oestradiol. LH = luteinising hormone and FSH = follicle-stimulating hormone. (1) LH binds to its receptor on the thecal cell and (2) stimulates androgen production. The androgen is aromatised to 17β-oestradiol in the granulosa cell by aromatase (4). Meanwhile, FSH binds to its receptor on the granulosa cell (3) and activates the aromatase. The 17β-oestradiol can enter the follicular fluid and enter the general circulation, and it can also bind to oestrogen receptors in the granulosa cell (5) to further activate more aromatisation of incoming androgen molecules (6).

still dependent on this hormone for their growth are no longer sufficiently stimulated and the process of their growth is arrested. The granulosa cells stop synthesising proteins and accumulate lipid droplets, the oocytes then die, and these follicles are destroyed by invading macrophages and leukocytes, the whole process being called atresia.

In humans, the largest antral follicle, which is no longer FSH dependent, can continue to grow under the influence of its own oestrogens, since it now has sufficient oestrogen receptors; this is a local (paracrine and autocrine) positive feedback effect. The oestrogen levels in the circulation, from days 6 to 13, increase exponentially to reach concentrations of 800 pmol.L^{-1} or more, while in the follicular fluid the concentration reaches levels up to 10000-fold higher. While some species may have multiple follicles reaching full maturity, in humans usually only one antral follicle reaches full size and development, and it is called the Graafian follicle. Towards the end of the follicular phase, if the very high concentrations of oestrogen in the circulation are maintained for a minimum of 36 hours, the feedback influence on the anterior pituitary gonadotrophs changes from negative to positive, and LH (and to a lesser extent FSH) production and release increase. This is known as the LH surge, and it is accompanied by a smaller surge in FSH. Within a few hours of the LH surge, various crucial changes occur within the Graafian follicle.

While LH cannot directly bind to the oocytes, it is nevertheless essential for initiating the final stages of the first meiotic division with the formation of the secondary oocyte (which has half the number of chromosomes and nearly all of the cytoplasm) and the first polar body (which has the remaining chromosomes and barely any cytoplasm), and the process of ovulation itself. At this late stage of development, usually by day 14, the outer granulosa cells synthesise LH receptors and in response to the high LH levels start to synthesise progestogen hormones, specifically progesterone and its precursor 17α-hydroxyprogesterone. The Graafian follicle is now quite large and is so close to the ovarian wall that it bulges out. The thin layer of thecal and outer granulosa cells closest to the ovarian wall degenerates, at least partly as a result of the LH-induced synthesis and release of proteases and other enzymes. This becomes the point (the *stigma*) from which the secondary oocyte and the first polar body are released at ovulation. The first polar body is phagocytosed, while the secondary oocyte is released directly into the peritoneal cavity. Here, it is caught by the *fimbria*, the finger-like projections around the opening of the Fallopian tube, which it then enters.

As a consequence of the eruption of the follicle with the loss of the secondary oocyte and the follicular fluid into the peritoneal cavity, the remaining cells collapse upon themselves, the *membrana propria* between the thecal and granulosa cells disintegrates and the follicular remnant becomes what is known as the corpus luteum, or luteal body. The original granulosa cells become large and are called luteal cells because they start secreting a yellowish carotinoid pigment called lutein. The remaining thecal cells become smaller luteal cells. All luteal cells respond to LH and FSH which stimulate the synthesis of the gonadal steroid hormones. The large luteal cells appear to be the source of 17β-oestradiol derived from androgenic precursors, while the small luteal cells are the main source of progesterone and, in humans, 17α-hydroxyprogesterone. The subsequent 14 days of the ovarian cycle following ovulation comprise the luteal phase, the corpus luteum becoming the important provider of the oestrogens and progestogens necessary for the maintenance of the secretory activity of the endometrium (Figure 7.4). As these steroid hormone concentrations increase in the circulation, they together exert a negative feedback on the hormones of the hypothalamo-adenohypophysial axis; consequently their production decreases. The corpus luteum now receives less stimulation by LH and FSH, and so it is no longer maintained and its production of steroid hormones diminishes. It gradually regresses and is absorbed into the ovarian stroma. As the oestrogen and progestogen levels decrease, their negative feedback influence on the hypothalamo-adenohypophysial system diminishes. LH and FSH levels then begin to rise, stimulating the beginning of the next cycle.

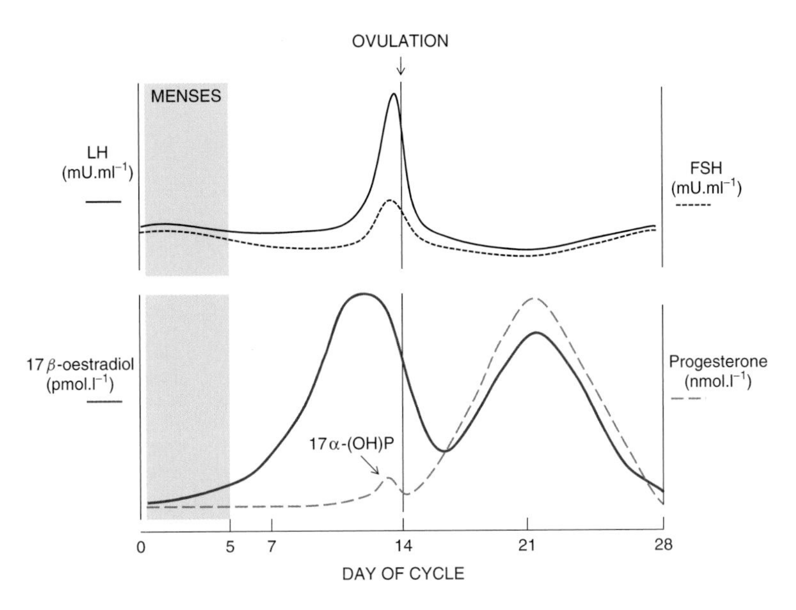

Figure 7.4 Diagram illustrating the changing pattern of gonadal steroids (17β-oestradiol and progesterone) and the gonadotrophins LH and FSH during the course of a typical menstrual cycle. 17α-(OH)P = 17α-hydroxyprogesterone.

The endometrial cycle

The female of all mammalian species normally provides not only half of the genetic material necessary for the creation of a new *conceptus*, but also the immediate environment within which the subsequent stages of development, from blastocyst to embryo, and ultimately to fetus, will occur. This environment is provided by the uterus, and the endometrial cycle refers to the changes which occur in the endometrial lining, beginning on day 1 with the loss of the two outer layers of the endometrial lining which had developed during the previous cycle, leaving only the basal layer.

The endometrial cycle consists of two phases, proliferative and secretory, which parallel the ovarian follicular and luteal phases. During the follicular phase, accompanying the growth and development of a mature Graffian follicle in the ovary, the endometrium recovers from the shedding of the previous outer layers of the lining produced over the previous cycle. The cells of the basal (*basalis*) layer proliferate and form the outer functional layer (*functionalis*), with the spiral arteries and secretory glands growing within the thickening uterine wall. This is the proliferative phase, and it is stimulated by the increasing levels of oestrogens being produced during the follicular phase of the ovarian cycle. Following ovulation at mid-cycle, the

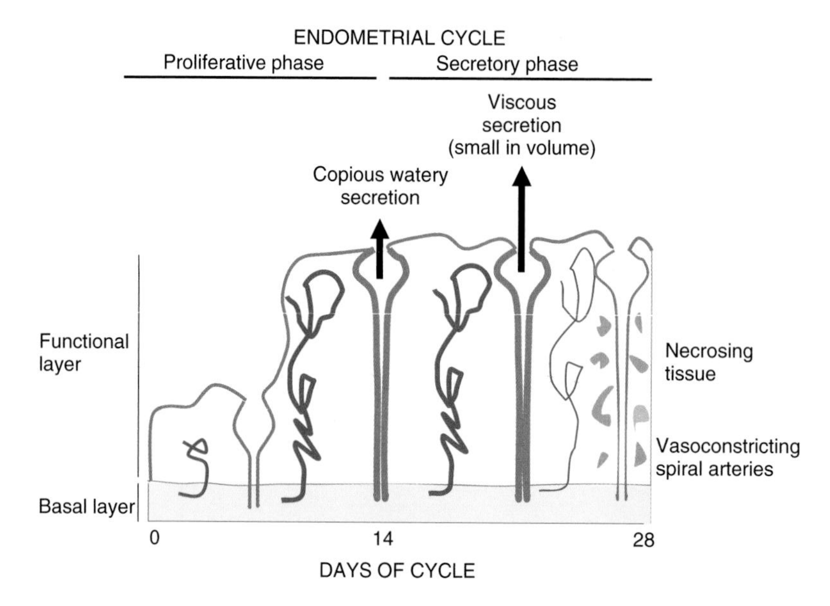

Figure 7.5 Diagram illustrating the endometrial changes occurring during the course of a typical menstrual cycle.

corpus luteum in the ovary now secretes not only 17β-oestradiol but also progesterone. The combination of these two hormones stimulates the new endothelial lining to secrete a fluid rich in nutrients and other molecules, in preparation for possible implantation should the ovum, released into the Fallopian tube, be fertilised by a spermatozoon. This part of the endometrial cycle is called the secretory phase (see Figure 7.5).

The relationship between the ovarian and endometrial cycles is clearly vital, and regulation of the latter depends on the hormones produced during the former in a sequence shown in Figure 7.6.

The ovarian hormones

Various polypeptide hormones are synthesised by ovarian cells, such as activins, inhibins and insulin-like growth factor 1 (IGF1). However, the principal ovarian hormones are steroids which can be grouped together as oestrogens and progestogens.

Oestrogens

Synthesis, storage, release and transport

By definition, oestrogens are molecules, natural or synthetic, which bind to specific oestrogen receptors and stimulate endometrial proliferation. The naturally occurring oestrogens are derived from androgenic precursors as

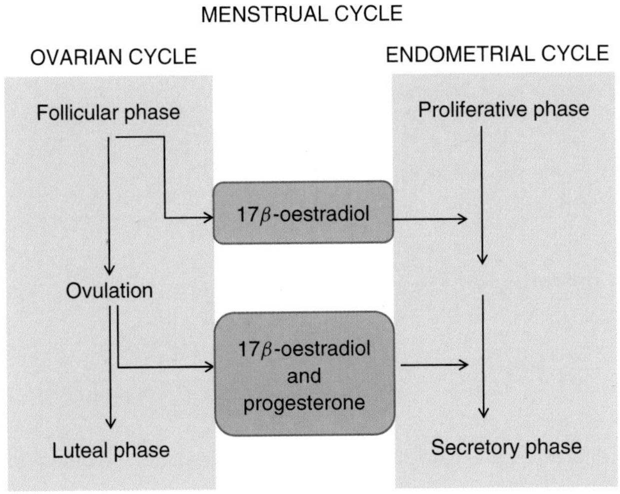

MENSTRUAL CYCLE

OVARIAN CYCLE

ENDOMETRIAL CYCLE

Follicular phase

Proliferative phase

17β-oestradiol

Ovulation

17β-oestradiol and progesterone

Luteal phase

Secretory phase

Figure 7.6 Diagram illustrating the key features of the ovarian and endometrial cycles, and the steroid hormones which link them.

mentioned elsewhere. During each menstrual cycle, the principal oestrogen in the circulation is 17β-oestradiol (often abbreviated to oestradiol), although some less potent oestrone is also produced. A third, even weaker, oestrogen called oestriol is the main oestrogen produced during pregnancy (see Chapter 9). As with all other steroid hormones, the initial precursor is cholesterol from which subsequent intermediary steroids, including progestogens, are produced. Oestrogens have 18 carbon atoms in their structure, and they are derived from androgens which have 19, so they are often called 18C and 19C steroids respectively. The androgens from which oestrogens are derived are androstenediol, androstenedione and testosterone (see Figure 7.7). Androstenedione is the substrate which can be converted to the more powerful testosterone which itself can be converted to 17β-oestradiol by the action of aromatase enzyme. Androstenediol can also be aromatised to the weak oestrogen oestrone which, in turn, can be converted to 17β-oestradiol. As explained earlier, the androgens are mainly synthesised in the thecal cells of the developing follicles during each cycle, and they are aromatised to oestrogens in the adjacent granulosa cells.

Oestrogens are also synthesised in other tissues of the body, particularly in adipose tissue, as well as in target cells of the arterial wall, skin and breast, in addition to the hypothalamus and other parts of the central nervous system (CNS). Oestrogens also play a major endocrine role during pregnancy, so not surprisingly the placenta is an important source. Body fat can become an important source of oestrogen in post-menopausal women, and this can be quite pronounced in obesity. The androgenic precursors in women come from not only the ovaries but also the adrenal glands.

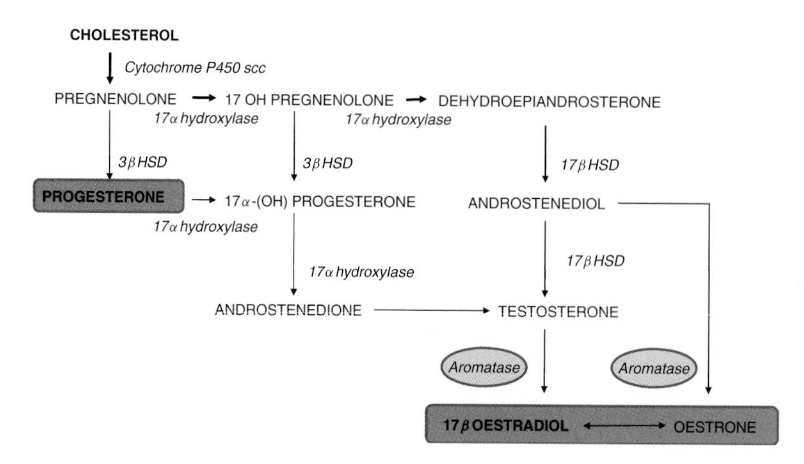

Figure 7.7 Diagram illustrating the main molecular stages involved in the synthesis of 17β-oestradiol (and progesterone).

Because steroids are lipophilic and can cross cell membranes with ease, they are not stored as such in the cells in which they are synthesised. The initial precursor cholesterol can be stored to a minor extent in the form of esters, but the oestrogens are synthesised only when androgens are available and aromatase enzyme activity is stimulated. Once they are synthesised they leave their cells of production presumably by diffusion across the cell membranes, and enter the fluid collecting in the follicular antrum, and the general circulation. In the circulation, in women, the plasma concentration varies throughout the cycle as indicated in Figure 7.4: there are two peaks, one prior to ovulation and the other approximately mid-way through the luteal phase. During pregnancy, plasma levels rise steadily to reach very high concentrations until parturition. In contrast, oestrogen levels in males do not vary very much and remain relatively low when compared to women at all stages of the menstrual cycle. Various tissues including Leydig and Sertoli cells in the testes, adipose tissue, the adrenal cortex and parts of the CNS contain the aromatase enzyme necessary for the conversion of androgens to oestrogens.

In the circulation, oestradiol is mostly transported bound to plasma proteins, in particular to the relatively specific sex hormone–binding globulin (SHBG, approximately 38%) which also binds androgens and progesterone, as well as to the more abundant and non-specific albumin (approximately 60%), with the bioactive (free, unbound) component comprising the remainder. The main metabolite of oestradiol is the weak oestrogen oestrone, much of it conjugated to the oestrone sulphate, produced mainly in the liver and excreted by the kidneys.

Oestrogen receptors and mechanism of action

There are two intracellular receptors for oestrogens called ERα and ERβ, both of which are located in the nuclei of target cells. The binding of the oestrogen to either of these nuclear receptors forms a complex which acts as a transcription factor, either activating or inhibiting specific target genes. Each ER consists of a transcriptional domain, a DNA-binding site, and a ligand- (usually the hormone) binding domain. Ligand binding to the ERα generally results in stimulated transcription, while binding to the ERβ leads to inhibition of transcription. Each receptor is found in a variety of target tissues, including breast and endothelium, and in some cells both receptors can be present. When this is the case, the ERβ inhibitory influence on transcription tends to be dominant. In the absence of oestrogen, the receptors form part of a heat shock protein complex which inhibits their potential activity. In the presence of oestrogen, however, the heat shock protein complex is disrupted, allowing the hormone to bind to the ligand-binding region of the receptor. When the oestrogen binds to its receptor, a number of other molecules called co-factors can either facilitate or repress ligand receptor complex-induced transcriptional activity. Once the hormone receptor complex has bound to its DNA-binding site and transcription of the relevant part of the DNA has occurred, there follows the usual sequence of reactions from mRNA formation to ultimate protein synthesis, the whole process taking up to an hour or more. Transcriptional activity associated with the ERs can also be influenced by molecules working through other intracellular signalling pathways which by-pass the hormone receptor interaction; this is ligand-independent receptor activation.

In addition, there is some evidence to suggest the existence of a much more rapid, non-genomic oestrogenic activity; for instance, changes in cell membrane properties occur within seconds or minutes of the hormone being applied to *in vitro* cell systems. Relatively little is known about such mechanisms of action. However, recent evidence suggests that such non-genomic mechanisms may play an important role in the development of various rapid membrane effects such as stimulation of transport systems (e.g. for ions).

Actions of oestrogens

The oestrogens are generally associated with actions in females (and are sometimes rather erroneously called the 'female' sex hormones) but it must be remembered that they are in fact also synthesised in males in whom they also have effects some of which are related to reproductive activity (see Chapter 6). However, they are produced in far greater amounts in females and it is in this gender that they therefore have the more pronounced effects (see Figure 7.8).

Metabolic effects

While oestrogens increase circulating triglyceride concentrations, they also increase the removal of low-density lipoprotein cholesterol (LDL-C) by stimulating the hepatic synthesis of LDL receptors. This is associated with the reduction in the circulating cholesterol level, which is beneficial for the cardiovascular system. These lipoproteins form particles which have a core of triglyceride and cholesteryl esters as well as the less hydrophobic cholesterol and phospholipids, surrounded by a surface of protein. They are derived from chylomicrons formed from the bile salts together with an inner core of hydrophobic lipids formed in, and absorbed from, the small intestine. These chylomicrons are subsequently reduced in size and density to very low-, then intermediate-, and finally low-density lipoprotein (VLDL, IDL and LDL respectively) fractions in the blood. Associated with the lipoproteins are smaller proteins called apolipoproteins which are found on their surface and which assist in directing the lipoprotein particles around the body and interacting with target cell lipoprotein receptors. The fat-containing lipoproteins can thus be directed to target cells where oestrogens stimulate their uptake and removal from the circulation.

In contrast, oestrogens stimulate the synthesis of high-density lipoproteins (HDL) which redistribute cholesterol and other lipids away from blood vessels to other parts of the body where they may be required, and particularly to the liver where they can be excreted in the bile.

In essence, the VDL and other low-density proteins transport cholesterol from the intestinal tract to the LDL receptors in target cells around the body, while HDL transports cholesterol from tissues where it is in excess to sites around the body where it is required, or to the liver from where it can be excreted via the bile (see Figure 7.8). Oestrogens increase the HDL:LDL ratio and are anti-atherosclerotic, thus being protective against coronary artery disease (CAD). Following the menopause, the decreased levels of circulating oestrogen are associated with a rise in atherosclerotic susceptibility and an increased risk of CAD.

Oestrogens also stimulate protein synthesis but are relatively weak anabolic steroids when compared with the androgens. Nevertheless, they are important particularly during puberty, when they participate in promoting the pubertal growth spurt. Furthermore, they stimulate hepatic protein synthesis which includes a number of the plasma binding protein globulins.

Endometrial effects

During the proliferative phase of the endometrial cycle, oestrogens stimulate cell division and the growth (proliferation) of the functionalis layer

PRINCIPAL ACTIONS OF OESTROGENS

REPRODUCTIVE	OTHER
Proliferation of endometrium	Metabolic effects
Stimulation of progesterone receptor synthesis	Cardiovascular effects
Breast development	Renal effects
Cervical secretion (copious and watery)	Skeletal effects
Female secondary sex characteristics	Interactions with other hormones
Central effects (e.g. negative and positive feedback, and behavioural)	Central effects

Figure 7.8 The principal actions of oestrogens (see section 'Actions of oestrogens').

of the endometrium. The radial branches of the arcuate arteries spiral up towards the surface as the layer thickens, and the ducts of the secretory glands also develop upwards from within the uterine stroma. While both ERα and ERβ are present, the main receptor mediating the effect of 17β-oestradiol in the endometrium is the ERα. It is likely that the growth-promoting effects of oestrogens are actually indirect, and are in response to the synthesis of epithelial growth factors which they promote.

Oestrogens are also necessary for promoting a secretion from the female reproductive tract which is favourable to the process of capacitation, an important maturation stage of the spermatozoa, should they have entered the tract following ejaculation (see Chapter 9).

Cervical secretion

During the follicular phase of the ovarian cycle, the increasing production of oestrogens stimulates the growth and maintenance of the cervix, and towards the end of the phase promotes the early secretion of a copious watery mucous containing various nutrients (sugars, proteins etc.) which is alkaline, and favourable to spermatozoa which can readily penetrate it.

General growth

Gonadal steroids have important effects on growth, and this is particularly marked at the time of puberty. While at least part of the effect is due to the enhanced stimulation of growth hormone from the adenohypophysis, oestrogens also stimulate local IGF1 production and so mediate more direct effects on growth themselves.

Musculoskeletal effects

Gonadal steroids have important effects on musculoskeletal growth and maturation, and this can be particularly evident during puberty. If excessively raised circulating androgens or oestrogens are present, they are associated with a rapid skeletal maturity, with fusion of the long bone epiphyses occurring earlier than normal and resulting in a reduced final height. Likewise, a deficiency in gonadal steroids will result in an enhanced final height due to the delay in fusion of the epiphyses of the long bones. Most of the effect of androgens is likely to be the result of aromatisation to oestrogens in the target tissues. Oestrogens promote bone mineralisation by stimulating osteoblasts (the bone cells that lay down the new bone matrix) and inhibiting some of the actions of osteoclasts (the bone cells that break down the bone matrix, releasing calcium salts). This 'protection' of bone structure is lost in women lacking normal circulating oestrogens, such as after the menopause, increasing their risk of developing osteoporosis (see Chapter 18).

Oestrogens also increase phosphate and calcium ion levels in the blood through intestinal and renal mechanisms, both effects enhancing the maintenance of bone matrix formation.

Renal effects

Oestrogens have effects on renal function. For example, they have been shown to stimulate renal sodium reabsorption in the proximal and distal tubules and this may indirectly lead to increased water reabsorption, at least partly through the increased plasma osmolality–vasopressin release mechanism (see Chapter 4). Part of the increased sodium reabsorption is probably indirect, and due to the action of oestrogens on stimulating hepatic protein synthesis, one such protein being angiotensinogen, the early precursor to angiotensin II (AII). Angiotensin II, apart from its other actions, stimulates sodium reabsorption directly, and indirectly by stimulating the synthesis of the adrenocortical mineralocorticoid aldosterone (see Chapter 10). On the other hand, 17β-oestradiol appears to decrease renin, angiotensin-converting enzyme and AII receptor synthesis, these all being actions which would counteract any effects associated with an increase in AII (see Chapter 10). Indeed, the overall effect of oestrogens on the renin–angiotensin system is inhibitory.

The effect of oestrogens on calcium handling by the kidneys is currently controversial. Reabsorption appears to be increased in the distal convoluted tubule since oestrogens stimulate luminal membrane calcium channel and intracellular transporter synthesis, but there is also evidence for a decreased reabsorption through other specific calcium channels in the luminal membranes. Oestrogens may also decrease the proximal tubular reabsorption of phosphate.

Furthermore, oestrogens appear to influence renal haemodynamics by stimulating nitric oxide synthesis, producing an afferent arteriolar vasodilation which will be associated with an increased renal plasma flow to the glomeruli and thus glomerular filtration rate.

Gastrointestinal effects

Oestrogens appear to stimulate calcium absorption in the small intestine by enhancing luminal calcium channels, and may enhance intestinal sodium-dependent phosphate absorption.

Central effects

Gonadal steroids are clearly associated with effects on behaviour, and some of them are the result of the establishment of neural networks at a very early age. Indeed, some behaviours described as particularly 'male', such as aggression and competitiveness, are likely to be due, at least in part, to oestrogens produced locally in neurones following aromatisation of androgens produced *in utero* or during early post-natal development. An explanation given to account for the absence of such 'male' behaviour developing in females is that the brain, during this critical stage of its development, is somehow protected from the direct influence of circulating oestrogens. For example, there is good evidence to suggest that oestrogens produced by the fetus are bound to some high-affinity protein such as α-fetoprotein which is present in high concentrations at that stage, thus preventing them from reaching central neurones. The feminising effects of oestrogens on brain development are likely to become manifest only as α-fetoprotein levels fall post-natally. Certainly, it seems clear that oestrogens are necessary for the development of certain female behaviours, such as lordosis (a crouching position adopted by a female rat in the presence of a male rat showing mounting behaviour), and the gender-specific neuronal development of the anterolateral nucleus in the hypothalamus. Other behaviours which are associated with oestrogens, in addition to reproductive behaviours, include cognition and social communication.

Another important central role of oestrogens, particularly 17β-oestradiol, is the feedback effect exerted on the hypothalamic GnRH neurones during the menstrual cycle (see section 'Control of the menstrual cycle').

Breast development

Breast tissue is comprised mainly of ducts and alveoli arranged in lobules. While various hormones and other factors are required for complete breast development and function (see Chapter 9), oestrogen plays a key role particularly in stimulating ductile growth. It is also involved in the prolific development of the alveolar lobules which occurs mainly during pregnancy.

Progesterone receptor synthesis

Oestradiol stimulates the synthesis of progesterone receptors in many target tissues, such as in the endometrium and breast.

Progestogens

Synthesis, storage, release and transport

Progestogens by definition are molecules, natural or synthetic, which bind to progestogen receptors and induce secretory activity in the endometrium. There are three naturally occurring progestogens: pregnenolone, progesterone and 17α-hydroxyprogesterone. By far the most important of these is the most potent one, progesterone. They have 21 carbon atoms in their structure and so are called C21 steroids. As shown in Figure 7.6, and as with the synthesis of corticosteroids (see Chapter 10), the initial precursor is cholesterol, with its characteristic cyclopentanoperhydrophenanthrene nucleus. This molecule is either synthesised within the production cells from acetate, or it reaches the cells from the general circulation. While the progestogens are usually precursors for other steroids, they are also clearly hormones in their own right.

Progesterone is mainly synthesised by the luteal cells of the corpus luteum during the luteal phase of the ovarian cycle. However, small amounts of progesterone and 17α-hydroxyprogesterone are also synthesised by the ovarian granulosa cells towards the end of the earlier follicular phase, just prior to ovulation. The progestogens are also synthesised in other parts of the body such as the adrenal glands and, importantly, the placenta during pregnancy (see Chapter 9). Adipose tissue is a useful store and source of progesterone.

As with the oestrogens (and androgens), the progestogens are lipophilic steroids and therefore are synthesised from precursors only when the appropriate enzymes are activated by an appropriate stimulus. Consequently, they are not stored, but released directly into the circulation as they pass readily through the lipid component of cell membranes. Once in the circulation, progesterone is transported mostly bound to plasma proteins: approximately 18% bound to corticosteroid-binding globulin (CBG), 80% to albumin with the remainder being unbound, this being the bioactive component. The circulating level of progesterone during the follicular phase of the ovarian cycle is normally low, but it rises considerably after ovulation to reach a peak approximately mid-way through the luteal phase (see Figure 7.4). As with oestrogens, should an ovum be fertilised during a menstrual cycle, levels of progesterone continue to rise throughout pregnancy. Again, normally levels of progesterone in men are reasonably steady, and considerably lower than in women.

Progesterone is rapidly cleared from the body, mainly by its hepatic conversion to the inactive derivative pregnanediol which is subsequently conjugated to its water-soluble glucuronide and excreted by the kidneys.

Progestogen receptors and mechanism of action

The progesterone receptor (PR) is a protein located in the nucleus of its target cells, and as with other nuclear receptors it has three main domains: a regulatory domain at the N-terminal, a DNA-binding domain next to a hinge section and a C-terminal ligand-binding domain. There are two isoforms of the human PR, PRA and PRB, with PRB having an extra 165 amino acids at the N-terminal.

Binding of hormone to a PR receptor increases its DNA-binding capability and its subsequent transcriptional activity. As a result, the synthesis of specific proteins is induced (or inhibited).

Actions of progesterone

Progestogens exert various effects associated particularly with the processes of reproduction, including the luteal phase of the menstrual cycle, pregnancy, parturition and lactation (Figure 7.9).

Endometrium

Progesterone plays a major role in stimulating the secretory activity of the endometrial glands during the secretory phase of the endometrial cycle. The endometrial secretion is particularly important should the ovum be fertilised because, particularly in the early stages, cell division is dependent on the nutrients provided in it.

PRINCIPAL ACTIONS OF PROGESTERONE

REPRODUCTIVE	OTHER
Stimulation of endometrial secretory activity	Renal
Breast development	Central (e.g. thermoregulation)
Cervical secretion (small volume and thick)	Precursor for androgen (and oestrogen) synthesis

Figure 7.9 The principal actions of progesterone (see section 'Actions of progesterone').

Breast development

Progesterone is another hormone necessary for the development of the alveoli of the breast lobules, particularly during pregnancy and lactation.

Cervical secretions

After ovulation, when circulating levels of 17β-oestradiol and progesterone both increase markedly, the progesterone influences the nature of the secretions produced by the glands lining the cervix. The secretion decreases in volume, and becomes more mucous. It becomes increasingly less 'sperm friendly' (i.e. it is more difficult for them to penetrate through the thicker secretion).

Renal effects

Progesterone can bind to mineralocorticoid receptors as an antagonist, thereby decreasing aldosterone-mediated sodium reabsorption in the distal nephron.

Central effects

Progesterone has various effects on the CNS. For example, the increase in basal body temperature (of up to 0.5° C) which normally occurs after ovulation, and is a useful indicator of this event, is associated with the increase in progesterone synthesis during the luteal phase. Another important role is to participate in the regulation of hypothalamic GnRH by exerting a dominant negative feedback effect, again during the luteal phase (see section 'Control of the menstrual cycle').

Inhibins

These hormones have already been considered briefly, with particular reference to their actions in the male. For the more general aspects of these hormones, the reader is therefore directed to that section in Chapter 6.

Synthesis, storage, release and transport

The granulosa cells of the follicles in the ovaries are the source of polypeptide hormones called inhibins. They exist as dimers, and belong to a super-family of molecules which include the activins, Mullerian Inhibiting Hormone (MIH, see Chapter 6) and transforming growth factor-β (TGFβ). Inhibins are also synthesised by the Sertoli cells in the testes, the hypothalamus and anterior pituitary, the placenta and other organs and tissues such as liver, skin and bone. There are two isomers of inhibin both of which have a common α subunit and either one of two β subunits, called A and B. Thus inhibin A is αβA and inhibin B is *a*βB.

The inhibins can be stored in secretory granules until released upon appropriate stimulation. It is likely that they are transported freely in the circulation; certainly their binding to other proteins in the blood is currently unknown. The anterior pituitary hormone FSH stimulates the synthesis of inhibin by the ovarian granulosa cells, which then exerts a negative feedback influence on the gonadotrophs in the anterior pituitary with a specific, but indirect, inhibitory effect on FSH production. It may also inhibit the release of GnRH from hypothalamic neurones since local production of inhibin is also possible.

Receptors and mechanisms of action

Inhibin actually binds to activin receptors, preventing the subsequent activation of activin-induced downstream events. However, it may also have a more potent effect by binding to a receptor for TGFβ called betaglycan which acts as a co-receptor for inhibin and greatly enhances its binding to the activin receptor (see Figure 7.10).

Actions of inhibin

Inhibin plays an important part in regulating the control of the ovarian cycle by specifically inhibiting FSH production, mainly through an indirect

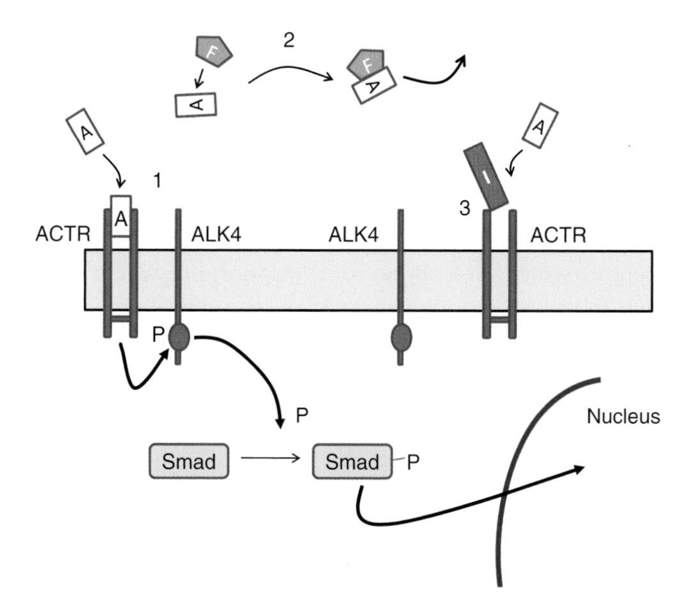

Figure 7.10 Diagram illustrating the membrane receptors for inhibin (I), activin (A) and follistatin (F) on a target cell (e.g. a gonadotroph). P = phosphorylation. (1) Activin receptor (ACTR) activation leads to transphosphorylation of an adjacent ALK4 receptor. (2) The consequence is phosphorylation of intracellular proteins such as Smads which can enter the nucleus and exert genomic effects. (3) Follistatin binds to the activin molecule and renders it inactive. Inhibin binds to the ACTR and blocks activin binding to that receptor (4). See text for further details.

mechanism involving the blocking of activin-induced stimulatory effects, at both pituitary and hypothalamic levels. It undoubtedly has other roles, as it is synthesised by other tissues, for example the placenta. While its specific physiological role remains unclear, it has a potential use as a diagnostic marker for the screening of disordered placental function during pregnancy.

Activins

Synthesis, storage, release and transport

Activins, like inhibins, are dimer polypeptides and members of the TGFβ super-family. The two sub-units comprising activin are both inhibin β subunits, so the three main bioactive dimers are activin A which consists of βAβA, activin B consisting of βBβB and activin AB consisting of βAβB. It is likely that activins, like inhibins, can be synthesised and stored within their cells of production, and then released when the cell is stimulated appropriately. As with inhibins, there is no knowledge about any binding that might occur in the circulation to other plasma proteins.

Activins are synthesised in many organs and tissues, including the granulosa and luteal cells of the ovaries, the placenta, the anterior pituitary and the hypothalamus. Indeed, activins seem to exert their effects mainly by local autocrine and paracrine actions within the tissues from which they are produced.

Receptors and mechanism of action

There are two types of activin receptors (ACTR) called types I and II in the plasma membranes of target cells. The ligand binds to the kinase domains of the type II receptors which then allows for transphosphorylation of the type I receptors which in turn phosphorylate downstream proteins such as the intracellular proteins called Smads. Phosphorylation of Smads enables these proteins to enter the nucleus where they can then act as transcription factors. There are two type II activin receptors, ActRII and ActRIIb, but only one type I receptor which is activin receptor–like kinase 4 (ALK4). Other ALKs exist, can use TGFβ as another ligand and could play a role in other signalling pathways.

Actions of activins

Activins are potent stimulators of FSH from the gonadotrophs in the anterior pituitary. It is quite possible that they also have central effects since subunits have been identified in neurones in various parts of the brain including the hypothalamus. Thus activins may have a stimulatory effect on the pulsatile release of GnRH. They also have effects in other tissues such as the placenta (see Chapter 9). Their activity is greatly reduced or abolished by inhibins, and follistatin when these hormones are present (see also Chapter 6).

Follistatins

Like inhibins, follistatins are inhibitors, or modulators, of activin action. Like the activins they are produced in many tissues including the anterior pituitary, and they have autocrine or paracrine effects.

Synthesis, storage, release and transport

Follistatins, like the activins and inhibins, are found in gonadal (e.g. follicular) fluid but are also synthesised in many other tissues in the body, including the anterior pituitary where they are found in various cell types including folliculostellate cells. They are proteins and as such may well be stored before their release upon suitable stimulation. Their mode of transport in the blood is likely to be free, unbound to other plasma proteins.

Receptors and mechanism of action

Follistatin acts as a binding protein for activin, masking its receptor-binding site and thereby essentially removing its bioactivity from the vicinity of its target cells (e.g. the gonadotrophs in the anterior pituitary). The effect is unlikely to be endocrine, but there is a lot of evidence pointing to a local autocrine or paracrine effect. Follistatin, by removing activin locally, acts as an FSH release inhibitor. Interestingly, it seems that activins stimulate its synthesis, providing a neat local negative feedback loop.

Actions of follistatin

As indicated in section 'Actions of inhibin', an important effect of follistatin is to modulate the release of FSH from the gonadotrophs by blocking the stimulatory activity of activins. It is likely that its other effects, in other tissues, may well also be associated with this interaction with the activins.

Control of the menstrual cycle

The events which occur during the proliferative and secretory phases of the endometrial cycle are regulated by the ovarian hormones, essentially 17β-oestradiol and progesterone. These steroids are produced by the growing antral follicles (ultimately by the Graffian follicle) and the corpus luteum, during the follicular and luteal phases of the ovarian cycle respectively. The production of these hormones is in turn controlled essentially by the hypothalamo-adenohypophysial axis. This consists of hypothalamic neurones producing the polypeptide gonadotrophin-releasing

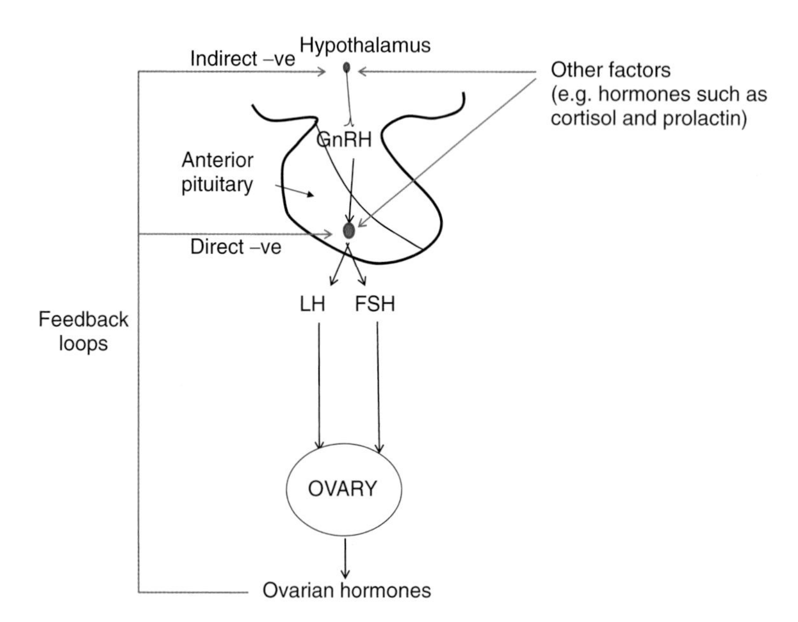

Figure 7.11 Diagram illustrating the principal components involved in the control of ovarian steroid production (see section 'Control of the menstrual cycle' for further details).

hormone (GnRH), and the two glycoprotein hormones produced by the gonadotrophs of the anterior pituitary, the gonadotrophins FSH and LH. All-importantly, there are feedback effects exerted by the gonadal steroids at both hypothalamic and adenohypophysial levels as well as other influences which influence gonadal steroid production (Figure 7.11).

The release of GnRH, as with other hypothalamic releasing hormone neurosecretions (see Chapter 2), is pulsatile and this is essential for effectively stimulating LH and FSH release since GnRH by continuous infusion initially stimulates, but then inhibits, release of the gonadotrophins. As indicated in Chapter 3, a gonadotrophin-inhibiting hormone called RFamide-related peptide-3 (RFRP-3) is likely to have direct inhibitory effects on the pulsatile release of LH (and to a lesser extent FSH). The influence of activins, inhibins and follistatin on GnRH, LH and FSH has already been mentioned.

At the beginning of each menstrual cycle (from days 1 to 5), consequent upon the decrease in gonadal steroid production, the spiral blood vessels vasoconstrict, blood flow to the endometrium is drastically reduced and necrosis of the outer endometrial layers results in their being shed. This is menstruation. Associated with the decreased gonadal steroid production, there is a decreased negative feedback exerted on the hypothalamo-adenohypophysial axis. Consequently, there are increased pulses of GnRH released from the hypothalamic neurones, and an increased release of LH and FSH from the anterior pituitary. These adenohypophsial hormones, particularly FSH, stimulate the development of up to 20 preantral follicles in the ovaries. These follicles begin to synthesise 17β-oestradiol (as well as the precursor androgens) and inhibin, which rise in concentration, not just in the growing follicles but also in the general circulation, exerting an increasing negative feedback on LH and FSH production (Figure 7.12a). As the FSH levels decrease, those antral follicles which are still growing under the influence of FSH are no longer sufficiently stimulated to continue developing so they stop growing and enter the regressive process called atresia. The usually lone surviving follicle (the Graafian follicle) continues to grow under the stimulatory effect of the oestradiol which it is now producing in large quantities. The granulosa cells have synthesised oestrogen receptors, so by a paracrine effect the follicle continues to grow and develop, producing even more oestradiol. As indicated in section 'The ovarian cycle', if the circulating concentration of oestradiol is sufficiently high for a long enough period of time (approximately 36 hours) it suddenly exerts a positive feedback effect on the hypothalamic

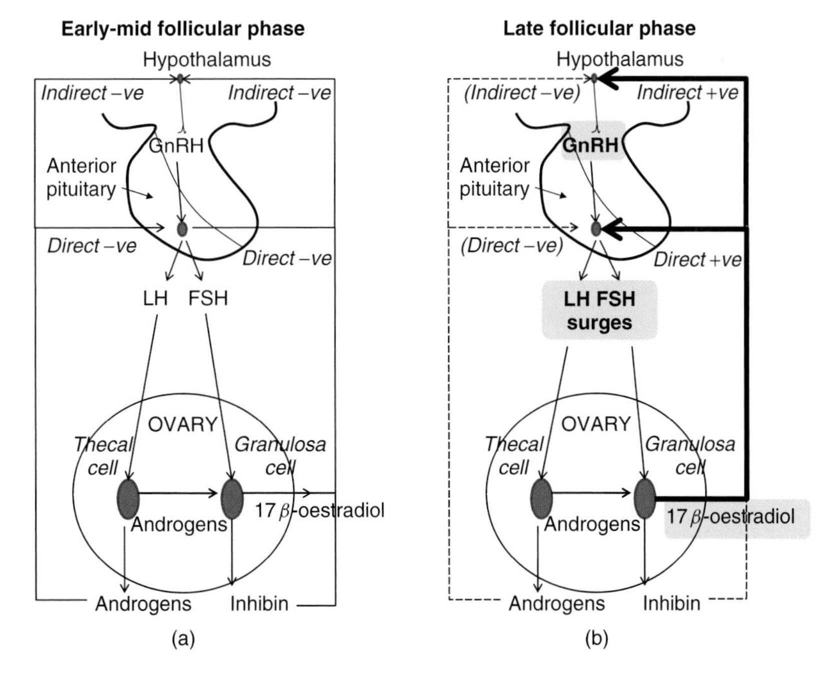

Figure 7.12 Diagrams illustrating control of ovarian steroid production during the menstrual cycle. (a) The early to mid-follicular phases, with increasing manifestation of negative feedback, and (b) the late follicular phase with the positive oestrogen-driven feedback, loops dominant, ensuring the production of the pre-ovulatory LH (and FSH) surges.

pulsatile release of GnRH, and also increases the sensitivity of the adenohypothalamic gonadotrophs to GnRH. This is partly brought about by the production of activins locally. Consequently, the production of LH, and to a lesser extent FSH, is stimulated, and their concentrations in the circulation surge (Figure 7.12b). The LH surge in particular is associated with the final maturation of the ovum and the final stages of follicular development, culminating in the release of the ovum at ovulation. This marks the end of the ovarian follicular phase during which the proliferation of the endometrium has taken place under the influence of oestradiol (the proliferative phase of the endometrial cycle). The oestradiol level now falls, the cells of the collapsed follicle become luteal cells and the corpus luteum begins to produce not only oestradiol but also progesterone under the influence of the rising LH and FSH levels. This is the luteal phase of the ovarian cycle and it coincides with the stimulation of the endometrial secretions by the increasing levels of both oestradiol and progesterone (the secretory phase of the endometrial cycle).

If the ovum is not fertilised, the rising levels of gonadal steroids now exert an increasing negative feedback on the hypothalamo-adenohypophysial axis. Consequently, there is a decrease in the pulsatile release of GnRH and a resulting decrease in the circulating gonadotrophin levels (Figure 7.13). As the LH and FSH levels fall, there is a decreased stimulation of the luteal cells and the oestradiol and progesterone levels then fall. The endometrium

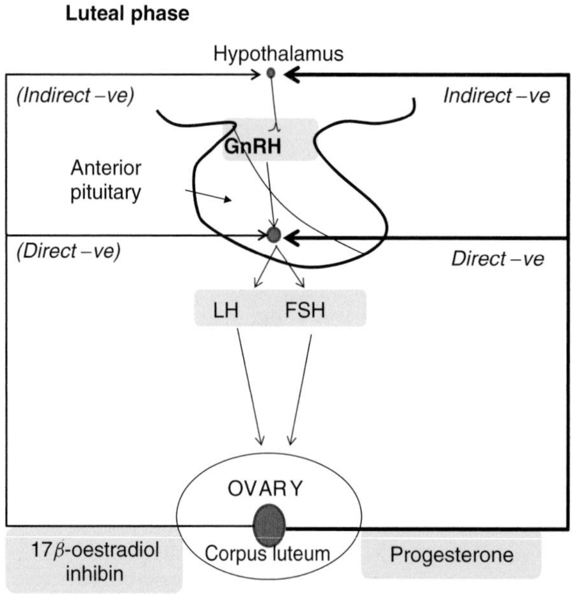

Figure 7.13 Diagram illustrating the control of ovarian steroid production during the luteal phase of the menstrual cycle.

is no longer maintained by the gonadal steroids, and the outer layers are shed, marking the beginning of the next cycle.

Interestingly, even though the oestradiol may reach the same high concentration as at the end of the follicular phase, it does not elicit a second LH (and FSH) surge because of the dominant inhibitory influence exerted by the high circulating concentration of progesterone, which was not present earlier. Again, the influence of other hormones also needs to be considered. Indeed, inhibin is likely to play an important negative feedback role during the menstrual cycle. It, too, is synthesised by the granulosa cells during the follicular phase of the ovarian cycle, and it exerts a specific negative feedback effect on FSH. It is probable that the decrease in FSH production during the mid-follicular phase is at least partly due to its effect on the gonadotrophs by blocking the action of locally produced activin, specifically with regard to FSH synthesis.

If the ovum released at ovulation is fertilised by a spermatozoon, then there is a continuing requirement for gonadal steroids since the secretory activity of the endometrium needs to be maintained and other tissues such as mammary also require these hormones for their development. Given that the increasing levels of ovarian gonadal steroids exert an inhibitory effect on the maternal hypothalamo-adenohypophysial axis, a new source of stimulation for their production is required. This is provided by the implanting blastocyst initially, as it starts to produce a hormone which can act as a ligand for LH receptors on the cells of the corpus luteum: human chorionic gonadotrophin (hCG; see Chapter 9).

Various other factors influence the production of gonadal steroids. Body weight is clearly a factor, since a decreased Body Mass Index (BMI) below the normal range (i.e. lower than 19) can be associated with oligomenorrhoea or amenorrhoea (disruption or loss of menstrual cycles respectively). Thus clinical conditions such as anorexia nervosa or even profound loss of weight associated with extreme, prolonged exercise are associated with disrupted menstrual cycles. Interestingly, the onset of puberty, at least in girls, is also associated with the attainment of a certain critical body weight. The hypothalamic polypeptide kisspeptin has been linked to the initiation of adult-like pulses of GnRH neurones at the time of puberty, for instance (see Chapter 8). Disrupted cycles are also associated with other endocrine disorders such as hypothyroidism and hypercorticalism (Cushing's syndrome) indicating that the iodothyronines are stimulatory, and glucocorticoids are inhibitory to the hypothalamo-adenohypophysial-ovarian axis. Hyperprolactinaemia is also associated with the loss of menstrual cycles due to an inhibitory effect on GnRH release, so this hormone too can have an inhibitory influence. Endorphins may also suppress gonadal hormone production. Many of these endocrine factors probably influence the pulsatile release of GnRH, but may also have effects on the pituitary gonadotrophs.

Clinical conditions

Polycystic ovarian syndrome (PCOS)

PCOS occurs in about 10% of women and is the most common cause of infertility in women. The symptoms of PCOS may begin with menstrual irregularities, or a woman may not know she has PCOS until later in life when symptoms or infertility occur. PCOS is associated with obesity and insulin resistance.

In patients with PCOS, the normal ovarian monthly cycle is disrupted. Instead of one oocyte developing, a number of cysts appear. Instead of the positive feedback described in Figure 7.3, followed by an LH surge and ovulation, the patient enters a metastable state with a fixed high LH:FSH ratio and no dominant oocyte, with follicular arrest. Instead of one follicle being dominant and releasing an egg, several follicles develop and then arrest at 5 mm to 7 mm in size. There is an increased secretion of thecal androgens, and insufficient aromatisation. Thus circulating testosterone levels are increased together with reduced 17β-oestradiol levels, and a high LH:FSH ratio. Ovulatory cycles are intermittent, and variable in length. The high levels of androgens sometimes increase acne and hirsutism.

There has been controversy in confirming the diagnosis, so various expert panels have been convened to agree on some standard diagnostic features. One such group has published the Rotterdam criteria for PCOS; to be diagnosed with PCOS using the Rotterdam criteria, a woman must have two of the following three manifestations:

1. Irregular or absent ovulation,
2. Elevated levels of androgenic hormones and/or
3. Enlarged ovaries containing at least 12 follicles (cysts) each.

Other conditions with similar signs, such as androgen-secreting tumours or Cushing's syndrome, must be ruled out. Polycystic ovaries with normal ovarian function and without hyperandrogenism should not be considered PCOS without further follow-up.

The treatment of PCOS depends on what the patient wants. In many cases, no treatment is required, and they may not even seek medical attention for irregular periods. Patients seek treatment either because they want fertility, or because they want to reduce the effects of androgens on the skin.

Patients who want fertility should be encouraged to exercise and lose weight. Metformin, a drug commonly used for patients with type 2 diabetes, can be helpful with regard to weight loss and ovulation. Clomiphene (a pituitary-specific oestrogen receptor antagonist) is more successful at achieving ovulation and pregnancy. Surgery to the ovary, including wedge resection (now rarely performed) and ovarian drilling, may result in spontaneous resumption of ovulation.

Patients who do not want a pregnancy but who want treatment for their acne and hirsutism can be given oestrogens, as in the combined oral contraceptive. The higher dose oestrogen-containing contraceptives such as Dianette and Marvalon are most popular with patients who suffer from PCOS and who are concerned about the appearance of their skin.

Primary ovarian failure

The normal menopause occurs between the ages of 45 and 55 years. At this point, it is normal for the ovary to fail, and for circulating gonadotrophin levels to increase dramatically. Premature menopause occurs when ovarian failure occurs early.

Early primary ovarian failure can occur due to autoimmune damage to the ovary. As with testicular failure, autoimmune ovarian failure occurs more commonly in individuals who have other autoimmune endocrine diseases, such as hypothyroidism, Addison's disease and pernicious anaemia. Patients who have premature ovarian failure with other autoimmune diseases have polyglandular autoimmune syndrome type 2. Patients may also suffer loss of libido, tiredness, hot flushes and osteoporosis. In general, patients who have a premature menopause may be started on hormone replacement therapy (HRT) as soon as they become menopausal, and continue till about the age of 50 years. Thus patients who have an oophorectomy for any reason may then be started on HRT. Patients who have either breast or endometrial cancer should not have HRT, as both cancers are often oestrogen receptor positive.

Postmenopausal osteoporosis

Oestrogen increases bone density and once the menopause occurs, the rate of bone loss increases. Figure 7.14 shows the normal rates of bone loss over time, and it is clear that bone loss accelerates at the onset of the menopause. This is one important reason to consider HRT.

Hormone replacement with oestrogen prevents bone loss at around the menopause, and it can reduce some of the symptoms that many women complain of, including hot flushes. Very long-term use of oestrogen can increase the risks of breast and endometrial cancer, so the use of HRT must be carefully considered on a case-by-case basis. Essentially patients with a family history of breast cancer are often advised to avoid HRT.

For women who have had a hysterectomy, oestrogen replacement alone can be considered. This will help both bone density and hot flushes, and as there is no uterus present, there is no risk of endometrial hyperplasia.

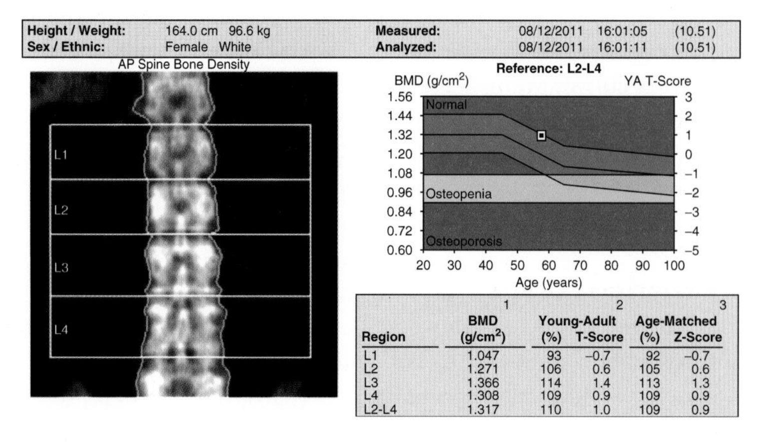

Region	BMD 1 (g/cm²)	Young-Adult 2 (%)	T-Score	Age-Matched 3 (%)	Z-Score
L1	1.047	93	−0.7	92	−0.7
L2	1.271	106	0.6	105	0.6
L3	1.366	114	1.4	113	1.3
L4	1.308	109	0.9	109	0.9
L2-L4	1.317	110	1.0	109	0.9

Figure 7.14 A dual-energy X ray absorptiometry report (DEXA). Low-dose X rays are used to determine the density of the bones in the spine. The density is compared to that of 'normal' women and plotted on a graph of patients, showing the usual loss of bone density over time of the menopause. The patient here has an excellent bone density at the age of 57.

For patients with an intact uterus, oestrogen is administered together with a progestogen. The progestogen is usually given in the second half of the cycle, and this is followed by an interval allowing for a menstrual bleed. This is known as combined HRT. This is because in women with an intact womb, oestrogen stimulates the growth of the womb lining (endometrium), which can lead to endometrial cancer if the growth is unopposed. A progestogen is given to oppose oestrogen's effect on the womb lining and reduce the risk of cancer, although it does not eliminate this risk entirely. A review of the available evidence suggests that in most patients, the risks of HRT exceed its benefits once patients reach the age of 50.

There are also a few 'bleed-free' products, such as Tibolone (which contains a steroid with a combination of oestrogenic, progestogenic and androgenic activity) or Kliofem (which contains both oestradiol and norethisterone). HRT is usually required only for short-term relief from menopausal symptoms and its use should be reviewed at least once a year with a doctor.

Selective oestrogen receptor modulators

There are slight differences in the oestrogen receptors in different parts of the body, so that it is possible to modify agents that are agonists at one of the oestrogen receptors such that they become antagonists at others. The well-known oestrogen receptor antagonist Tamoxifen has long been used as a drug to prevent or treat patients with breast cancer. It transpires that patients on Tamoxifen also have improved bone density, so that although it is an antagonist of the oestrogen receptor in the breast, Tamoxifen is an agonist in bone. Tamoxifen is also an agonist in the uterus, so it does increase the risk of endometrial hyperplasia.

Raloxifene is another oestrogen receptor modulator that has been designed as an agonist in bone (so is licenced as a treatment for osteoporosis), but it is also an antagonist in both breast tissue and endometrium. It thus does not increase risk of cancer in either site.

Oral contraceptives

Higher dose oestrogen-containing pills administering at least 20 micrograms of ethinyl oestrodiol can be used to inhibit ovulation. These work by negative feedback (see section 'Control of the menstrual cycle') suppressing the release of anterior pituitary gonadotrophins to undetectable levels. In the absence of gonadotrophins, the ovary does not ovulate.

Ovarian cancer

Because the ovaries are small, deeply situated internal organs, ovarian cancer presents late, unlike testicular cancer. Thus by the time the patient has any symptoms, the cancer has often spread. Presenting features include abdominal bloating with malignant cells causing ascites, increased abdominal girth and an urge to pass urine. The cause of ovarian cancer is unclear. Some cases have a family history, and genetic factors are likely. However, there are environmental associations that are difficult to explain. There seems to be a lower prevalence of ovarian cancer amongst women who have had several pregnancies, or spent time on the oral contraceptive. Early age at first pregnancy, and being older with your final pregnancy, have also been thought to have a protective effect. These factors are not independent, and their effects might all be due to bias. For example it is possible that women who are well exposed to medical care and are of higher social class have a lower risk of ovarian cancer. The disease is more common in industrialised nations. Women have a 2% lifetime risk of developing ovarian cancer. Older women are at highest risk. More than half of the deaths from ovarian cancer occur in women between 55 and 74 years of age.

Once women present with the features of ovarian cancer, it will already have spread, and the prognosis is thus poor. There is no good screening test for ovarian cancer.

Appendix: Clinical Cases

Clinical Scenario 7.1

A 35-year-old woman is referred by her GP because she is having difficulty in trying to conceive a baby. She stopped taking the oral contraceptive 2 years ago, and still has not conceived. She has been having irregular periods for the last 12 months and has a BMI of 29 kg/m^2. On direct questioning, it transpires that she has had only four periods in the last year.

Questions

Q1. What would be your initial action in the clinic?
Q2. What are the likely causes of infertility in this patient?
Q3. What investigations would you do?
Q4. What treatment options are available?

Answers to Clinical Scenario 7.1

A1. What would be your initial action in the clinic?

The diagnoses in this chapter should be carefully considered.

Take a detailed history and examine the patient carefully, in particular determining what your initial action in the clinic would be. Enquire when her periods first started (menarche), whether they have always been irregular and the number of days of bleeding during a menstrual cycle. One should also enquire whether she has had any previous pregnancies including terminations, and if so were these with the same partner? Has the partner had any children before, if he has had a previous partner?

Clearly if she has had any pregnancies, or her husband has fathered any children, this would suggest a newly acquired cause of infertility.

Examination should be general but then specifically examine the secondary sexual characteristics (e.g. breast development and axillary and pubic hair growth). One should look in particular for features of excess facial hair. The patient may tell you that she has had excess hair, but has used local therapy such as laser treatment to remove it, so in addition to looking for it, one should also ask about it.

Also her partner should be asked to attend to ensure that he is fertile.

A2. What are the likely causes of infertility in this patient?

One should consider the whole hypothalamo-pituitary-ovarian axis, and consider diseases in the pituitary chapter in addition to ovarian causes.

- Polycystic ovarian syndrome (diagnosed by having features of the Rotterdam criteria, see section 'Polycystic ovarian syndrome (PCOS)') is the likeliest diagnosis with this BMI.
- Gonadotrophin failure:
  - Hypothalamic or pituitary disease.
- Hyperprolactinaemia:
  - Prolactinoma.
  - Primary hypothyroidism (low thyroxine levels lead to reduced negative feedback causing a rise in TRH which increases prolactin release).
- Androgen excess produced by a gonadal tumour causes negative feedback resulting in low GnRH and low LH and FSH.
- Premature menopause also needs to be excluded.
- Post-pill amenorrhoea: It may be normal to not have a menstrual cycle for up to a year when stopping the oral contraceptive pill as the oestrogens in the pill will have caused negative feedback resulting in low GnRH, LH and FSH. It can then take up to a year for the axis which has been down-regulated to recover. However, most patients recover their axis more quickly.

A3. What investigations would you do?

Blood sample analyses would include:

- LH, FSH and oestradiol.
- Low oestradiol and high LH and FSH suggest primary gonadal failure.
- Low oestradiol and low LH and FSH suggest hypothalamic or pituitary disease.
- High testosterone (3.0 nM) with a high LH:FSH ratio (approximately a 3:1 ratio) would suggest PCOS.
- Progesterone measured on day 21 of her menstrual cycle – should be raised if ovulation is occurring. This assumes a 28-day cycle. Thus you really need to get a sample 7 days before the next period, and one way to do this is to obtain a sample every 3 days from day 21, so that one catches the peak in progesterone in someone who has irregular periods.
- Prolactin and thyroid function tests.
- Androgens (testosterone, androstenedione and DHEAS).
- Ultrasound scan of ovaries and/or uterus.

A4. What treatment options are available?

- Treat the cause (e.g. hypothyroidism or hyperprolactinaemia).
- If PCOS is confirmed, then the patient should lose weight until her BMI is 23 (this is the ideal weight for ovulatory cycles). Metformin can also help to make the periods regular (the mechanism by which it does this is not known). If she does not conceive despite these measures, then she should be referred to an infertility specialist for treatment, and a combination of clomiphene and metformin is most suitable. Recent evidence suggests that metformin is really effective only when clomiphene is added.

Further Reading

Bilezikjian, L.M., Blount, A.L., Donaldson, C.J. & Vale, W.W. (2006) Pituitary actions of ligands of the TGF-b family: activins and inhibins. *Reproduction*, 132, 207–15.

Brunette, M.G. & Leclerc, M. (2001) Effect of estrogen on calcium and sodium transport by the nephron luminal membranes. *Journal of Endocrinology*, 170, 441–50.

Cornil, C.A. & Charlier, T.D. (2010) Rapid behavioural effects of oestrogens and fast regulation of their local synthesis by brain aromatase. *Journal of Neuroendocrinology*, 22, 664–73.

CHAPTER 8

The Endocrine Control of Puberty

Introduction

From the moment of fertilisation, the new life form starts to grow and develop. Initially, there will simply be a number of cell divisions resulting in a morula (or ball) of cells, nutrients reaching the dividing, growing cells from the immediate environment by diffusion. However, once implantation has taken place, subsequent growth of the embryo into a fetus is extremely rapid. Indeed, the fastest period of growth during our entire lives occurs *in utero*, when the peak growth rate can be measured in $mm.day^{-1}$. At the time of birth, the growth rate (e.g. in head-to-toe length) is still rapid but decreasing from its peak, and measured now in $cm.year^{-1}$ (see Figure 8.1). For the next few years, growth remains substantial, but the rate decreases steadily until the age at which certain changes to the body begin to take place. These changes are associated with the preparation of the body for future sexual intercourse with the potential for procreation, this being the basic, instinctive drive for the survival of the species. In females, particularly, these changes are associated with the ability to successfully carry a developing fetus during pregnancy. In addition to the physical changes taking place in our bodies at this time, there are many other developments taking place. For instance, our brains are also maturing, and these changes will be associated with our developing cognitive and psychosocial behaviours, and a growing awareness of the world in which we live. The period of time in our lives when these changes occur is called adolescence, and the particular stage associated with the physical, sexual changes is called puberty. Puberty is also the time when there is a sudden increase in linear growth called the pubertal growth spurt.

Pubertal changes

The age at which puberty begins varies from individual to individual, but in general it occurs approximately 2 years earlier in girls than in boys. It can begin as early as 7–8 years in girls, and 9–10 in boys, but can be later

Integrated Endocrinology, First Edition. John Laycock and Karim Meeran.

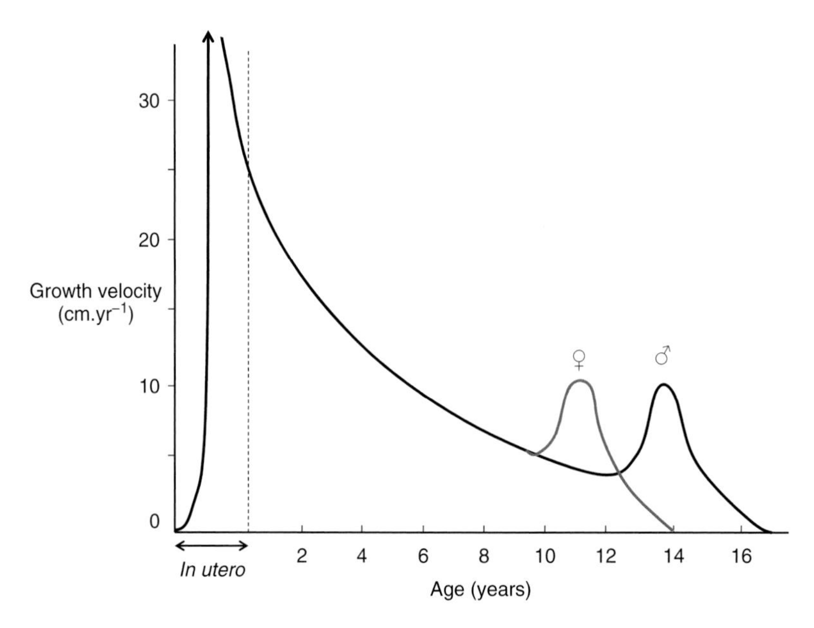

Figure 8.1 Diagram illustrating growth velocity curves, which are highest *in utero*, with gender-differentiated pubertal growth spurts.

in both genders, so that the ranges are 7–13 and 9–15 years in girls and boys respectively. The best indicator of the start of puberty is in girls, with the occurrence of their first 'period', or menstruation (menarche). In boys, a similar indicator of sexual maturity is not so obvious but is likely to be a spontaneous erection (and maybe ejaculation) occurring one night, during sleep. Associated changes include the development of the secondary sexual characteristics, physical growth and mental development, and all of these can occur over a period of 3–4 years or longer.

Secondary sexual characteristics

In females these are the breasts, the pubic (in the genital region) and axillary (armpit) hair, the widening of the hips and the distribution of fat, while in males they include the growth of the testes and penis, pubic and axillary hair, facial and body hair, greater muscle development and generally narrower hips than girls. For the breasts, pubic hair and testicles, the various stages of development were first clearly described by a paediatrician, James Tanner, in the 1960s. These are known as the Tanner stages and provide a useful guide to classify the normal sexual development of a boy or girl. It is important to appreciate how much normal variation there is in male penile size and the shape and size of female breasts. The Tanner staging for normal sexual development is as follows:

In boys: Male genitalia

Stage 1: Prepubertal stage.

Stage 2: Testes and scrotum enlarge; the scrotum changes texture and reddens.

Stage 3: Further enlargement of testes and scrotum; penis lengthens.

Stage 4: Further increase in penile length and circumference, with the glans developing; testes and scrotum continue to grow, with the colour of the scrotum darkening.

Stage 5: Adult genitalia.

Girls: Breast development

Stage 1: Prepubertal stage.

Stage 2: Breast bud with raised papilla (nipple) and enlargement of areola.

Stage 3: Further enlargement of breast and areola, which remain together in contour.

Stage 4: Areola and papilla separate from breast contour forming a secondary mound.

Stage 5: Adult stage, with projection of papilla only.

Pubic hair in boys and girls

Stage 1: Prepubertal stage (any hair at this stage is similar to rest of abdominal hair).

Stage 2: Sparse growth, long, straight or curled, along the labia or around the base of the penis.

Stage 3: Increased growth of darker, coarser and curlier hair, spreading in area but still relatively sparse.

Stage 4: Hair adult in type, but still restricted in area.

Stage 5: Adult in type and amount.

Physical growth

The pubertal growth spurt starts approximately 2 years earlier in girls than in boys. Linear growth ceases when the epiphyses of the long bones fuse, after which no further such growth can occur. For this reason, girls enter the growth phase from a lower starting point and reach the stage when the epiphyses have fused earlier than boys with the consequence that, in general, they end up shorter in stature. The growth spurt is related to the pubertal stages described earlier so that the growth rate in boys starts at 5–6 cm.year^{-1} at stage 1 and gradually increases to 7–8 cm.year^{-1} by stage 3, reaching 10 cm.year^{-1} by stage 4, with the final height being reached around 17–18 years. In girls the increase is similar, although the maximal growth rate of approximately 8 cm.year^{-1} is reached by stage 3, with final height being attained at around 15–16 years (Figure 8.1).

In addition to the measure of linear growth, body weight also increases during puberty. The body mass index (BMI) which relates the two (body weight in kg/height in m^2) has a normal range for adults of 20–25, with overweight being 25–30, obese 30–35 and 'clinically obese' 35 and higher. A BMI lower than 20 is usually taken as indicating someone who is underweight for his or her height. Interestingly, during the prepubertal years the BMI distribution is usually lower than for adults, and from age 2 to 6 it actually falls because the rate of increase in weight is surpassed by that of height.

It is also interesting to appreciate the gender difference in body composition that occurs during puberty. In girls as they reach adulthood, their lean body mass (muscle with respect to body fat) tends to decrease because of an increase in body fat deposition, while in boys it is likely to increase because of increased muscle formation.

Mental changes

The period during which the various physical changes take place during puberty is also the time for mental development. Characteristic changes include the development of abstract reasoning, logical thought, the appreciation of consequences, the establishment of identity and relationship to (and with) others. This is the time for the acceptance of increased autonomy and responsibility, and is often a period of experimentation (e.g. with drugs such as alcohol). There is also an increased intensity of mood states, and depression is relatively high from this stage.

These mental developments are associated with brain growth (e.g. increased synaptic formation) and the endocrine changes at puberty (gonadal steroids in particular). A genetic influence is also present.

Adrenarche

From its peak at birth, the circulating level of the weak androgen dehydroepiandrostenedione sulphate (DHEAS) decreases so that by the end of the first year of life, the plasma concentration is low. It then remains low until approximately 2 years before the onset of puberty when, in both boys and girls, there is a sudden increase in weak androgen production by the adrenal glands (known as adrenarche) which predates the pubertal increase in gonadal steroid hormone (androgen and oestrogen) production (see Figure 8.2). The stimulus for triggering the onset of adrenarche is unknown, but reaching a certain body weight may be a key factor. Furthermore, the purpose of adrenarche is unclear, other than to indicate that puberty is 'on the way'. The only identifiable related effect of the consequent increase in circulating DHEAS is the initial production of axillary and pubic hair, the development of the apocrine glands in the skin

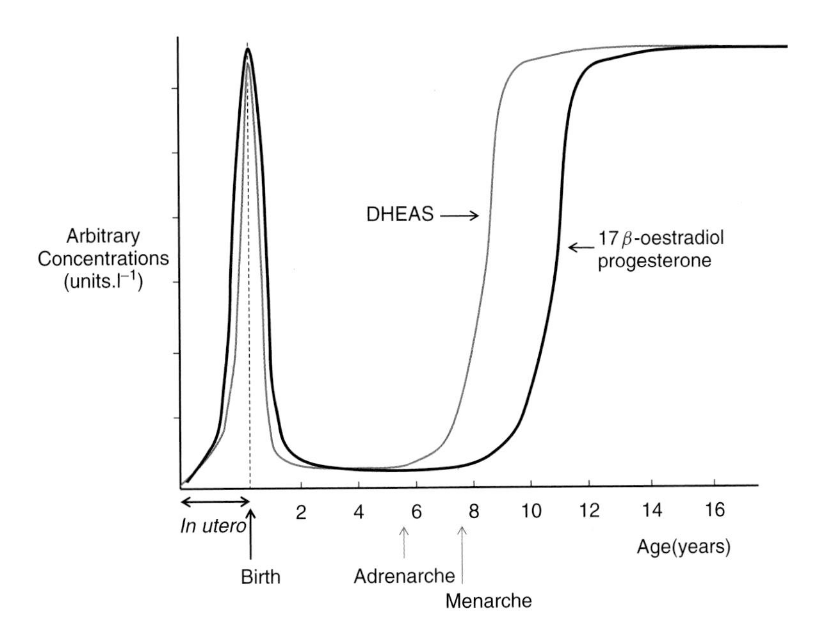

Figure 8.2 Diagram illustrating the changing levels of specific steroid hormones in girls: dehydroepiandrosterone sulphate (DHEAS) from the adrenals, followed by 17β-oestradiol and progesterone from the gonads approximately 2 years later. The same pattern is seen in boys, with adrenarche and the beginning of puberty delayed by approximately 2 years.

of these regions, and the associated secretion of sweat. Indeed, DHEAS can be converted to the more powerful testosterone (which can in turn be converted to the even more powerful dihydrotestosterone) in local tissues such as the armpit and groin, and it is probably this local production that is effective. The adrenocortical steroids such as aldosterone and cortisol show no such rise in production at this stage.

Consequences of adrenarche will be the early production of body odour due to the increased apocrine secretion, and the increased production of sebum by the skin with the potentially problematic accompanying acne which can occur.

The endocrine basis for the pubertal changes

The growth spurt, the development of the secondary sex characteristics and the mental changes which occur at puberty are all associated with hormonal effects. Basal levels of various metabolic hormones such as insulin, glucagon, the iodothyronines and adrenaline are necessary, together with the determining feature of the onset of increased production of somatotrophin (growth hormone) and the gonadal steroids. Gonadal steroids, which are at adult levels at birth before decreasing to low prepubertal levels, suddenly begin to be produced in increasing amounts at the time of puberty, some 2 years after adrenarche (Figure 8.2). Gonadal steroid production in both genders is regulated mainly by the adenohypophysial gonadotrophins luteinising hormone (LH) and follicle-stimulating hormone (FSH), and these hormones also are produced in low basal amounts from year 1 until the onset of puberty. Both these hormones, as well as somatotrophin, are to a large extent controlled by hypothalamic hormones (see Chapters 2 and 3). Thus, gonadotrophin-releasing hormone (GnRH) mainly regulates LH and FSH release, and both growth hormone–releasing hormone (GHRH) and the inhibitory hormone somatostatin regulate somatotrophin release.

The key to the onset of puberty would therefore seem to be linked to the attainment of some developmental stage of the hypothalamus relating to the increased pulsatile release of specific hypothalamic hormones, particularly GnRH.

What is the signal for the onset for puberty?

There has been much debate about what actually triggers the onset of puberty. Indeed, there are some interesting observations that need to be considered. For instance, there is a clear secular trend showing that puberty has been starting earlier and earlier since the beginning of the Industrial Revolution in the 19th century in the Western world, when records of puberty first began, in all nations where such information has been gathered. Over the past 170 years or so, the age at which puberty is first recorded, not surprisingly by using the marker of menarche in girls, has decreased from approximately 16 down to 10 years, levelling out over the last 20–30 years (Figure 8.3). What has happened over that period of time? The Industrial Revolution has been associated with many changes including improved civic sanitation, improved lighting (gas and then electric) and improved nutrition. One intriguing possibility is, in fact, to do with the increased exposure to lighting that has taken place over that period, from being limited to daylight hours (with the possibility of limited use of candles) to the extended day with artificial lighting being provided by the advent of gas and then electricity. Many animals have a fertile season during each year governed by the length of day (e.g. rutting in deer), so maybe the longer exposure to light we have nowadays is relevant? In humans, evidence for this possibility is scanty; for example studies of girls born with impaired sight fail to show a delayed advent of

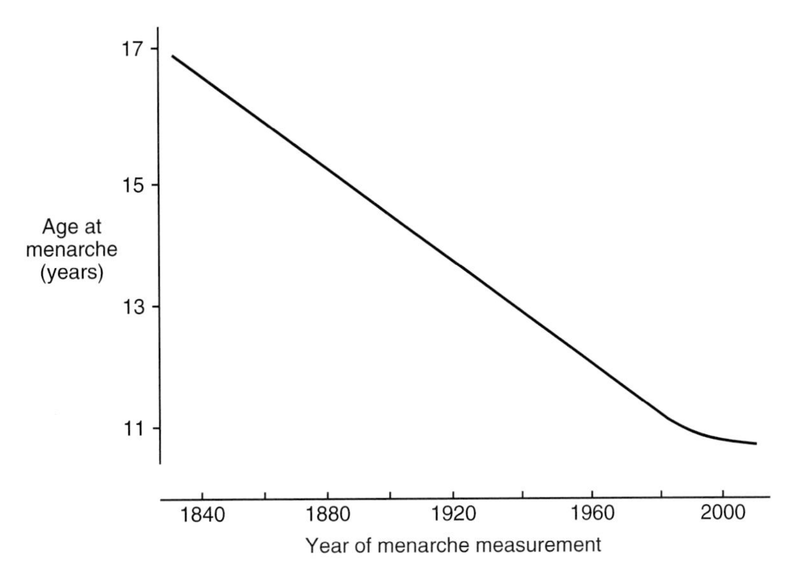

Figure 8.3 Diagram illustrating the secular change in the age of menarche in Western countries over the last 170 years, corresponding with the advent of the Industrial Revolution up to the present day.

puberty although the specific influence of factors such as wavelength and intensity has not so far been studied.

Another, potentially more likely, contender is the level of nutrition. While it is clear that there has indeed been a decrease in the age of onset of puberty, population studies indicate that the body weight at which puberty begins has remained remarkably constant, at 47.5 kg in girls (see Figure 8.4). Clearly there will be a range of body weights relating to menarche in individual girls and this value of 47.5 kg is a (study) population average, but nevertheless an influence of body weight on reproductive capability would appear to be reasonably logical. For a girl to be able to reproduce successfully, and carry a fetus to term, she would require a certain metabolic status indicating capability. Indeed, a poor diet and low nutritional status are clearly factors which are associated with premature births and less favourable outcomes for successful delivery.

At present, the general consensus is that the onset of puberty is linked to the switching on of hypothalamic GnRH neurones by factors associated with the metabolic status of the child. The nature of the pulsatile release of GnRH then switches from the low prepubertal to an adult-level production pattern, and this then mediates the release of gonadotrophins from the anterior pituitary and the various changes associated with puberty itself.

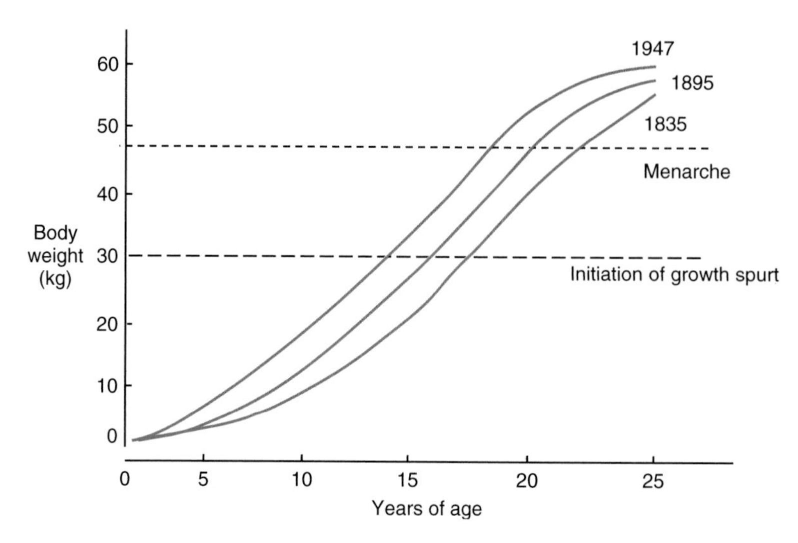

Figure 8.4 Diagram illustrating how the trend of an earlier menarche (and earlier initiation of the growth spurt) is nevertheless associated with the attainment of specific body weights in population studies.

The regulation of GnRH neurones

The GnRH neurones are located particularly in the medial basal hypothalamus, an area which includes the arcuate nucleus, a key area involved in the regulation of feeding and appetite. These GnRH neurones receive innervations from both stimulatory and inhibitory fibres which release a variety of neurotransmitters. Stimulatory fibres include neurones releasing glutamate while inhibitory fibres include those releasing gamma-aminobutyric acid (GABA) and neuropeptide Y (NPY). There is some evidence to suggest that at birth the inhibitory neurones are silent and the stimulatory neurones are active so that the pulses of GnRH are similar to those produced in the later adult stage. Then, the inhibitory fibres become active and GnRH pulses become reduced and prepubertal. This stage seems to last until puberty when, once again, the stimulatory glutamatergic neurones are activated and become the dominant influence on GnRH release which now becomes established in the adult pattern (see Figure 8.5). Undoubtedly, there will be other modulating influences which also influence the activity of the GnRH neurones; for instance, nearby neurones which release naturally occurring opioids such as enkephalin may also be inhibitory on the glutamatergic neurones during the prepubertal stage.

What is it that acts as the switch for re-activating the glutamatergic neurones at puberty? It is likely that metabolic hormones provide the key; in addition to hormones such as insulin-like growth factor 1 (IGF1),

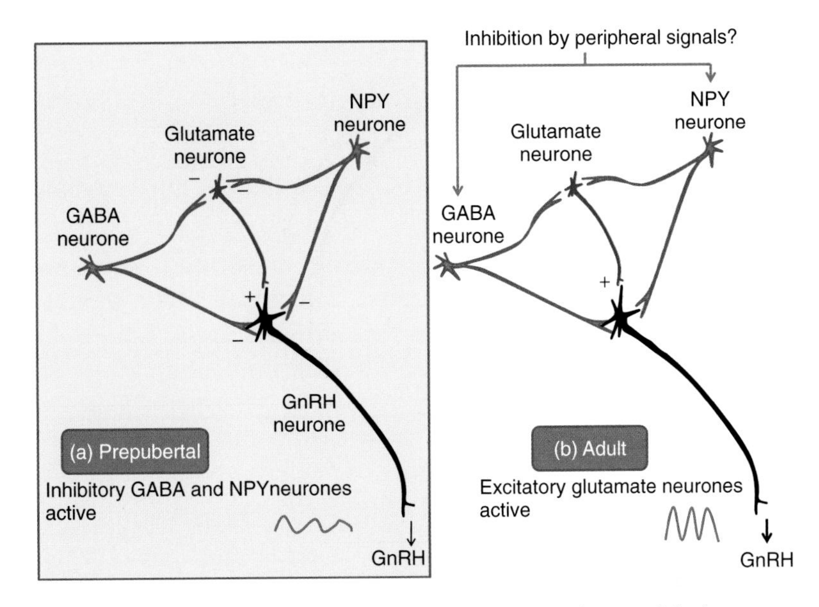

Figure 8.5 Diagram illustrating the key GnRH neurones and some of the known interacting inhibitory and excitatory neurones which influence their activity, in (a) the prepubertal state and (b) the adult state. The potential peripheral signals which inhibit the inhibitory GABA and NPY neurones, allowing the glutamatergic neurones to activate the GnRH neurones, are likely to include molecules such as leptin, insulin-like growth factor 2 (IGF2) and the gonadal steroids.

iodothyronines and the gonadal steroids themselves, one likely factor may well be leptin. This is a hormone produced by adipose tissue; its concentration in the plasma increases in direct proportion to the amount of body fat. This would be an ideal intermediate between the nutritional status of the body and the hypothalamic puberty trigger, the GnRH neurones.

The kiss gene and its products

There is yet one more interesting aspect to the search for the onset of puberty trigger: the KiSS1 gene. This is a gene that was identified at the State University of Pennsylvania Medical College which is located in the town of Hershey in the United States, source of the famous chocolate kisses, hence the name apparently. The KiSS1 gene product is a 121–amino acid protein which is cleaved to produce a 54–amino acid polypeptide called kisspeptin. Kisspeptin is a ligand for an originally identified orphan G protein–coupled receptor called GPR54 (Figure 8.6).

As mentioned earlier, puberty is initiated by the onset of production of adult-level gonadal steroids (gonadarche) which follows the increased

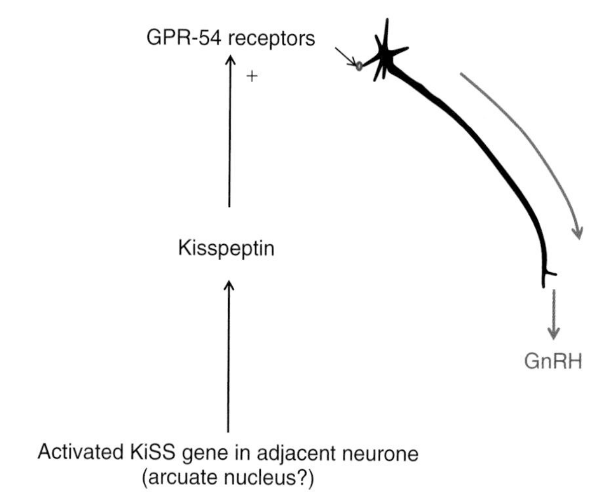

Figure 8.6 Diagram illustrating how adjacent neurones expressing the KiSS gene within the arcuate nucleus produce the polypeptide kisspeptin which activates GPT54 receptors located on the GnRH neurones.

production of anterior pituitary LH (and FSH), in turn initiated by the resurgence of pulsatile GnRH from the hypothalamus. What evidence is there for a role of the KiSS-1 gene, kisspeptin and its receptor in timing the onset of puberty? Both the KiSS-1 gene and GPR54 are located in the medial basal hypothalamus in rodent models, particularly in the arcuate nucleus where many GnRH neurones are located. Inactivating mutations in the GPR54 gene are associated with hypogonadotrophic hypogonadism and absent puberty, whilst activating mutations in the GPR54 gene cause precocious (early) puberty. GPR54 mRNA is actually expressed in GnRH neurones in the arcuate nucleus. Furthermore, puberty is associated with increased KiSS-1 expression, and the intravenous administration of pulses of kisspeptin induces maintenance of raised pulsatile LH initiated by priming doses of GnRH, suggesting that the kisspeptin is stimulating the GnRH neurones via the GPR54 receptor.

Thus the KiSS gene, kisspeptin and its GPR54 receptor seem to be intimately involved with gonadal function and the onset of puberty (Figure 8.6). How these molecules fit into the overall initiation of puberty is the next stage to identify in this fascinating story.

Further Reading

Hameed, S., Jayasena, C.N. & Dhillo, W.S. (2011) Kisspeptin and fertility. *Journal of Endocrinology*, 208, 97–105.

CHAPTER 9

The Hormones of Pregnancy, Parturition and Lactation

Introduction

Pregnancy is the consequence of successful fertilisation of an ovum by a spermatozoon. In humans, the time it takes for that initial formation of a diploid cell to develop into a fully grown fetus ready to be delivered from the mother's uterus at parturition is normally approximately 260–270 days. This is generally divided into three trimesters. The whole process involves the formation of a placenta which will act as a vital 'intermediary' between the mother and growing fetus. One particularly important function of the placenta is to provide hormones which are essential for the maintenance of normal growth and development of the fetus itself. Indeed, because of the important collaboration which exists between placenta and fetus, they are generally considered together as a fetoplacental unit, as least with regard to their endocrine function.

While it is inappropriate to cover the subject of pregnancy in any detail, and the reader is directed to textbooks on reproduction, there are certain aspects that can be mentioned here because they emphasise the importance of hormones in the whole regulatory process.

Fertilisation

In order to successfully fertilise an ovum, the spermatozoa need to have reached a certain stage of development, which includes the ability to swim effectively in their containment fluid. Spermatozoa are released from their association with the Sertoli cells of each testis into the seminiferous tubules together with the seminiferous fluid, at which stage they are quiescent and incapable of fertilising an ovum. Nutrients such as glycoproteins and fructose necessary for maintaining spermatozoon viability are produced by the Sertoli cells under the influence of androgens. From the seminiferous tubules the spermatozoa pass into the rete testis and epidydymis, and it is here that much of the fluid is reabsorbed under the influence of oestrogens

Integrated Endocrinology, First Edition. John Laycock and Karim Meeran.

synthesised from androgen precursors in, and released from, the Sertoli cells. The spermatozoa are concentrated by the removal of much of the fluid, and stored within the muscular vas deferens. A contraceptive method for males is the tying off of the vasa deferentia, preventing the passage of spermatozoa into the urethra. The spermatozoa that accumulate above the ligature are removed by phagocytic leukocytes.

When the appropriate stimulus activates the sympathetic innervations of the vasa deferentia, the muscular walls contract and the stored spermatozoa are propelled via the ejaculatory duct into the urethra, along with the fluid released from the seminal vesicles which usually represents two-thirds of the total ejaculate volume. More fluid (and useful molecules such as nutrients) is provided by other accessory sex glands including the prostate which provides most of the remainder. The seminal fluid containing spermatozoa is called semen, and it is this which is normally released into the female vagina at copulation by the process of ejaculation. The volume of the semen in human males is of the order of 2–5 ml, but the volume in other species can be much greater. In human semen the concentration of spermatozoa is normally $15–120 \times 10^6 ml^{-1}$. Semen contains fibrinogen and fibrinogenase as well as fibrinolytic enzymes, so that when it is deposited within the vagina it rapidly clots, and then subsequently gradually returns to the fluid stage as spermatozoa are released through the cervix into the uterus itself.

While in the vas deferens, the spermatozoa develop a limited capability to move by means of whiplash-like activity of their tails (activation), but they are still unlikely to be able to fertilise an ovum. Full capability to move effectively and to fertilise an ovum are properties which are attained only within the female reproductive tract in the presence of oestrogens, in a process called capacitation. This involves the loss of the glycoprotein protective 'coat' which was deposited onto the spermatozoa in the vas deferens, the establishment of fully developed whiplash movements and changes to the surface membrane over the spermatozoon head which ultimately lead to the acrosome reaction. Of the many spermatozoa released in the ejaculate, very few actually enter the uterus and survive the journey up to the Fallopian tube where the ovum is usually to be found after ovulation. However, in the presence of the ovum, the first spermatozoon to reach it binds to a glycoprotein called ZP3, on the zona pellucida membrane around the ovum. In the presence of progesterone, calcium ions enter the spermatozoon head, and the inner surface of the membrane and the large enzyme-containing vesicle, called the acrosome, become exposed to the exterior: this is the acrosome reaction. Another binding site, on the now exposed inner membrane of the spermatozoon head, is presented to the zona pellucida surface where it binds to another glycoprotein called ZP2. This is associated with the penetration of the spermatozoon through the membrane by the simultaneous release of hyaluronidase and other (proteolytic) enzymes from the ruptured acrosome which digest their way through the zona, allowing it to reach the ovum surface. The ovum then immediately undergoes a zonal reaction during which the second polar body is expelled and the ZP2 and ZP3 molecules are degraded by enzymes released from the cortical granules in the oocyte. The genetic material from the (haploid) spermatozoon now enters the (haploid) ovum, and a new diploid cell is produced.

In this highly simplified description of events leading to the formation of a diploid zygote, it is clear that the maternal ovarian steroids play an essential role (see Figure 9.1).

Implantation and creation of the placenta

Fertilisation usually takes place in a Fallopian tube, and the first cell divisions take place over the following 3–4 days as the dividing cells drifts towards, and then into, the progesterone-primed uterus (albeit in the simultaneous presence of 17β-oestradiol in humans). Initially the cells, called blastomeres, divide, each cell receiving its necessary nutrients from the endometrial secretions by simple diffusion. However, once the number of cells has reached approximately 16, they compact to form a

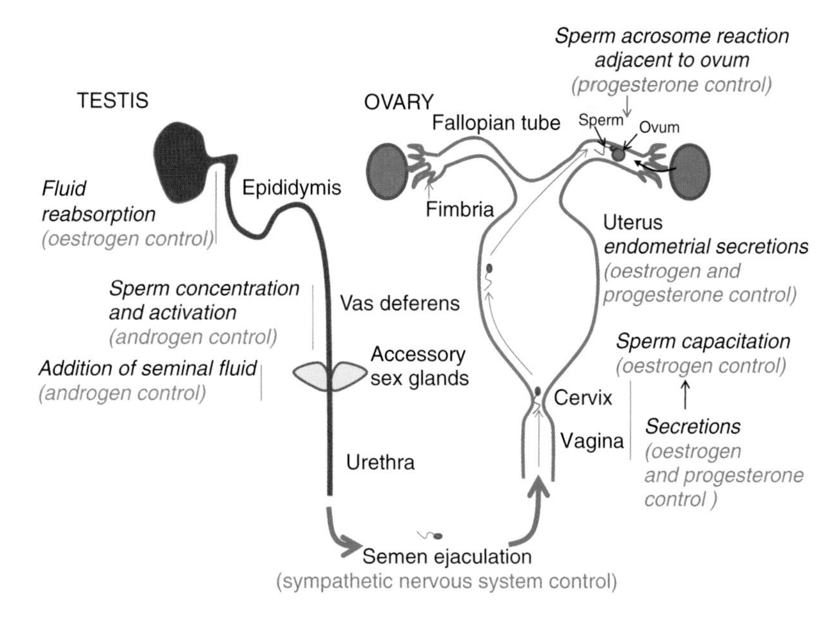

Figure 9.1 Diagram illustrating the hormonal background to key events which are necessary for the spermatozoon to become fully capable of fertilising the released ovum.

spherical morula, at which stage the inner cells would have difficulty receiving sufficient nutrients by simple diffusion alone. The cells then begin to specialise into an inner cell mass and an outer layer of trophoblast cells surrounding a fluid-filled blastocoelic cavity, forming a blastocyst. From initial fertilisation to the formation of the blastocyst, the group of developing cells is called the conceptus and it remains surrounded by the zona pellucida. As the morula becomes a blastocyst, it moves from the Fallopian tube into the uterus. The inner cell mass develops into first the embryo and later the fetus, while the trophoblast cells becomes extra-embryonic tissue ultimately becoming part of the placenta. For the next few days, as the blastocyst develops within the uterus, its cells are still getting their necessary nutrients from the endometrial secretions, but by days 6–9 it is ready to implant. In humans implantation is invasive, meaning that it will penetrate into the endometrium where subsequent development takes place.

The first stage of implantation is the attachment phase when the trophoblast cells attach to the endometrial surface (see Figure 9.2). Before this can happen, the zona pellucida is removed by digestive enzymes secreted by the trophoblast cells and/or the endometrial glands. For

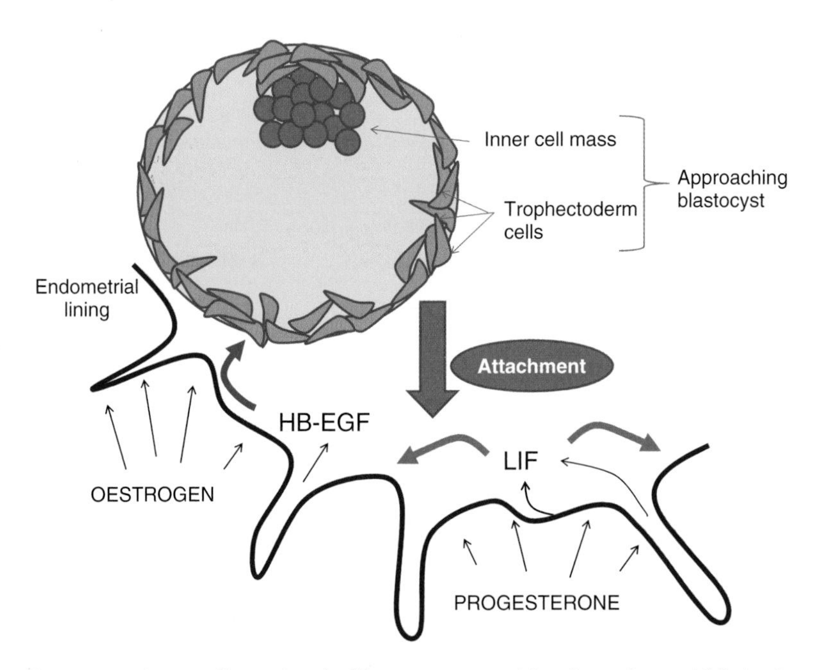

Figure 9.2 Diagram illustrating the blastocyst approaching the endometrial lining just prior to its attachment. LIF = leukaemia inhibitory factor and HB-EGF = heparin-binding epidermal growth factor (for further details, see section 'Implantation and creation of the placenta').

attachment and subsequent implantation to occur successfully requires a particular endocrine environment of oestrogens superimposed upon the dominant progestogen background. Various locally produced factors are involved in the processes of attachment and implantation. One key molecule involved in bringing about the apposition of the outer trophoblast cells with the epithelial cells lining the endometrium is the cytokine glycoprotein Leukaemia Inhibitory Factor (LIF). Release of this molecule by the endometrial glands and epithelial cells lining the uterus is believed to be stimulated by progesterone, and it reaches its peak in the mid- to late luteal phase, and in early pregnancy. LIF promotes general receptivity of the endometrial lining to the approaching blastocyst, and without it implantation does not occur. The more localised attachment process is brought about by various molecules including the oestrogen-stimulated release of heparin-bound epidermal growth factor (HB-EGF), the cytokine interleukin-11 (IL-11) and leptin.

Within hours of the outer trophoblast cells touching the endometrial lining, they begin to infiltrate the epithelial cells and penetrate the stroma. The stromal cells during the late part of the luteal phase are transformed into decidual cells, and these cells are phagocytosed by the invading trophoblast cells which lose their membranes and interconnect, forming a syncytiotrophoblast. The decidua becomes a source of nutrients for the now implanting blastocyst, in a process called decidualisation. The epithelial cell layer gradually re-forms over the implanting blastocyst as it invades the endometrium. IL-11, which is released by the invading trophoblast cells and by decidual cells, is believed to play a role in the decidualisation process.

Subsequent growth of the decidualised tissue and the development of maternal arteries and arterioles provide the basis for the formation of the placental interface between mother and fetus.

Hormones of pregnancy

Human chorionic gonadotrophin (hCG)

It is already clear that the ovarian steroid hormones, oestrogens and progesterone, are essential for the early process of endometrial preparation and for the process of implantation, and at this very early stage of pregnancy their source is the corpus luteum in the ovary. During the first 6 weeks of pregnancy, loss of the ovaries results in the termination of pregnancy, but thereafter ovariectomy has no effect on it. This is because of the crucial role of the ovaries, specifically the corpus luteum, during those first few weeks by providing the necessary quantities of oestrogens and progesterone. Subsequently, the fetoplacental unit takes over this role.

There is an apparent paradox which requires resolution with regard to the continuing provision of steroid hormones by the ovaries during the first few weeks of pregnancy: what stimulates the luteal cells once the increasing circulating levels of both 17β-oestradiol and particularly progesterone have inhibited the anterior pituitary production of luteinising hormone (LH) and follicle-stimulating hormone (FSH) by their direct and indirect negative feedback effects on the maternal hypothalamo-adenohypophysial axis? The answer is another glycoprotein hormone very similar to LH, with the same α sub-unit, called human chorionic gonadotrophin (hCG), which is produced by the implanting syncytiotrophoblast cells. This molecule can be detected by the 10th day of pregnancy, and indeed a standard pregnancy test depends on the measurement of a fragment of this hCG molecule. It binds to the LH receptors on the luteal cells in the corpus luteum and thereby stimulates these cells to continue synthesising 17β-oestradiol and progesterone, despite the now diminished maternal LH (Figure 9.3). Its concentration in the maternal blood increases to reach maximum levels around the 10th to 12th week of pregnancy after which levels decline, although it is present (in varying small amounts) throughout the rest of pregnancy.

Oestrogens

From the sixth week of pregnancy, the placenta is the source of most of the oestrogens being produced, but both mother and fetus provide the necessary precursor molecules. The largest quantity of oestrogen now

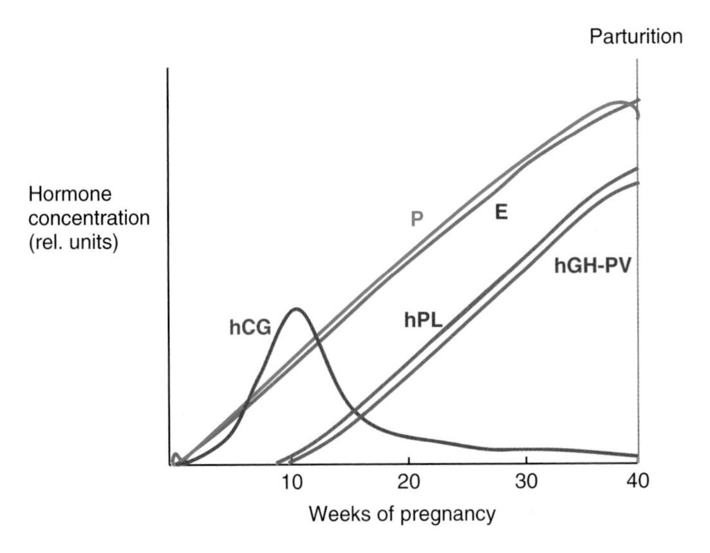

Figure 9.3 Graph showing the changing levels of various hormones throughout pregnancy. E = oestrogen; hCG = human chorionic gonadotrophin; hGH-PV = human growth hormone placental variant; hPL = human placental lactogen and P = progesterone.

produced is no longer 17β-oestradiol but the less potent oestriol, with the metabolite oestrone (which also has weak oestrogenic activity) also being produced in small amounts. One androgenic precursor provided by both mother and fetus is dehydroepiandrosterone sulphate (DHEAS) which is aromatised to oestrone and 17β-oestradiol by the placenta. The oestriol is synthesised by the placenta purely from the androgenic 16α-hydroxy-DHEAS which is produced by the fetus (Figure 9.4). By late pregnancy, oestrogen levels are normally very high.

Measurement of urinary oestrogen excretion levels can be a useful indicator of fetal well-being. The oestriol concentration is likely to be more adversely affected than the other two oestrogens when the fetus is distressed, since its precursor is almost entirely provided by it. In this case, the ratio of oestriol to the sum of the other two oestrogens (17β-oestradiol and oestrone) would be decreased.

Oestrogens are vital for the normal growth and development of the fetus throughout the pregnancy, and are necessary for preparing the mammary glands by stimulating their growth and development (see Chapter 7). They are also necessary for the preparation of the smooth muscle of the

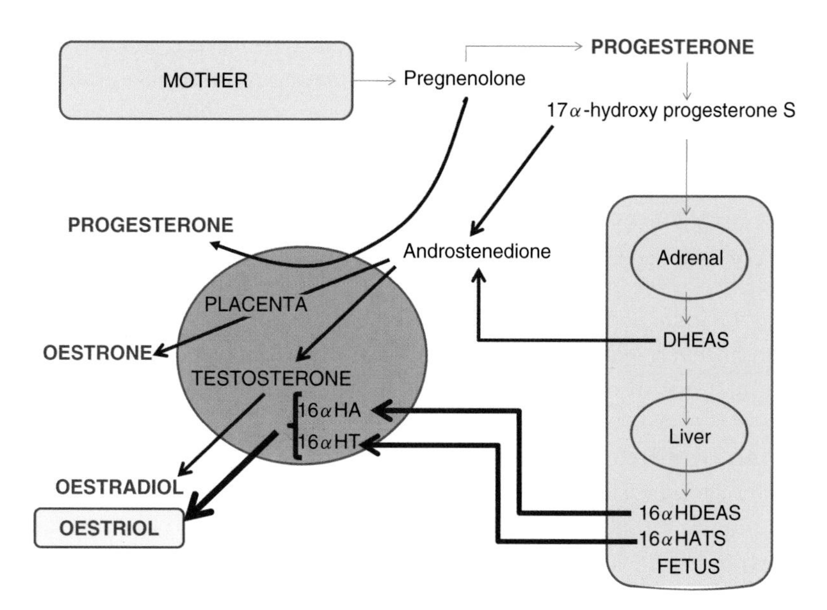

Figure 9.4 Diagram illustrating the important relationship which exists between the mother regarding the provision of precursors, the fetus which converts these precursors to dehydroepiandrosterone sulphate (DHEAS) and then androstenedione in the adrenals, and the conjugated 16α-hydroxydehydroepiandrosterone sulphate (16αHDEAS) and 16α-hydroxyandrostenetriol sulphate (16αHATS) molecules which are converted respectively to the deconjugated forms 16α-hydroxyandrostenedione (16αHA) and 16α-hydroxytestosterone (16αHT) in the placenta.

myometrial lining of the uterus. They stimulate the synthesis of oxytocin receptors and promote the synthesis of prostaglandins. In addition, they stimulate the production of prolactin from the maternal adenohypophysis. The reason why some of these effects of oestrogens do not actually occur noticeably until the approach of parturition is at least partly due to the counteracting inhibitory effects of progesterone.

Progestogens

During pregnancy, progesterone is also necessary for the normal growth and development of the fetus as well as other, maternal tissues such as the mammary glands. Its source is chiefly the placenta, once this tissue has taken over the synthesis role from the corpus luteum. Its concentration in the maternal blood increases steadily throughout pregnancy reaching concentrations of the order of 120–200 $ng.ml^{-1}$, a 10-fold increase compared with the peak during the luteal phase of the menstrual cycle.

Progesterone counterbalances the generally stimulatory effects of the oestrogens, for instance on the synthesis of oxytocin receptors and prostaglandins. It also appears to stimulate maternal appetite.

Relaxin

This polypeptide hormone is produced by the corpus luteum and then the placenta throughout pregnancy. In women, plasma relaxin concentrations are highest towards the end of the first trimester after which they fall to a steady lower level throughout the rest of pregnancy. It is a cytokine with a structure similar to that of insulin. It stimulates angiogenesis, so this effect on the vasculature may be particularly important during the establishment of the maternal-fetal interface (i.e. the placenta). At least in other mammals, when levels reach their highest at term, it seems mainly to have a role in the induction of cervical 'softening', a pelvic and cervical expansion due to an effect on associated ligaments, to help passage of the fetus at delivery. It may also have a relaxing effect on uterine smooth muscle.

Human placental lactogen (hPL)

Human placental lactogen, like hCG, has structural and other similarities with anterior pituitary hormones, in this case with growth hormone and prolactin. As hCG levels decline from about the 10th week of pregnancy, hPL levels begin to rise. It is synthesised in increasing quantities from the fetoplacental unit from the end of the first trimester onwards. It is a protein hormone of 119 amino acids, which has mainly growth hormone and some prolactin activity, increasing lipolysis from the progesterone-stimulated maternal fat deposits. It increases insulin resistance as pregnancy develops, and in susceptible women the rising levels of glucose in the blood can lead to gestational diabetes mellitus.

Human growth hormone placental variant (hGH-pV)

This hormone is also produced by the fetoplacental unit and as with hPL its levels rise from the end of the first trimester onwards. Like GH it is a protein, with 217 amino acids. It seems to have such similarity to growth hormone itself that as its levels rise, maternal GH levels fall, indicating an effective negative feedback on the mother's adenohypophysis. By acting on GH receptors, it is likely to provide GH-like effects on the mother's metabolism, increasing lipolysis and protein synthesis. In the liver, a major role is to stimulate the synthesis of insulin-like growth factor 1 (IGF1).

Other maternal hormones

As pregnancy progresses, many of the maternal endocrine glands increase in size and consequently hormone production rises. Thus there is an increased production of all the anterior pituitary hormones except for the gonadotrophins LH and FSH (inhibited by the rising levels of gonadal steroids) and GH (which appears to be inhibited by the increasing levels of the GH placental variant). Likewise the thyroid increases its output of iodothyronines, and the adrenals increase their output of glucocorticoids. At least part of the latter effects is likely to be due to the oestrogen-stimulated hepatic production of the plasma-binding proteins, raising total hormone levels.

Parturition

Delivery of the fetus at parturition depends at least partly on the change in the ratio of oestrogen to progesterone as this will now allow oestrogenic effects to predominate, for instance in the myometrium. Evidence in most animals indicates that this is precisely what occurs just before parturition, although evidence in humans is still controversial. Oestrogen at this approaching end stage of pregnancy is associated with increasing contractions of the uterine smooth muscle, the myometrium, preparatory to the successful expulsion of the fetus from the maternal uterus, combined with the softening of the cervical ligaments maybe through the mediation of relaxin.

Evidence also indicates that it is the fetal hypothalamus which initiates the process of parturition by regulating the production of corticotrophin-releasing hormone (CRH). This hormone controls the production of corticotrophin (ACTH) (see Chapter 3), and this hormone in turn diverts the steroid synthesis pathway in the fetal adrenals from one producing predominantly progesterone to one which now synthesises more of the oestrogens. Consequently, when maternal oxytocin synthesis is increased in the oxytocinergic neurones of the hypothalamus just before parturition, there is also an oestrogen-driven increase in oxytocin receptor numbers in the myometrial cells. Release of oxytocin is actually triggered by the stimulus of stretch of the cervix and uterine wall. This is the neural component of a neuroendocrine reflex arc which has the release of oxytocin as the efferent response (see Chapter 4). Oxytocin increases the inward movement of calcium into the myometrial cells and stimulates prostaglandin (e.g. $PGF_{2\alpha}$) synthesis which is also assisted by the oestrogen-stimulated synthesis of the precursor arachidonic acid. These prostaglandins increase the cytoplasmic calcium ion concentration within the cells by stimulating their release from intracellular calcium stores such as the sarcoplasmic reticulum and microsomes, while oxytocin directly stimulates the uptake of calcium ions from the extracellular fluid via calcium channels (see Figure 9.5).

In the myometrial cell, this increase in cytoplasmic calcium ion concentration is crucial in bringing about the contraction of smooth muscle.

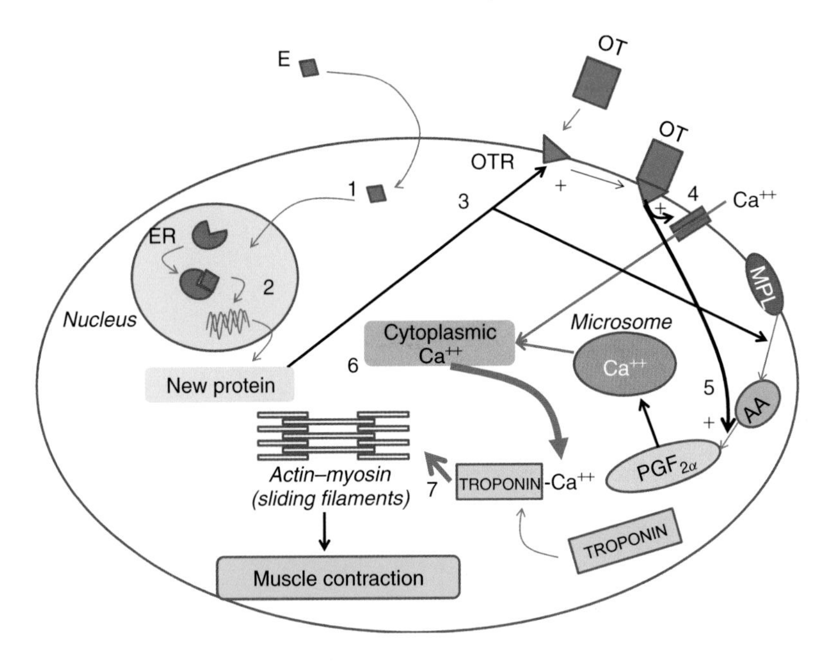

Figure 9.5 Diagram illustrating how oestrogen domination of the myometrium at parturition influences the process of smooth muscle contraction. (1) Oestrogen (E) binds to its intranuclear receptor (ER) and the complex then acts as a transcription factor on the DNA, with the resultant synthesis of new proteins (2). These include (3) the synthesis of arachidonic acid (AA), and oxytocin receptors (OTR) which, in the presence of oxytocin, results in (a) an increase in movement of calcium ions (Ca^{++}) from extracellular to intracellular compartments (4) and (b) following the oestrogen-stimulated synthesis of arachidonic acid (AA) from membrane phospholipid (MPL), the subsequent synthesis of prostaglandin $PGF_{2\alpha}$ (5) which stimulates the release of Ca^{++} from intracellular stores such as the microsomes into the cytoplasm (6). Calcium ions then bind to the troponin proteins activating myosin kinase which stimulates the formation of actin-myosin filaments (7) resulting in the contraction of the smooth muscle by the sliding filament theory.

The calcium ions bind to the intracellular protein calmodulin which then activates myosin kinase which in turn stimulates the formation of the actin–myosin bonds which comprise the sliding filament machinery of the myometrial cell. The consequence of all these effects initiated by oestrogens and mediated by oxytocin and the prostaglandins is that the uterine wall contracts and the increasingly rhythmic contractions ultimately lead to the delivery of the baby.

Lactation

Once the baby is born it requires feeding, and the natural source of that food is provided by the mother's milk. Throughout pregnancy, mammary tissue is stimulated to grow and develop not only under the influence of oestrogens and progesterone, but also in the presence of insulin, IGF1, iodothyronines, growth hormone (and the placental GH variant) as well as prolactin and hPL. As the hormone levels generally increase during pregnancy, as parturition approaches there is an increasing stimulation

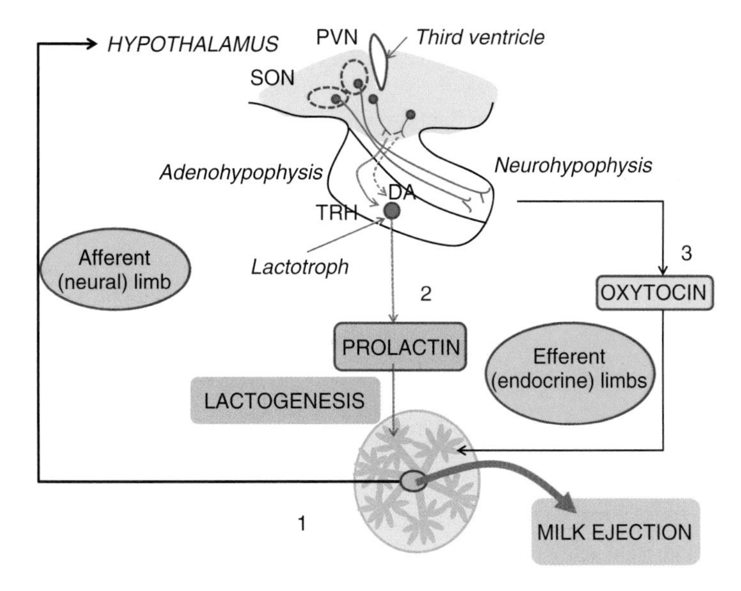

Figure 9.6 Diagram illustrating the two neuroendocrine reflex arcs relating the stimulation of nerve endings around the nipple and the activation of afferent neural pathways (1) with the stimulation of prolactin release from the lactotrophs of the adenohypophysis (2) as a consequence of the hypothalamic inhibition of dopamine (DA, inhibitory) release and the stimulation of thyrotrophin-releasing hormone (TRH), and (3) the stimulation of the oxytocic neurones originating in the paraventricular and supraoptic nuclei (PVN and SON respectively) with the release of oxytocin from the nerve endings in the neurohypophysis.

of the alveolar enzymes involved in lactogenesis (i.e. milk production). Prolactin (and hPL) is a key hormone in this respect.

However, after parturition, another hormone becomes important regarding the actual delivery of milk from the alveoli, and along the ducts to the nipple where the milk is expressed. As indicated in Chapter 4, oxytocin is a potent constrictor molecule, particularly with reference to the myometrial lining of the womb and the myoepithelial cells lining the mammary alveoli and ducts. What makes the process of lactation particularly interesting is that its regulation provides a beautiful example of compatible regulation by neuroendocrine reflex arcs, one involving prolactin, and the other oxytocin.

When the baby suckles, it stimulates mechanoreceptors around the nipple resulting in increased nerve impulses up an afferent pathway to the brain. Here, there are neural pathways to the hypothalamus which (i) stimulate the oxytocinergic neurones which then release pulses of oxytocin from the nerve terminals in the neurohypophysis, and (ii) activate an inhibitory pathway to the dopaminergic neurones and a stimulatory pathway to the thyrotrophin-releasing hormone (TRH) neurones, which together control prolactin production and release. As a consequence of the baby suckling at the nipple, the breast is stimulated to eject previously synthesised and stored milk from the alveoli by oxytocin (the milk ejection reflex), and at the same time is stimulated to produce more milk by prolactin-induced lactogenesis for the feeding to follow. Both of these reflexes have the same shared neural afferent limb, and two different efferent endocrine limbs involving the release of two very different hormones (Figure 9.6).

CHAPTER 10

The Adrenal Glands (1): Adrenal Cortex

Introduction

Life is incompatible with the absence of adrenal function, so the hormones produced by the adrenal glands clearly play a vital role in the regulation of the body's homeostasis. The adrenals were originally called the supra-renal glands, which clearly describes their anatomical location, being associated with the superior poles of the two kidneys. Each of the two adrenal glands is composed of an outer part called the cortex (from the Latin for the bark of a tree) and a central medulla (from the Latin for marrow, or core). These two parts are quite different from each other and will be considered separately, although there is some relationship between the two. Only the adrenal cortex is vital for life.

Adrenal embryology and general structure

The fetal adrenals are detectable from around the sixth week *in utero*. The adrenal cortex develops from clusters of mesenchymal cells along the developing coelomic cavity wall close to the urogenital ridge. During fetal development, and during the first year after birth, each adrenal cortex consists initially of an inner fetal, and an outer permanent, zone. During that first year post-partum, the inner fetal zone regresses while the permanent zone proliferates and differentiates into outer glomerulosa and inner fasciculata zones. An additional, even more inner zona reticularis develops subsequently. During the fetal development stage, certain neural crest cells migrate to the coelomic cavity lining and become surrounded by the developing cortex. These neural crest cells form the medulla at the centre of the adrenal gland. Thus the adult adrenal gland consists of a cortex comprising an outer zona glomerulosa, a middle zona fasciculata and an inner zona reticularis which surrounds the cells of the adrenal medulla. The zona fasciculata can usually be clearly differentiated from the outer glomerulosa and inner reticularis cells, being made up of 'chains'

Integrated Endocrinology, First Edition. John Laycock and Karim Meeran.

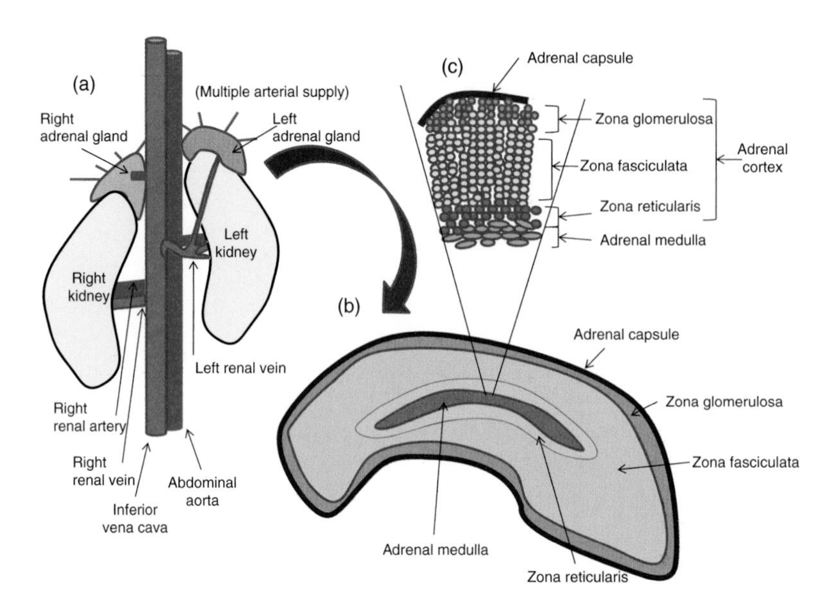

Figure 10.1 Diagram illustrating (a) the general anatomy of the adrenal glands, (b) a cross-section of a single adrenal with the three zones of the outer cortex and the inner medulla identified and (c) an exploded view illustrating the cell formations comprising the three zones.

of cells giving the effect of bundles (Latin, *fasciculata*) (see Figure 10.1). The cells of the adrenal medulla are called chromaffin cells because they can be readily stained with chromic acid salts.

Each adrenal gland lies within its own capsule attached to the upper pole of its respective kidney. The adrenal gland weighs about 5 g and is pyramidal (left) or crescent shaped (right), both measuring approximately 5 × 2 cm long, and 1 cm thick, in the adult. Blood reaches each adrenal gland by a number of small arterial branches from the abdominal aorta, as well as the inferior phrenic, the renal and the intercostal arteries. The blood enters an arteriolar network just below the capsule and then flows down radial capillaries which pass through the three cortical zones to reach the adrenal medulla. Blood then enters a central vein in the middle of the medulla. Blood passes through the short right adrenal vein to drain directly into the inferior vena cava. Blood from the longer left adrenal vein usually drains into the left renal vein, or sometimes into the left inferior phrenic vein, before entering the inferior vena cava. Most of the arterial blood reaches the adrenal medulla indirectly, having passed down through the three cortical zones, and is partially deoxygenated by the time it reaches the chromaffin cells. Less than 10% of the medullary arterial blood reaches the chromaffin cells directly, through small arterioles penetrating through the cortex.

Synthesis, storage, release and transport of the adrenocortical hormones

As described earlier, the cells of the adrenal cortex form three distinct layers, or zones. The zona glomerulosa represents the outer 15% of the cortex, the main part (approximately 75%) being the radial rows of cells comprising the zona fasciculata. The glomerulosa cells are small and grouped together in spherical clusters. The fasciculata cells are larger and full of lipid and extend towards the innermost reticularis cells which are irregularly distributed and contain little lipid. The chromaffin cells of the medulla are clearly different and separate, being typical secretory cells with many granules.

The three cortical zones are associated with steroid hormones called corticosteroids. Because steroids are lipophilic, the adrenocortical hormones cannot be stored to any great extent as they would simply cross the lipid components of membranes and move out of the cells into the general circulation, once synthesised. In other words, they are generally synthesised 'on demand' when the adrenocortical cells are stimulated appropriately. A small amount of cholesterol is present (stored) in the cells, however, in an esterised form which is readily available for hormone synthesis. Historically the hormones have been classified according to their principal physiological activities: (i) hormones having a major role in mineral (Na^+, K^+ ions) regulation, called mineralocorticoids; (ii) hormones having a major role in metabolic regulation, called glucocorticoids and (iii) hormones more commonly associated with the gonads, the androgens and oestrogens, commonly called sex hormones (see Figure 10.2). The mineralocorticoids and glucocorticoids have 21 carbon atoms in their chemical structures, as does their important precursor progesterone, while androgens and oestrogens have 19 and 18 carbon atoms respectively.

All the corticosteroids are synthesised according to cell-dependent, enzyme-induced conversions of substrates to the final end products via various intermediaries. The original precursor molecule is cholesterol which can reach the cells from the blood mainly as low-density lipoprotein (LDL) cholesterol, or be synthesised from acetyl coenzyme A within the cells. The basic structure of cholesterol is known as the tongue-twisting cyclopentanoperhydrophenanphrene nucleus (see Figure 10.3).

The first important step in corticosteroid hormone synthesis is the rate-limiting transport of intracellular cholesterol across the outer to the inner mitochondrial membrane where, in the presence of the enzyme P450scc (cholesterol side chain cleavage enzyme) it is converted to pregnenolone. A specific steroidogenic acute regulatory (stAR) protein mediates this transport process, which is controlled by the adenohypophysial hormone corticotrophin (discussed elsewhere in this chapter). The pregnenolone

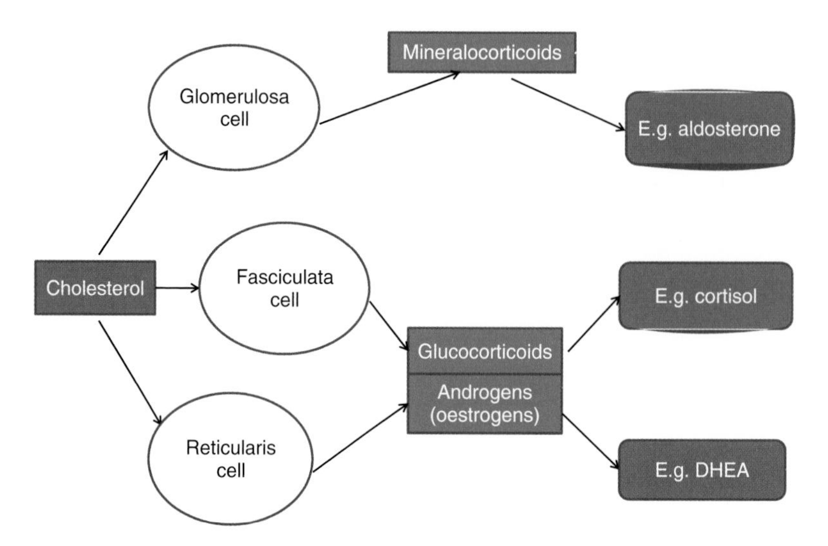

Figure 10.2 Diagram illustrating the three groups of corticosteroids from the adrenal cortex (minerocorticoids, glucocorticoids and the sex hormones) and the principal products in humans (aldosterone, cortisol and androstenedione respectively). DHEA = dehydroepiandrostenedione.

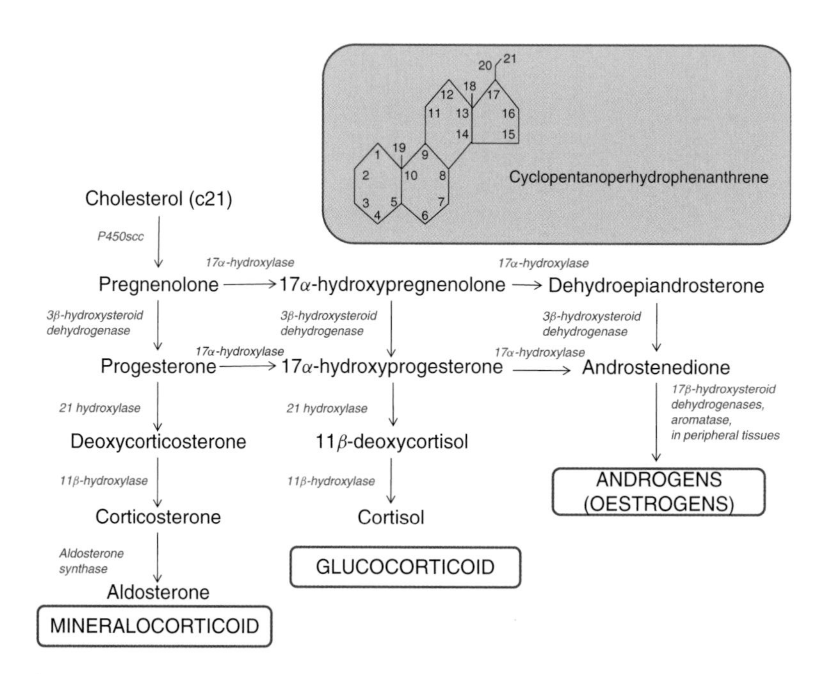

Figure 10.3 Diagram illustrating the synthesis pathway for the three groups of adrenal corticosteroid hormones (mineralocorticoids, glucocorticoids and the sex hormones). Enzymes are given in italics.

enters the cytoplasm where it is converted to progesterone by the enzyme 3β-hydroxysteroid dehydrogenase. Subsequent conversions are dependent on the specificity of enzymes present in the cells of the three zones; indeed, they also depend on their specific intracellular location (e.g. cytoplasm or mitochondrion).

In the glomerulosa cells, progesterone is converted to deoxycorticosterone by 21-hydroxylase, which is found in the microsomes and endoplasmic reticulum. Deoxycorticosterone is then converted to corticosterone by the mitochondrial enzyme 11β-hydroxylase. In humans corticosterone, a weak mineralocorticoid, is converted to the far more potent molecule aldosterone (via an intermediate molecule, 18-hydroxycorticosterone) by aldosterone synthase. The fasciculata and reticularis cells do not contain aldosterone synthase and so cannot produce aldosterone. However, they do contain 17α-hydroxylase, which is absent from the glomerulosa cells. This is another enzyme located in the microsomal-endoplasmic reticulum fraction, and this enzyme converts both pregnenolone and progesterone to their 17α-hydroxy metabolites respectively. As can be seen from Figure 10.3, the 17α-hydroxypregnenolone can also be converted to the 17α-hydroxyprogesterone by the 3β-hydroxysteroid dehydrogenase enzyme. Mainly in the fasciculata cells, but also to some extent in the reticularis cells, the 17α-hydroxyprogesterone is converted to 11-deoxycortisol by the enzyme 21-hydroxylase, and this in turn is converted by 11β-hydroxylase to the final end product in humans, the powerful glucocorticoid cortisol. Both 17α-hydroxypregnenolone and 17α-hydroxyprogesterone can also be converted to the weak androgens dehydroepiandrosterone (DHEA) and androstenedione respectively by the action of 17α-hydroxylase, and these actions take place mainly in the reticularis cells, but also to some extent in fasciculata cells.

In humans, the principal minerocorticoid produced by the zona glomerulosa is aldosterone, while the principal glucocorticoid is cortisol (or hydrocortisone). Cortisol can be converted to a biologically inactive form called cortisone by various target tissues including the liver and kidneys, which contain the enzyme 11β-hydroxysteroid dehydrogenase (11β-HSD) which exists in two isoforms. One, $11\beta HSD_1$, converts inactive cortisone to biologically active cortisol and is found mainly in the liver, while the other is $11\beta HSD_2$, which converts cortisol to cortisone. The latter isoform is found not only in the liver but also in the kidneys where it plays an important inactivating role. Of the weak androgens produced, they are mostly DHEA and androstenedione, both of which can be converted to more potent androgens or oestrogens in peripheral tissues depending on whether reductase or aromatase enzymes are present in the cells. When the cells are stimulated appropriately, the parent precursor molecule cholesterol rapidly undergoes the various enzyme-catalysed conversions described in this section, the final hormone end product depending on the presence or absence of specific enzymes in particular cell types, namely, 17-hydroxylase which is present only in fasciculata and reticularis cells, and aldosterone synthase which is present only in glomerulosa cells.

The adrenal androgens represent approximately 50% of the circulating androgens in females, the remainder coming from the thecal cells of the developing ovarian follicles, and later in the menstrual cycle the corpus luteum. The adrenals are therefore an important source of these hormones and their subsequent metabolites, the oestrogens. In males, the adrenal androgens are normally of little significance since the majority of the circulating androgen content is provided by the more potent testosterone (and dihydrotestosterone) produced by the testes (see Chapters 6 and 7).

The corticosteroid hormones, being lipid soluble, are believed to simply diffuse across the lipid components of membranes once they have been synthesised. Once in the circulation, over 90% of the corticosteroids are bound to plasma proteins, providing a protected circulating 'store' of these hormones. The total circulating concentration of cortisol is far greater than that of aldosterone. Like all other hormones which bind to plasma proteins, a dynamic relationship exists between the plasma protein, and the bound and free hormone fractions (see Chapter 1). The two main plasma proteins involved in corticosteroid transport are a specific (high-affinity) α_2-globulin synthesised in the liver called transcortin (or corticosteroid-binding globulin (CBG)) and the non-specific (low-affinity) but high-capacity (because there is so much more of it) albumin (see Table 10.1). Transcortin binds most of the cortisol and corticosterone in the circulation, while albumin 'mops up' much of the remainder of both of these hormones. The biologically active, free, hormone component is of the order of 10% for cortisol and 40% for aldosterone. CBG is also found in the brain, in areas such as the hypothalamus where it is co-localised with vasopressin and oxytocin in the paraventricular neurones projecting to the median eminence, the neurohypophysis and other parts of the brain.

Table 10.1 Normal plasma concentration ranges and approximate % of protein-bound and free components for cortisol and aldosterone in humans. Note the 1 000-fold difference in circulating cortisol and aldosterone concentrations.

Hormone	Total concentration (plasma) nmol/l	Transcortin (%)	Albumin (%)	Free (%)
Cortisol	9:00 AM: 200–700 12:00 AM: < 250	75	15	10
Aldosterone	Standing: 0.2–0.8 Supine: 0.1–0.45	20	40	40

It should be noted that there is a 1000-fold greater amount of cortisol than aldosterone in the circulation normally, and even greater quantities of weak androgens. Regarding the latter hormones, they act mainly as precursors for the peripheral synthesis of more potent molecules such as the androgens testosterone and dihydrotestosterone, and oestrogens such as 17β-oestradiol. They are transported in the blood mainly bound to sex hormone–binding globulin (SHBG) which is synthesised by the liver. They are considered in far more detail in Chapters 6 and 7 where the gonadal steroids are considered.

The corticosteroids have relatively long half-lives in the circulation, the $t^{1/2}$ for aldosterone being of the order of 15 minutes and for cortisol considerably longer at approximately 90 minutes.

While aldosterone concentrations in the blood will vary to some extent depending on posture, for instance, cortisol concentrations vary continually, being released in pulses driven by the pulsatile release of corticotrophin (adrenocorticotropic hormone, or ACTH) from the anterior pituitary. Furthermore, there is normally a clear circadian rhythm superimposed on this pulsatile release profile (see section 'Regulation of corticosteroid production').

Corticosteroid hormones during fetal and neonatal development

As mentioned earlier, during fetal growth and during the first year of life, the adrenal glands develop cortices containing just two zones: a larger inner fetal zone and an outer definitive zone. The fetal zone regresses after birth, while the definitive zone gradually differentiates and proliferates into outer glomerulosa and inner fasciculata zones. By one year of life, the innermost reticularis zone has developed. During fetal development, it is the large fetal zone which predominantly produces steroids, and these are mainly the weak androgens DHEA and androstenedione, which are the precursors of the fetoplacental oestrogens. The production of oestrogens during pregnancy by the fetoplacental unit, in collaboration with the maternal provision of precursor molecules, is considered in Chapter 9.

Pregnancy and corticosteroids

Pregnancy is a normal physiological condition associated with a hypertrophy of the adrenals and an increased production of corticosteroids. This is accompanied by a two- to threefold increase in the hepatic synthesis of plasma proteins, stimulated by the greatly increased production of oestrogens by the fetoplacental unit. The consequence is that there is an increase in transcortin and SHBG synthesis, at least partly accounting for the overall rise in the total corticosteroid concentrations in the blood. The same situation can arise following treatment of women with oestrogenic compounds.

Corticosteroid receptors and mechanisms of action

All the corticosteroids readily cross the plasma membranes of their target cells and enter the cytoplasm where they bind to their specific receptor molecules. Aldosterone and cortisol bind to mineralocorticoid and glucocorticoid receptors (MR and GR), which share considerable homology; particularly, cortisol binds equally readily to both receptors. While the binding of aldosterone to the GR is an unlikely event physiologically, given the 1000-fold difference in their circulating concentrations, the same is not true for the binding of cortisol to the MR. The glucocorticoid receptor has two splice variations both of which are found in numerous tissues of the body including brain tissue. The MR is more discretely distributed in the body, the main target organs where they are located being the kidneys, as well as the gastrointestinal tract and the sweat glands.

The mechanism of action for cortisol has been studied in some detail, and it is likely that many aspects are similar for aldosterone. The first stage is the binding of cortisol to its cytoplasmic receptor, and this appears to involve the receptor's initial dissociation from specific heat shock proteins which are also present in the cytoplasm. The hormone receptor complex is then translocated to the nucleus where it acts as a transcription factor exerting specific genomic effects. Gene transcription can be either stimulated or repressed, and it is quite likely that various other factors can influence the final effect by conferring tissue specificity on the type of response produced. The end result of glucocorticoid action is the regulation of new protein synthesis. It is now quite clear that glucocorticoids influence many hundreds of different genes by stimulation or repression, accounting for the diverse actions associated with them.

Aldosterone has much more specific actions, being directed at the regulation of ion transport across epithelial cells in its target tissues. It binds to its cytoplasmic MR, and the hormone receptor complex is translocated into the nucleus where it acts as a transcription factor.

Interestingly, the MR has an equal affinity for cortisol and aldosterone. Given that there is far more cortisol in the circulation in humans than aldosterone, one might expect the MRs to be continuously bound by the far more abundant cortisol with a permanently 'switched-on' mineralocorticoid activity. It is therefore of much physiological (and clinical) interest to appreciate that the enzyme 11β-hydroxysteroid dehydrogenase type 2 (11βHSD_2), which is present in these cells, converts the active cortisol to the biologically inactive cortisone molecule, thus protecting the integrity of MR regulation by aldosterone under normal circumstances

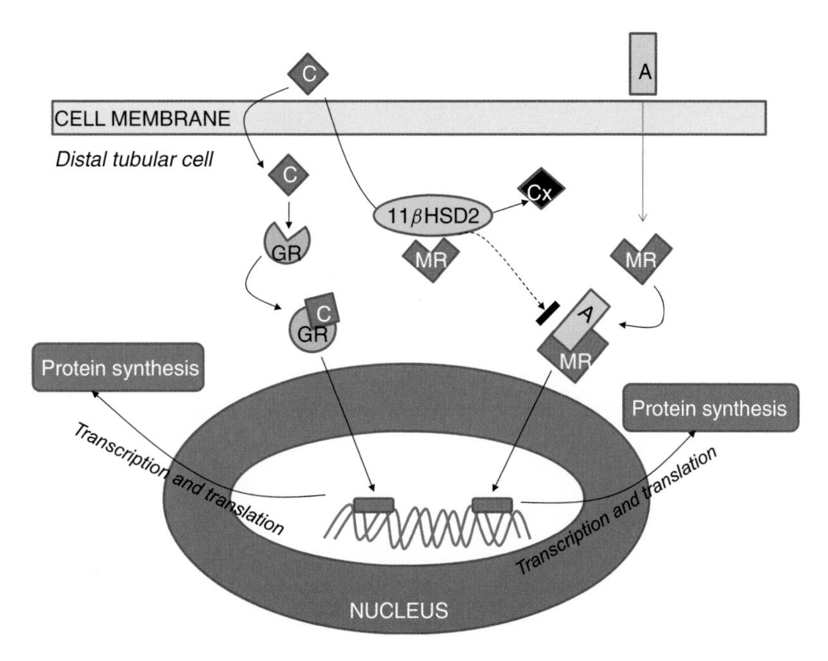

Figure 10.4 Diagram illustrating the important action of enzyme 11β-hydroxysteroid dehydrogenase type 2 which inactivates cortisol (C) by converting it to the inactive cortisone (Cx), allowing the mineralocorticoid aldosterone (A) to bind to the mineralocorticoid receptor (MR). Cortisol can still act on its glucocorticoid receptor (GR).

(see Figure 10.4). The MR and $11\beta HSD_2$ are co-localised in target cells, including in non-renal target tissues such as the brain. It is quite possible, however, that in the presence of excess glucocorticoids such as in Cushing's syndrome, the $11\beta HSD_2$ enzyme system becomes saturated and increased mineralocorticoid activity occurs (see later).

The weak androgens and derived oestrogens also bind to their intracellular (nuclear) androgen and oestrogen receptors (AR and ER) respectively and the formed hormone-receptor complexes act as transcription factors in their target cells. Their genomic actions are discussed more fully in Chapters 6 and 7.

All these steroid hormones, because their mechanisms of action involve the processes of transcription and translation with the ultimate formation of new proteins, have relatively long latent periods, with effects taking up to 24h to begin being manifested. However, over the last few years there has been increasing evidence to support the likelihood that the corticosteroids not only have their well-recognised genomic actions but also have some actions which are observed within minutes of their application to cell preparations, indicating more rapid non-genomic effects. The receptors and precise mechanisms for these actions generally remain elusive currently.

Physiological actions of corticosteroids

Mineralocorticoids (aldosterone)

Aldosterone, the principal mineralocorticoid in humans, exerts its physiological actions on epithelial cells in the kidneys (its main target organs), as well as in the intestinal tract, eccrine sweat glands and salivary glands. In the cells of the distal convoluted tubule and cortical collecting duct, aldosterone regulates the synthesis of an apical membrane sodium channel and also subunits of the basolateral membrane Na^+-K^+-ATPase. The enhanced activity of the basolateral Na^+-K^+ ATPase maintains the low intracellular Na^+ concentration which allows Na^+ ions to enter the cell down their electrochemical gradient through the aldosterone-regulated apical Na^+ channels from the tubular fluid. The overall effect of aldosterone is therefore to stimulate the reabsorption of sodium ions from the tubular fluid into the circulation.

Aldosterone also stimulates the synthesis of a potassium channel in the apical membranes of its target cells in the distal nephron, and plays a major role in the regulation of potassium balance by promoting the secretion of this ion into the tubular fluid, and ultimately its excretion in the urine. The increase in sodium reabsorption across the apical membrane produces a negatively charged tubular lumen which provides the driving force for enhancing the secretion of potassium ions into the tubular fluid. Therefore, aldosterone is an important, direct physiological regulator for controlling renal potassium excretion, and hence the plasma potassium concentration. Furthermore, aldosterone increases H^+ excretion which explains why primary hyperaldosteronism (Conn's syndrome) is associated with a metabolic alkalosis (see Figure 10.5). In the intestinal tract (colon), a Na^+/H^+ exchange transporter is stimulated by aldosterone, increasing the absorption of Na^+ (and Cl^-) ions across the cells and into the general circulation in exchange for H^+ (and HCO_3^-).

While aldosterone clearly plays an important role in regulating the amount of sodium in the extracellular fluid, a consequence is that it also indirectly regulates the extracellular fluid volume. The increased plasma sodium ion concentration stimulates the central osmoreceptors, which results in increased vasopressin release from the neurohypophysis. Vasopressin stimulates distal tubular water reabsorption and thus, under normal conditions, the extracellular fluid volume (ECFV) is increased (see Chapter 4). This is one long-term mechanism by which the arterial blood pressure is maintained. Not surprisingly therefore, a likely sign of increased or decreased mineralocorticoid activity is hypertension or hypotension, respectively.

In the presence of chronically raised aldosterone production, the expansion of the ECFV is limitied to a maximum of 15% by means of various

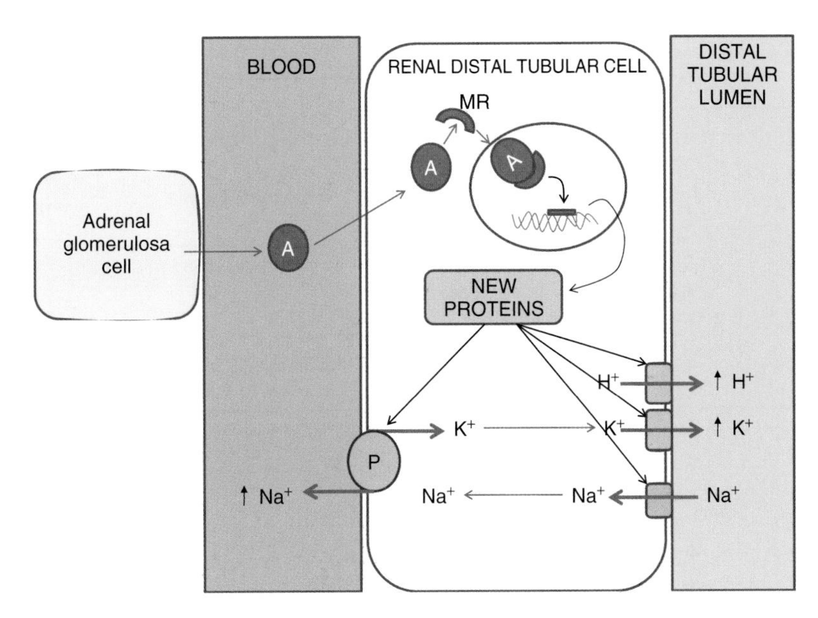

Figure 10.5 Diagram illustrating the mechanism of action of aldosterone in a renal distal tubular cell, with the resulting stimulation of sodium, potassium and hydrogen ion channel synthesis directed to the apical (tubular) membrane, and the increased activity of the Na^+–K^+ pump in the basolateral membrane.

compensatory mechanisms which come into play and which together are sometimes referred to as the 'mineralocorticoid escape' phenomenon. These mechanisms include an increased release of atrial natriuretic peptide (ANP) particularly from the left atrium of the heart. ANP, as its name indicates, exerts a compensatory renal natriuresis (see Chapter 12).

It is clinically useful to appreciate that the precursor to aldosterone in the synthesis pathway is another, weaker, mineralocorticoid called corticosterone (see Figure 10.3). In fact this molecule is the principal mineralocorticoid in rodents. In cases of 11β-hydroxlase deficiency in humans, the production of aldosterone and cortisol is reduced or absent. Consequently the negative feedback normally exerted by cortisol on the hypothalamo-pituitary axis is reduced and corticotrophin production increased, resulting in the stimulation of the biosynthesis pathway for the corticosteroids (see regulation of cortisol, below). In this case, there will be an increased production of the precursor molecules, including corticosterone which mitigates somewhat the loss of aldosterone. Hence in this instance the glucocorticoid deficit can be severe but the minerocorticoid loss minimal.

Glucocorticoids (cortisol)

Cortisol is a far more ubiquitous hormone in its actions than aldosterone. Being a glucocorticoid, as the name suggests, it has important metabolic

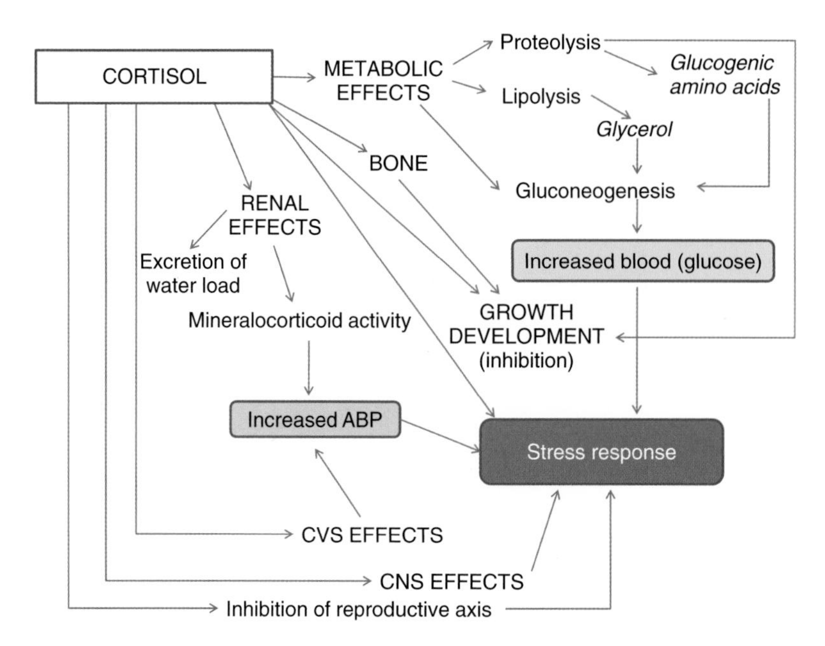

Figure 10.6 Diagram illustrating the principal effects of glucocorticoids such as cortisol.

effects in addition to a variety of other actions some of which become important when the hormone is present in large concentrations in the circulation (see Figure 10.6).

Thus glucocorticoids are considered to have 'physiological' and 'pharmacological' actions which are simply effects within the same spectrum but which become more noticeable at different circulating concentrations.

Physiological actions

Carbohydrate metabolism and the blood glucose concentration

Cortisol stimulates the synthesis of 'new' glucose from non-carbohydrate precursors (gluconeogenesis) mainly in the liver, but also in the kidneys and small intestine under certain circumstances. Stimulation of the hepatic neoglucogenic enzymes results in an increase in the blood glucose concentration. The initial non-carbohydrate precursors are certain amino acids derived from protein breakdown which produce pyruvate and oxaloacetate, and glycerol from lipid catabolism which produces glyceraldehydes. The pyruvate, oxaloacetate and glyceraldehydes are all converted to glucose on stimulation of the appropriate enzymes. A separate indirect effect of cortisol on maintaining an increased blood glucose concentration is the decreased sensitivity of the peripheral tissues to insulin that it induces. This is at least in part produced by the increased concentration of circulating free fatty acids which decrease tissue responsiveness to insulin. Thus in peripheral tissues such as muscle and fat, the uptake of glucose and its cellular utilisation are decreased. Furthermore, cortisol has a permissive effect on the actions of hormones which increase glucose production, such as glucagon and adrenaline. A common consequence of prolonged raised levels of circulating glucocorticoids, as in Cushing's syndrome, is therefore an increase in blood glucose concentration manifested by a raised fasting blood glucose level and a prolonged glucose tolerance test. A high proportion (up to 50%) of Cushing's syndrome patients will actually develop secondary diabetes mellitus.

Protein catabolism

Cortisol stimulates the catabolism of protein, which provides the amino acids necessary for the process of gluconeogenesis. Glucogenic amino acids which can be utilised for gluconeogenesis include glycine, arginine, glutamine and isoleucine. A consequence of this action, when the circulating cortisol concentration is chronically raised, is muscle wasting which is particularly evident in the arms and legs (see Cushing's syndrome).

Lipid catabolism

The physiological action of cortisol on fat is to stimulate its breakdown by the process of lipolysis, providing free fatty acids and glycerol, the latter molecule being a substrate for gluconeogenesis. This effect is in contrast to the lipogenesis observed in patients with chronic, raised circulating glucocorticoid levels. The usual explanation to account for this apparent switch-over is that the raised blood glucose concentration will stimulate insulin release. Insulin stimulates lipogenesis, so this is an indirect consequence of the raised cortisol concentration in the blood. Interestingly, in Cushing's syndrome patients, the increased distribution of fat is particularly prominent in the visceral and centripetal areas, and at the upper back region (called a 'buffalo hump'), rather than in subcutaneous areas. The explanation for this differential distribution, which is characteristic of the condition, is still unclear although it may be associated with a specific increase in GR numbers in these regions.

Renal effects

Glucocorticoids appear to be necessary for the normal excretion of a water load. This is at least partly because they increase the glomerular filtration rate (GFR) by inducing the vasodilation of the afferent (but not the efferent) arterioles. Cortisol also has an inhibitory effect on the release of vasopressin from the neurohypophysis decreasing its activity in the collecting ducts, and this consequently promotes an increase in the excretion of free water.

Bearing in mind that the MR is equally sensitive to cortisol and aldosterone, and that normal circulating levels of cortisol are some 1000-fold greater (as mentioned earlier), the principal renal effect of glucocorticoids might conceivably be a profound increase in sodium reabsorption (and potassium and hydrogen secretion) in the mineralocorticoid-sensitive section of the renal nephron. This is the relatively water-impermeable distal convoluted tubule as well as the vasopressin-sensitive cortical collecting duct. The increased sodium reabsorption would be associated with an increased plasma osmolality, and osmotically sensitive cells in certain circumventricular organs in the brain would be stimulated, resulting in an increased release of vasopressin from the neurohypophysis. Vasopressin in turn would then exert its antidiuretic effect on the principal cells of the collecting duct (see Chapter 3). Thus the overall increased mineralocorticoid effect on stimulating sodium reabsorption is an expansion of the extracellular fluid volume which ultimately results in a raised arterial blood pressure (see Figure 10.5). However, under normal circumstances, this is not the case because of the presence of the enzyme 11β-hydroxysteroid dehydrogenase (11β-HSD type 2) which converts the biologically active cortisol to the biologically inactive cortisone (see Figure 10.4). This 'protects' the MR from cortisol, allowing it to preferentially bind aldosterone. When this enzyme's activity is reduced or absent, either because of a genetic defect or because its action is blocked by compounds such as glycyrrhizinic acid or glycyrrhetinic acid (both found in licorice extract) in susceptible people, this can result in cortisol-induced sodium retention, hypokalemia and hypertension. This condition is known as apparent mineralocorticoid excess syndrome.

Cardiovascular effects

Chronically raised circulating cortisol concentrations such as are seen in Cushing's syndrome patients are commonly associated with hypertension. It is generally assumed that this is likely to be due to the manifestation of the cortisol-induced mineralocorticoid activity which might occur if the 11β-HSD_2 protective system is overwhelmed (see section 'Renal effects'). However, some synthetic glucocorticoids such as dexamethasone also increase blood pressure despite the fact that they have very little affinity for the renal MRs. Thus other cardiovascular actions are also likely, including the stimulation of phenylethanolamine N-methyl transferase (PNMT) in the adrenal medullary chromaffine cells which methylates noradrenaline to adrenaline, the enhanced sensitivity of vascular tissue to pressor molecules such as the catecholamines and angiotensin II, and the decreased synthesis of vasodilator prostaglandins. Glucocorticoids also stimulate hepatic angiotensinogen synthesis and this increased substrate production may result in enhanced circulating angiotensin II levels. This potent vasoconstrictor molecule also stimulates aldosterone production from the zona glomerulosa cells (see section 'Aldosterone').

Bone

Glucocorticoids inhibit osteoblast and increase osteoclast activities and this is probably the reason for the osteoporosis associated with chronic raised circulating levels of these substances. Furthermore, the glucocorticoid-induced increase in GFR and an inhibitory effect on renal calcium reabsorption are associated with an increased excretion of calcium ions in the urine resulting in a negative calcium balance.

Growth and development

Cortisol stimulates lung maturation in the late stages of fetal development by enhancing the synthesis of various surfactant proteins produced by type II alveolar cells which line the alveoli and small bronchioles. Surfactant acts by reducing the surface tension of the lung fluid making the organ more compliant and expandable. Premature babies born before cortisol has exerted this important effect are susceptible to breathing difficulties at birth and without appropriate treatment would be prone to die of asphyxia.

Glucocorticoids are also associated with a depressed linear growth in children when they are present in excess. This is likely to be due at least in part to the increased protein breakdown in muscle, connective tissue and bone matrix, together with a lack of normal bone development. Glucocorticoids inhibit the production of osteoprotegerin, an inhibitor of osteoclast function, resulting in increased osteoclast activity, breaking down the bone matrix. They also stimulate the production of RANKL which is an osteoclast differentiation factor necessary for the development of osteoclasts from precursor cells (see Chapter 19).

The gonads and reproduction

Glucocorticoids inhibit the pulsatile release of hypothalamic gonadotrophin-releasing hormone (GnRH) thereby decreasing the consequent pulsatile release of luteinising hormone (LH) and follicle-stimulating hormone (FSH) from the anterior pituitary. Consequently the gonads are less stimulated by these gonadal controlling hormones and sexual function is reduced (e.g. loss of regular menstrual cycles in women). There may also be other effects on the brain, either directly due to increased glucocorticoids or secondarily due to decreased gonadal steroids, which result in decreased desire for sexual activity (i.e. decreased libido).

Central nervous system (CNS)

GRs and MRs are present in various areas of the brain including the prefrontal cortex, cerebellum, amygdala, hippocampus and hypothalamus.

Corticosteroids, being lipophilic, can probably cross the blood–brain barrier without the need for specific transporters. However, there is also evidence that these steroids can actually be synthesised *de novo* in the brain, for instance in certain neurones and astrocytes. The role of these locally synthesised 'neurosteroids' is unclear, particularly given the small amounts actually synthesised here compared with the adrenal output of these hormones, but since they are generally in close proximity to the GRs and MRs, it is quite possible that they have paracrine or autocrine effects.

Glucocorticoids in physiological concentrations have beneficial effects on the CNS. For example, they act as transcription factors for selective gene expression (e.g. synthesis of adhesion molecules) in the endothelial cells which form the blood–brain barrier, providing protection for the neurones from the deleterious effects of toxic molecules and inflammatory cells in the blood. In contrast, high circulating glucocorticoid concentrations are associated with certain neurological disorders such as depression, euphoria and memory loss. Disruption to synaptic terminals and connections in the hippocampus has been observed when glucocorticoid levels are raised, for example, and these effects together with other neural changes could be linked to memory loss. Clinical depression is one disorder which is also linked to high circulating levels of cortisol. However, the causal relationship between circulating cortisol levels and depression remains controversial at present since depression is not an inevitable consequence of raised glucocorticoid concentrations.

Cortisol also has an important indirect negative feedback influence on the hypothalamic neurones which synthesise and release corticotrophin-releasing hormone (CRH) and also on neurones which produce vasopressin which also acts as a corticotrophin-releasing factor.

The stress response

Cortisol is essential in regulating the body's homeostasis in response to stressors. This is the one key role which makes glucocorticoids vital for the maintenance of life. Indeed, cortisol is the one defining hormone of the true (chronic) stress response. Stressors can be from either the internal (e.g. hypoglycaemia) or external (e.g. infection) environments, and the responses that they elicit are due to the ubiquitous nature of cortisol's actions on the body. These actions would include the metabolic effects culminating in a raised blood glucose concentration, the maintenance of the blood pressure and systemic circulation and other effects generally associated with raised circulating levels of the hormone such as depressing the immune response. The body's resources are diverted in order to deal with the particular stressor, and consequently non-essential body functions are depressed. In particular, reproductive activity can be considered as non-essential in these circumstances, and the glucocorticoids switch off (inhibit) the hypothalamo-adenohypophysio-gonadal axis. However, when glucocorticoids remain raised in the circulation chronically, their long-term effects can become detrimental. Clinically, these excessive effects are associated with Cushing's syndrome (see section 'Cushing's syndrome'), and include manifestations such as secondary diabetes mellitus and hypertension.

Pharmacological actions

Some of the physiological actions of glucocorticoids actually manifest themselves more profoundly when circulating concentrations are raised, and these are generally called 'pharmacological' effects. These actions are of clinical relevance and form the basis for their well-established use in medical practice, and are also common features seen in patients with Cushing's syndrome. These actions result in immunosuppressive, anti-inflammatory and anti-allergic effects all of which involve glucocorticoid interactions with the synthesis of a wide variety of cellular mediators.

Immunosuppression

Glucocorticoids have an important physiological effect controlling the natural immune response, whereby foreign or abnormal molecules, and tissues such as bacterial, viral or fungal infections, are destroyed and removed from the body.

Briefly, the immune system includes cells called T helper (Th) cells which express a specific surface cluster differentiation protein called CD4 (hence known as $CD4^+$ Th cells). The normal immune response begins when $CD4^+$ Th cells respond to an antigen presented to them by antigen-presenting cells (ATC) such as monocytes, B cells or dendritic cells, and 'help' the body by initiating the immune response (see Figure 10.7). These cells then synthesise interleukin-2 (IL-2) receptors and release IL-2 which causes an auto feedback proliferation of these activated T cells. Subsequently, this clone of identical cells differentiates into two subgroups of T helper cells called Th1 and Th2. The Th1 cells also produce IL-2, which causes further cell proliferation, and an interleukin called interferon gamma (IFNγ), which has direct effects of its own (e.g. prevents viral replication) and causes a further differentiation into cells expressing a different surface protein called CD8 ($CD8^+$ T cells). These cells in turn produce various other cytokines which act in different ways such as chemotactins and activators of those cells that mediate the cellular component of the immune response (e.g. phagocytic monocytes). In contrast, the Th2 cells produce interleukin-4 (IL-4) which stimulates further cell proliferation and acts on B cells which also proliferate and produce antibodies which mediate the humoral immune response.

The physiological effect of glucocorticoids is to depress this normal immune response, keeping it in check. This is achieved by inhibiting

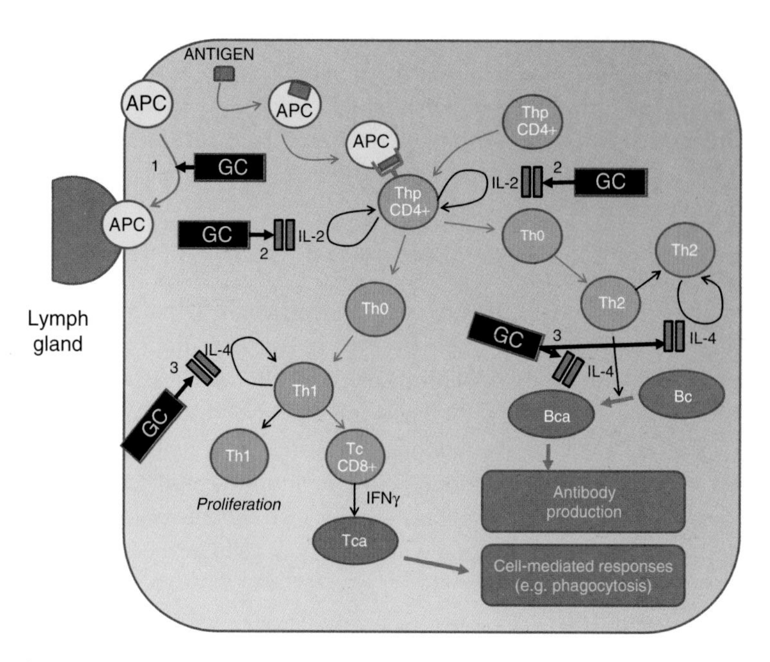

Figure 10.7 Diagram illustrating the effects of glucocorticoids (GC) such as cortisol on cells of the immune system. APC = antigen-producing cells; Bc = B cells; Bca = activated B cells; IL-2 = interleukin-2; IL-4 = interleukin-4; IFNγ = gamma interferon; Th CD4+ cells = T helper CD4+ (naive) cells; Th0 = activated T cells; Th1 and Th2 = T cell subsets; TcCD8+ = differentiated T cells and Tca = activated T cells. (1) Initially GC stimulate the uptake of APC into lymph tissue; (2) GC inhibit the autocrine proliferating effect of IL-2 on the ThCD4+ cells and (3) GC also inhibit the autocrine proliferating effect of IL-4 on the TH1 cells.

the production of IL-2 from the $CD4^+$ Th and Th1 cells of the immune system, thus decreasing the proliferation and differentiation of these cells in response to the initial antigen presentation. When glucocorticoids such as cortisol are present chronically in high circulating concentrations, the overall effect becomes one of immunosuppression. This initially beneficial effect can now become problematic: patients with Cushing's syndrome have a poor response to infections, for example. On the other hand, these steroids now have an important clinical role, precisely when it is beneficial for the immune response to be depressed (e.g. preventing rejection of a transplanted organ).

Anti-inflammatory response

Glucocorticoids also play a physiological role in regulating the normal inflammatory response, preventing it from becoming excessively activated.

The normal acute inflammatory response consists of three clear phases, in response to any form of tissue damage: local vasodilatation, increased leakage of capillaries and increased migration of phagocytes and other leukocytes from the circulation into the damaged area. For example, the initial response to a scratch on the skin is a rapid activation of nearby mast cells and macrophages which promptly release nitric oxide, histamine and various other inflammatory molecules such as bradykinin into the surrounding area. These substances induce an immediate local vasodilatation which is associated with a reddening of the skin around the damaged area. This is followed by an increase in the intracellular gaps between the endothelial cells of the capillary wall, resulting in their increased 'leakiness'. Plasma subsequently moves out of the capillaries into the surrounding tissue which swells. Various chemotactic proteins are also released and these attract leukocytes in the blood, including neutrophils, eosinophils and basophils, which pass between the endothelial cells by a process called diapedesis. These cells enter the tissue where they engulf and digest invading organisms and damaged cells and release toxic molecules such as superoxide, hydrogen peroxide, lysosyme and acid hydrolases which participate in the destruction of bacteria and other invading organisms. Other systems are also activated including the complement and immune systems. The complement system involves approximately 30 proteins which are normally present in the circulation in an inactive form and which, when activated (e.g. by the binding of an antibody to its antigen), enter a cascade of reactions forming molecules which promote a variety of actions aimed at destroying and removing invading cells and other harmful debris. These actions include opsonisation (the binding of an opsonin such as an antibody or complement protein, necessary for phagocytosis), chemotaxis (attraction of phagocytic cells), the lysis of target (e.g. bacterial) cell membranes and increased vascular permeability.

Glucocorticoids are effective anti-inflammatory agents which block all components of the normal inflammatory response. They act partly by inhibiting the release of inflammatory molecules such as histamine and bradykinin from leukocytes. They also stabilise the activation of the complement and immune systems. These steroids play an important role not just in the physiological regulation of the normal inflammatory response, but also because they are important pharmacological agents which are used widely clinically to treat disorders such as rheumatoid arthritis where the normal inflammatory response is out of control.

Anti-allergic action

An allergen is an antigenic molecule which can promote an excessive immune response in certain individuals, with the release of cytokines, histamine and other molecules which then induce the allergic reaction. Normally, the allergen will be 'recognised' by the immune system and appropriately dealt with via the activation of Th cells as part of the usual immune response. However, in some individuals there may be a genetic susceptibility to reacting excessively to certain invading organisms or molecules (the allergens), with potentially serious consequences. There is also the belief that early exposure during childhood promotes a normal immune response. If a child is raised in an excessively hygienic environment, it is possible that the normal controlled response to that allergen is not produced, with the consequence that a later exposure to that allergen then induces an excessive reaction. Pollens (the grains of plant seeds) or specific proteins in food products (e.g. cheese or shellfish) are typical candidates for inducing the allergic response in susceptible individuals.

Molecules such as histamine are released from mast cells and other leukocytes, causing the itching, sneezing and more serious respiratory distress associated with an allergic reaction. Glucocorticoids are useful treatments in individuals suffering an allergic reaction because they regulate the normal immune response including the release of histamine.

Androgens

While the production of androgens is an important feature of the fetal adrenal gland, it is relatively minimal in adults. This is certainly true normally in males, since the testes produce vastly greater quantities of the more potent androgen testosterone. However, in females, the adrenals are an important source of circulating androgens, and while circulating concentrations are usually considerably lower than in males, they do have various physiological actions such as the stimulation of axillary hair and certain effects on the brain such as the maintenance of libido. Androgens are also the precursor molecules for oestrogens, so conversion of weak androgenic molecules to oestrogens in tissues such as adipose is clearly an important role.

Regulation of corticosteroid production

Aldosterone

As with many other hormones, the principal control mechanism of aldosterone production is linked to its main physiological action, albeit indirectly. The principal action of aldosterone is to stimulate the reabsorption of sodium in the renal distal nephron. Consequently, via stimulation of central osmoreceptors and vasopressin release, this is associated with increased distal tubular water reabsorption and therefore expansion of the extracellular fluid volume (ECFV). As indicated earlier, this would be associated with a long-term increase in mean arterial blood pressure were aldosterone to be produced in excess chronically such as in Conn's syndrome.

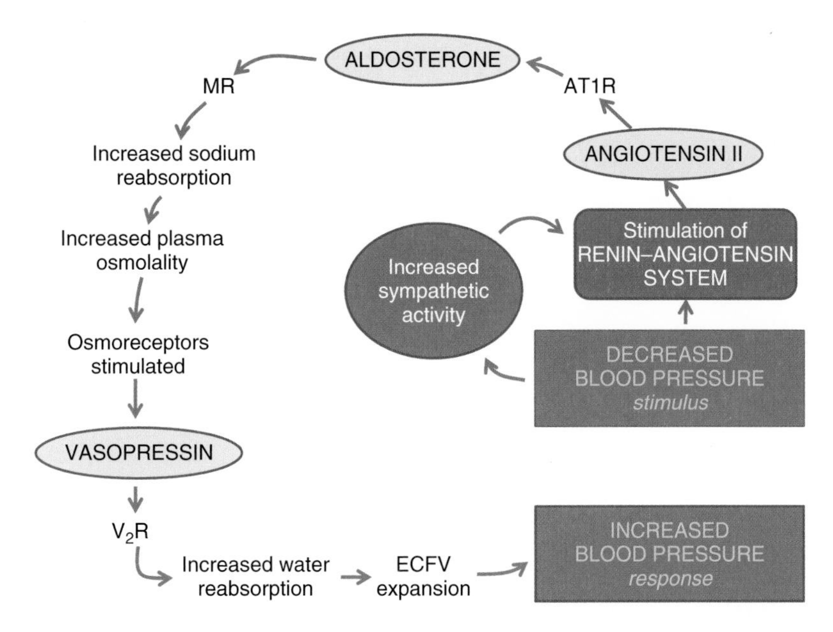

Figure 10.8 Flow diagram illustrating the basic pathway driving the renin–angiotensin–aldosterone system in response to a decrease in arterial blood pressure with a resultant increase in the blood pressure. AT1R = angiotensin II receptor type 1, ECFV = extracellular fluid volume and V2R = vasopressin type 2 receptor.

The principal control system linking blood pressure to the production of aldosterone is called the renin–angiotensin system, and the sensor for this system is located in the kidneys (Figure 10.8).

The renin-angiotensin system (RAS)

The main sensor involved in aldosterone release is a specialised cell in the kidney, called the juxtaglomerular (JG) cell, which responds to decreases in renal arterial blood pressure. As the name suggests, these cells are close (*juxta*) to the glomerulus of the renal nephron, and form part of the walls of the afferent arterioles through which blood reaches the glomerular capillaries in Bowman's capsule. Each adult human kidney has approximately one million renal 'units', or nephrons, and each one of these nephrons has a filtering component called Bowman's capsule which contains capillaries through which blood passes and is filtered. The blood reaches these capillaries via an afferent arteriole and leaves them via an efferent arteriole. The JG cells are adjacent to specialised sodium-sensing cells located at the top of the ascending limb of the loop of Henle which is part of the renal nephron, in a region called the macula densa (Figure 10.9). These macula densa cells detect changes in the sodium ion concentration of the tubular fluid as it reaches the top of the loop of Henle.

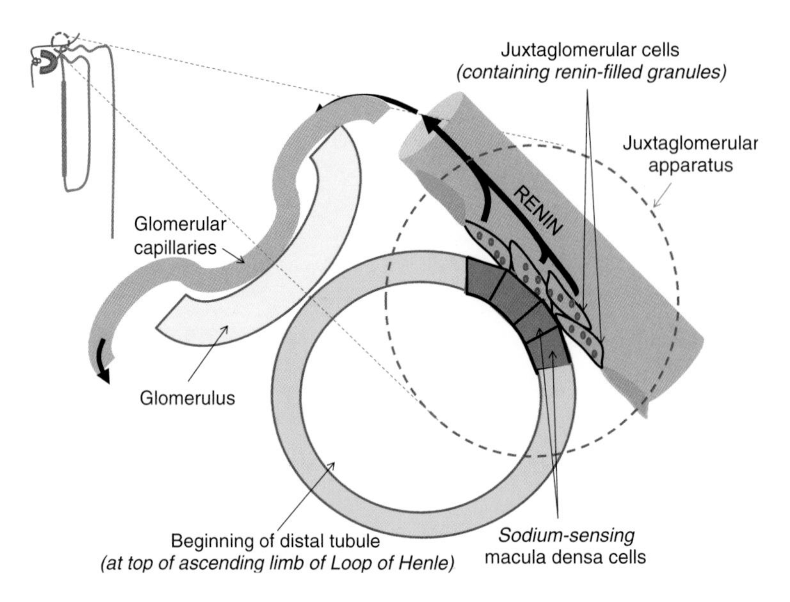

Figure 10.9 Diagram illustrating the juxta-glomerular apparatus, comprising the renin-secreting juxtaglomerular cells of the afferent arteriole and the sodium-sensing cells of the distal convoluted tubule where it arises from the top of the thick ascending limb of the loop of Henle.

The juxtaglomerular cells respond directly to changes in renal arterial blood pressure, but they are also innervated by sympathetic nerve fibres. When the blood pressure falls, not only do these cells respond directly, but also the sympathetic nerve fibre activity will be increased, further stimulating the JG cells. The 'ensemble' of JG cells is sometimes referred to as the JG apparatus. When stimulated, the JG cells release an enzyme called renin into the general circulation, and this protease enzyme acts on a circulating plasma protein called angiotensinogen which is synthesised in the liver. A decapeptide called angiotensin I (AI) is split off the angiotensinogen precursor. This, in turn, is cleaved by yet another enzyme called angiotensin-converting enzyme (ACE) produced by endothelial cells (particularly in the lung vasculature) which removes another two amino acids leaving an octapeptide called angiotensin II (AII). The sequence of catalytic events beginning with the renal release of renin and the final production of angiotensin II is known as the renin–angiotensin system (RAS). Yet another product is produced from AII and this is a septapeptide called angiotensin III (AIII), the physiological significance of which remains unclear.

Yet another stimulus for the production of renin from the JG cells is a decrease in the sodium ion concentration in the glomerular filtrate as it reaches the top of the ascending limb of the loop of Henle where it is detected by the specialised macula densa cells which somehow communicate with the adjacent JG cells. Various hormones such as vasopressin and catecholamines also exert an influence on the RAS. Vasopressin decreases renin release, and while there is evidence for a direct effect on the JG cells it is likely that it can also exert an indirect negative influence through its vasopressor activity.

AII is a hormone with various physiological actions: it is a potent pressor molecule, it increases renal proximal tubular sodium reabsorption and it also stimulates the zona glomerulosa cells of the adrenal cortex to synthesise aldosterone. There are two G protein-coupled angiotensin II receptors, AT1R and AT2R, distributed in the cell membranes of the glomerulosa cells of the adrenal cortex, and in other tissues including neurones in the brain, cardiac and vascular smooth muscle. In the zona glomerulosa cells, AII binds to the AT1R. Indeed, AII is the principal controlling stimulus for the production of aldosterone, and the whole regulatory pathway is sometimes called the renin–angiotensin–aldosterone system (RAAS). As mentioned, aldosterone also has a chronic potentially pressor effect; consequently the RAS is an important target for antihypertensive drugs, with ACE inhibitors being one of the mainstays of treatment for hypertension. In addition, AII has effects on the CNS including a stimulatory action on the 'thirst centre' in the lateral hypothalamus, and a stimulatory controlling influence on vasopressin production. There is much evidence to suggest that the various components of the RAS including renin exist within the CNS. AII is actually synthesised in specific neurones which release it as a neurotransmitter. AII may influence cardiovascular regulation as well as salt and fluid balance by central actions.

Sodium and potassium ion concentrations

The actual physiological actions of aldosterone are to stimulate the distal tubular reabsorption of sodium and the secretion of potassium. As might be expected, changes in the circulating concentrations of these two ions also have direct effects on zona glomerulosa cells and aldosterone release. A decrease in [Na^+] and an increase in [K^+] concentrations, both of the order of 10%, directly stimulate this hormone's release, indicating that they are not of major physiological importance. The principal controlling mechanisms for aldosterone release are shown in Figure 10.10.

Corticotrophin (ACTH)

Since the inner zones of the adrenal cortex are very much under the control of the hypothalamo-adenohypophysial system, it is interesting to note that corticotrophin has only a permissive effect on aldosterone production, stimulating enzymes involved in the synthesis of common precursors early in the steroid synthesis pathway. Corticotrophin (ACTH) is considered in more detail in Chapter 3.

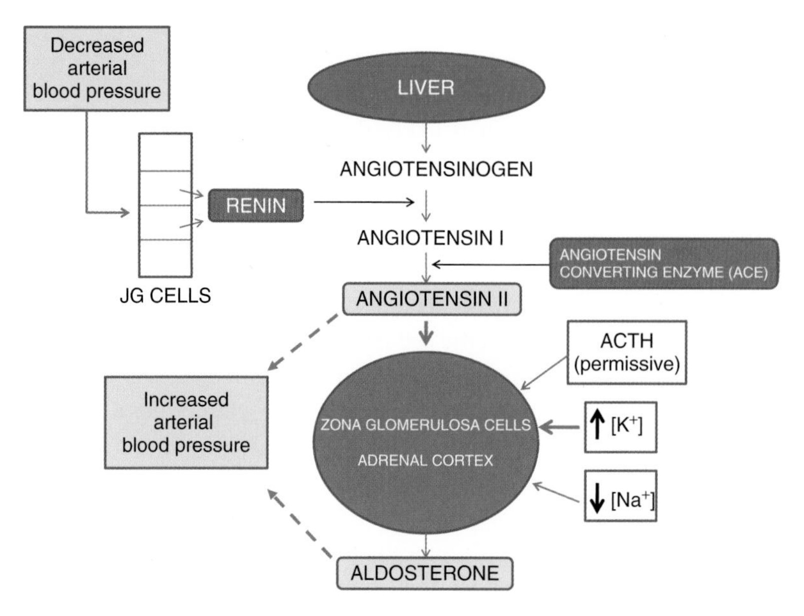

Figure 10.10 Diagram illustrating the control of aldosterone production, with the main drive being provided by the renin–angiotensin system (see text).

Cortisol

By far the most important controlling influence over the production of cortisol is from the hypothalamo-adenohypophysial axis. The anterior pituitary produces corticotrophin (ACTH) which directly stimulates various enzymes in the cells of the two inner adrenocortical layers, those of the zonae fasciculata and reticularis which then produce the glucocorticoids, mainly cortisol (in humans) as well as the adrenocortical androgens. Corticotrophin is itself under the direct control of hormones produced in the hypothalamus. These are CRH and vasopressin which are produced in the cell bodies of parvocellular neurones from the paraventricular nuclei of the hypothalamus. These hormones reach the corticotroph cells of the anterior pituitary via the hypothalamo-adenohypophysial portal system (see Chapters 2 and 3). While most of these neurones produce either CRH or vasopressin, there are some neurones which produce both (see Figure 10.11).

Many factors influence the release of CRH and vasopressin, and it is likely that different stimuli for corticotrophin release operate through either, or both, of these releasing hormones from the hypothalamus. For example, CRH neurones receive a neural input from catecholaminergic fibres from the brainstem and they are stimulated in response to stressors such as decreases in blood glucose concentration (glucoprivation), infections or inflammatory processes, and decreased body temperature. In addition to

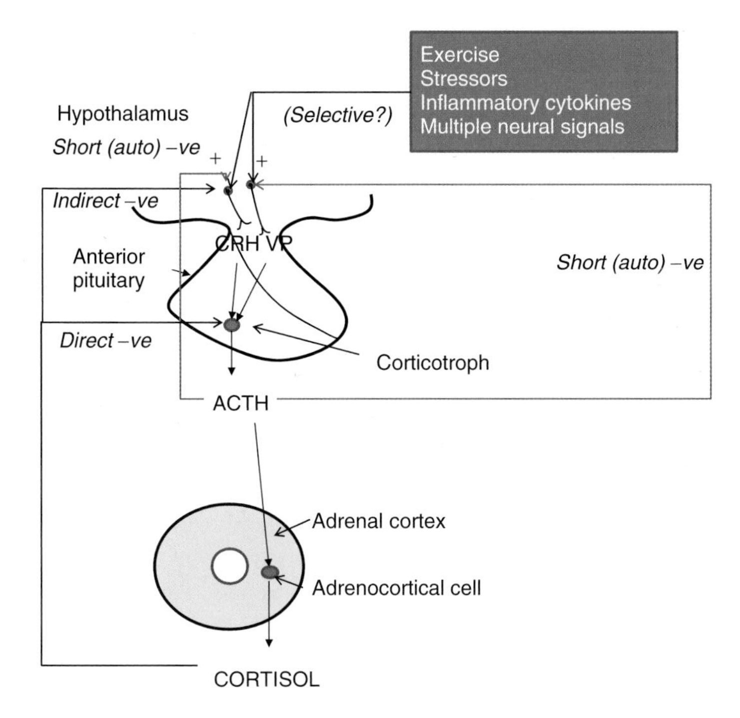

Figure 10.11 Diagram illustrating the hypothalamo-adenohypophysial axis exerting important control of the production of cortisol from the adrenal cortex, and some factors which influence the regulatory process.

releasing either adrenaline or noradrenaline as the neurotransmitters, other molecules co-localised with them such as neuropeptide Y and glutamate may have more specific (currently unknown) controlling functions.

Various environmental stressors such as extreme exercise, excessive cold or heat or prolonged water deprivation also stimulate the release of glucocorticoids. For example, extreme exercise is likely to influence glucocorticoid production at least partly by the glucoprivation that can be encountered in this situation, probably acting via the brainstem catecholaminergic fibres.

There are also important negative feedback loops which operate between the adrenal cortex and the hypothalamo-adenohypophysial axis. Firstly, cortisol directly inhibits the production of corticotrophin from the anterior pituitary, and, secondly, it also indirectly inhibits it by decreasing the activity of the hypothalamic CRH neurones. Thus when the adrenals are stimulated to produce more cortisol under normal circumstances, this glucocorticoid acts back on the hypothalamus and the anterior pituitary to reduce production of CRH and ACTH respectively, thereby decreasing further stimulation of the adrenals and restoring circulating cortisol levels back to normal. Furthermore, it seems likely that ACTH also has a short (or auto) negative feedback loop on the hypothalamic production of CRH. Thus cortisol production is precisely regulated under normal conditions.

Because of the pulsatile nature of CRH release from hypothalamic neurones, ACTH production is also produced in pulses, as is cortisol. Furthermore, there is a very clear circadian rhythm that is superimposed on the production of ACTH and therefore cortisol. This normal circadian rhythm is such that peak circulating levels of cortisol can be measured at around 7:00–8:00 AM, with the lowest levels occurring around 11:00–12:00 PM, and this can be of use clinically (see Figure 10.12).

Adrenocortical disorders

Adrenocortical failure (Addison's disease)

Adrenal cortical failure can be due either to destruction of the adrenal gland, or to a congenital abnormality. The adrenal glands can be destroyed by tuberculosis, by cancer invading the adrenal glands or by the immune system causing autoimmune destruction of the adrenals. Tuberculosis of the adrenal glands is still the commonest cause worldwide, and it used to be the most common cause in the United Kingdom although it is quite rare nowadays. Adrenal failure was first described by Thomas Addison almost 200 years ago, and acquired adrenal failure is thus known as Addison's disease.

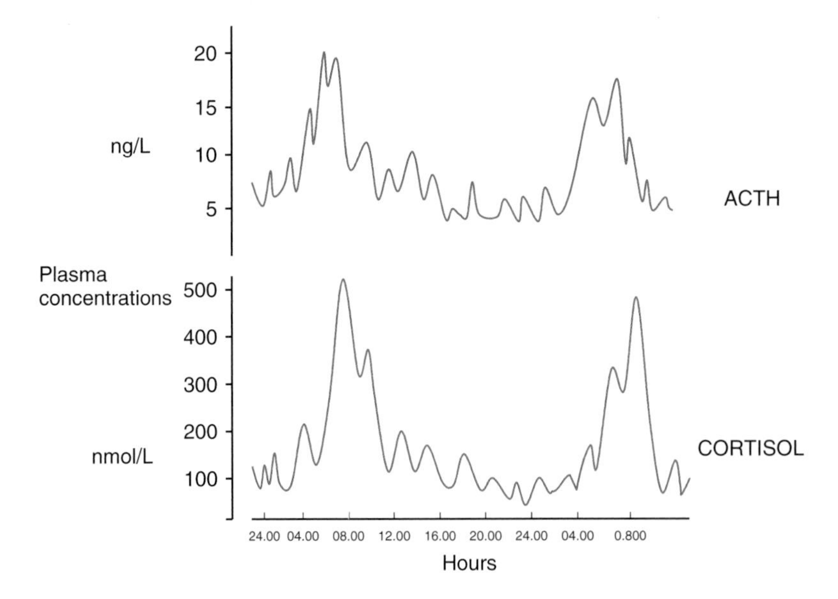

Figure 10.12 Diagram illustrating the normal circadian variation in plasma ACTH and cortisol concentrations. Note the peak cortisol in the early morning, and the low around midnight.

Symptoms and signs of primary adrenal cortical failure

When the adrenal gland fails to make cortisol and aldosterone, the blood pressure starts to fall and the electrolyte concentrations in the plasma change, with a fall in plasma sodium and a rise in plasma potassium. As the circulating cortisol concentration falls, the lack of negative feedback on the hypothalamo-pituitary axis results in an increase in precursor POMC synthesis, and consequently an increase in circulating ACTH. Thus there is an increase not only in ACTH, but also in melanocyte-stimulating hormone (MSH) which is also derived from POMC (see Chapter 3). This in time will stimulate hyperpigmentation in patients with adrenal failure, and the appearance of a "particularly good tan" is one way of confirming the diagnosis.

In addition to the hypotension, the lack of cortisol can cause non-specific tiredness and exhaustion. If basic electrolytes are measured, patients may have slightly low sodium and raised plasma potassium levels. Because cortisol levels are normally highest at 9:00 AM, the standard basic investigation is a 9:00 AM blood sample for cortisol measurement. A very low cortisol level should precipitate referral to a specialist endocrine centre for further investigations of cortisol deficiency. It is important to realise that cortisol levels vary with time of day, so that a low level of cortisol at 9:00 AM is abnormal, while a low level in the late evening or afternoon might simply be perfectly normal. In addition, the diurnal rhythm may be affected in shift workers, who are asleep during the day. Therefore, if the diagnosis is suspected, the gold standard is a dynamic investigation. The dynamic investigation of choice for Addison's disease is the short synacthen test.

Short synacthen test

A baseline sample of blood is collected for cortisol and ACTH measurements. In Addisonian patients, the baseline cortisol might be low, and the ACTH will be high. A subcutaneous or intramuscular injection of 250 mcg SYNthetic ACTH (hence SYNACTHen) is administered, and blood samples taken 30 and 60 minutes later to measure plasma cortisol concentrations. In patients with Addison's disease, there will be only a minimal rise in cortisol despite the injection of the huge dose of ACTH. In normal individuals, the cortisol will rise by at least 170 nM to at least 450 nM (these values vary slightly between different hospitals depending on the assay used in that hospital). A full list of updated tests can be found on the Imperial Centre for Endocrinology website by looking up the Hammersmith Hospital Endocrine Bible (http://www.imperialendo.com).

Once Addison's disease has been confirmed, a patient will then need lifelong treatment with both a glucocorticoid and a mineralocorticoid. Usually hydrocortisone and fludrocortisone are used as replacement drugs, respectively. Dose adjustment should be made in a specialist unit, and patients should be advised that they might need extra steroid cover at times of illness. They should always carry a steroid card, so that in the event of injury, medical staff know that extra steroids need to be administered. Many patients are on a daily dose of hydrocortisone of 20 mg (taken as 10 mg on awakening, with 5 mg at about lunchtime and 5 mg in the early afternoon). There should *not* be a night-time dose, as this has the effect of awakening the patient. The final dose should be given at about 5:00 PM, or perhaps 6:00 PM at the latest, if the patient is hoping to sleep normally at night. The patient will also need between 50 and 100 mcg fludrocortisone daily.

Addisonian crisis

Patients who are not treated with hydrocortisone and fludrocortisone (e.g. if the Addison's disease is not diagnosed or the patients inadvertently stop their treatment) are at risk of a salt-losing, hypotensive Addisonian crisis, from which patients may not recover if the diagnosis is not made in time and steroids are not administered. Nausea and vomiting are prominent features, and patients may find they are unable to keep down their steroids, which only worsens the problem. An Addisonian crisis is thus a medical emergency. Patients need to be urgently transferred to hospital for parenteral (intravenous or intramuscular) administration of 100 mg of hydrocortisone. In an emergency, this will have both glucocorticoid and sufficient mineralocorticoid activity to save the person's life. They will also require rehydration with saline as they are likely to be very salt depleted due to a combination of the vomiting and renal sodium loss in the absence of mineralocorticoid. Once patients are stabilised, and have a normal blood pressure, they can be recommenced on their normal hydrocortisone and fludrocortisone replacement drugs.

Congenital adrenal hyperplasia (CAH)

If a neonate is born without one of the enzymes that should normally be in the adrenocortical cells, there might be failure of cortisol and/or aldosterone synthesis. This will result in increased hypothalamo-pituitary-adrenal (HPA) axis activation *in utero*, and there is thus profound stimulation of the adrenal glands before birth. The consequence is that the adrenal glands become very large indeed. This is known as congenital adrenal hyperplasia (CAH). The clinical features of this condition can be worked out by looking at the enzyme that is missing (see Figure 10.3 and Table 10.2). The most common cause of CAH is deficiency of the enzyme 21-hydroxylase. If the deficiency is complete, as in one-third of all cases, the adrenals will not synthesise any aldosterone or cortisol. The adrenal glands will thus be stimulated by large amounts of ACTH, but the only outcome of the raised ACTH will be increased levels of precursors to cortisol and aldosterone, such as 17-hydroxyprogesterone. High levels

Table 10.2 Principal enzyme deficiencies which result in CAH with the main diagnostically useful criteria in each case.

Enzyme deficiency	Blood pressure	Potassium	Genitalia	Notes
21-hydroxylase	Hypotension	Hyperkalaemia	Virilisation	90% of cases
11-hydroxylase	Hypertension	Hypokalaemia	Virilisation	10% of cases
17-hydroxylase	Hypertension	Hypokalaemia	Absent puberty	Rare
Other enzymes				Very rare

of this molecule will be channelled through other pathways, and consequently one other effect will be an increased secretion of androgens, in particular testosterone. This will result in a virilised fetus, and if the child is a girl, then ambiguous genitalia may be present. Just adequate levels of maternal cortisol and aldosterone will have crossed the placenta in*utero* and kept the fetus alive, but as soon as the child is born, this endocrine provision will no longer be available. Within 24 to 48 hours, the signs of adrenal failure will become apparent, and the child is at risk of a salt-losing Addisonian crisis. Children may become floppy, hypotensive and unresponsive, and without treatment will not survive. If the child is a girl, then the diagnosis will be suspected at birth, because of the ambiguous genitalia. Those children can be given hydrocortisone, and a salt-losing crisis will be averted. Boys, on the other hand, will have normal male external genitalia, even if there is an excess of testosterone, so that the diagnosis may not be suspected at birth until other signs of adrenal failure become apparent.

The second most common cause of CAH is 11-hydroxylase deficiency. This condition will also cause virilisation of a female fetus, but does not cause a hypotensive crisis. This is because the immediate precursor 11-deoxycorticosterone, which is now produced in excessive quantities, binds to the aldosterone receptor and is biologically active. Thus children with this condition present with hypertension and hypokalaemia due to the excessive mineralocorticoid activity, as well as virilisation.

Children with 17-hydroxylase deficiency also present with childhood hypertension because of high circulating levels of aldosterone, but are cortisol and sex steroid deficient. These children become unwell at times of stress, and will remain prepubertal without treatment, as the enzyme will be missing from the gonads as well as the adrenals.

CAH is an autosomal recessive condition, so that couples who have a child with the condition are at risk of having further children also suffering from the condition.

To confirm the diagnosis of congenital adrenal hyperplasia, a synacthen test can be used. In this situation, in addition to measuring cortisol levels in the blood, the steroid precursors are measured. When ACTH is administered, a patient with CAH will exhibit an exuberant rise in the circulating concentrations of steroid precursors, but a minimal or no increase at all in cortisol. Thus in 21-hydroxylase deficiency, there will be a large increase in plasma 17-hydroxyprogesterone on administration of synacthen, but no rise in cortisol.

As with Addison's disease, the treatment of this condition is lifelong replacement of the relevant missing steroid. However, unlike patients who have Addison's disease, the dose and type of steroid used are difficult to manage and should be arranged by a specialist centre. It is important to give adequate doses and frequency of a steroid to suppress the production of the precursors of cortisol to minimise the effects of other steroids (such as testosterone) being produced.

Tumours of the adrenal cortex

Non-functioning tumours

Non-functioning adrenal tumours are usually of no clinical consequence. Guidelines suggest that the risk of malignancy becomes significant when they have grown to 4 cm, so lesions larger than 4 cm should be removed.

Functioning tumours

Functioning tumours of the adrenal gland can cause hypertension; adrenaline, aldosterone and cortisol when produced in excess can each cause hypertension. When hypertension occurs in young people, secondary causes such as an adrenal tumour have to be considered. There may be clues in the case history to make one suspect the correct diagnosis.

Conn's syndrome (excess aldosterone)

Clinical features

Patients present with hypertension, and when measured, the plasma potassium concentration may be low. The hypertension gradually worsens, but patients may not have a diagnosis for several years. Rarely the hypokalaemia may make the patient feel unwell or develop a cardiac arrhythmia which brings them to medical attention.

Diagnosis

Conn's syndrome usually occurs from a unilateral adrenal tumour, but sometimes patients have a bilateral adrenal hyperplasia. Tumours originating in the zona glomerulosa are not under the control of the renin–angiotensin system and consequently can secrete large amounts of unregulated aldosterone. Once the diagnosis is biochemically confirmed, by repeated measurement of raised plasma aldosterone concentrations in the presence of suppressed renin levels (usually induced by a saline infusion), then imaging should be undertaken. A computerised tomography (CT) scan of the adrenals will demonstrate the presence or absence of a tumour. When found, tumours are usually smaller than 1 cm in diameter. Even in the presence of a tumour, one should then proceed with selective venous sampling of both adrenals, to ensure that the source of excess aldosterone is unilateral, and to confirm that a unilateral adrenalectomy will be effective. The aldosterone causes an increase in blood pressure, which in turn suppresses the renin. Conn's syndrome should be suspected in any patients who have hypertension and are hypokalaemic, as aldosterone increases the renal loss of potassium. In some cases, a hypokalaemic alkalosis may be present. If on the angiogram the aldosterone is found to be coming from both adrenals, this would suggest bilateral adrenal hyperplasia.

As indicated, the diagnosis should be confirmed by measuring the plasma aldosterone and renin concentrations. Patients with Conn's syndrome will have raised aldosterone and suppressed renin levels. It is important to make these measurements on patients who have been taken off any drug treatments for hypertension because these can alter the aldosterone-to-renin ratio. In particular, beta blockers cause a fall in renin so patients on beta blockers can have a high aldosterone-to-renin ratio even if they do not have Conn's syndrome. Patients with suspected Conn's syndrome should be referred to a clinical centre where Conn's syndrome is commonly managed.

Treatments

Patients with Conn's syndrome may often be treated with several anti-hypertensive agents because it can be difficult to adequately control their raised blood pressure. They are also often on potassium supplements in view of the profound continuous loss of potassium in this condition.

It is also important for patients to know about the different operations that are available when an adrenal tumour needs to be removed. An alternative to laparoscopic adrenalectomy is the retroperitoneoscopic adrenalectomy. The technique has the advantage of avoiding the visceral innervation of the peritoneal cavity and is used by many large centres around the world. In a comparison of laparoscopic versus retroperitoneoscopic adrenalectomies at Hammersmith Hospital with a typical range of indications for surgery −27% for non-functioning adenomas, 25% for Conn's syndrome, 11% for phaeochromocytomas, 16% for Cushing's and 21% for other benign pathologies – the retroperitoneoscopic approach was found to be associated with an earlier oral intake and lower analgesic requirement. Most significantly, the length of hospital stay was reduced by a mean of 43 hours when the operation was performed retroperitoneoscopically.

Following surgery, if successful, patients can usually stop potassium supplements and some or all of the anti-hypertensive drugs.

If bilateral adrenal hyperplasia is diagnosed, removing both adrenals would be effective but is not recommended, because that would result in the patients requiring lifelong cortisol and fludrocortisone (a mineralocorticoid) replacement, and missing occasional doses can result in an Addisonian crisis, which may be fatal. Thus patients are usually managed medically. Aldosterone antagonists such as spironolactone or epleronone are effective in these patients.

Cushing's syndrome

The clinical features of Cushing's syndrome are the same as Cushing's disease and are given in Chapter 5.

Diagnosis

Tumours of the zona fasciculata of the adrenal cortex may secrete unregulated amounts of cortisol. This will result in Cushing's syndrome, which clinically is very similar in appearance to Cushing's disease, when the cause is specifically a pituitary tumour, except that there is no increase in pigmentation because, unlike in Cushing's disease, the ACTH is suppressed (also see Chapter 5). The diagnosis is confirmed by the dexamethasone suppression test as described in Chapter 5, where the adrenal tumour continues to secrete a relatively fixed amount of cortisol, in the presence or absence of dexamethasone.

Treatments

Once confirmed, a cortisol-secreting adrenal tumour should be removed. Because cortisol suppresses peripheral protein synthesis, rates of wound healing after surgery are slowed, so that large scars can result in prolonged hospital stays. The advent of laparoscopic and retroperitoneoscopic surgery has made a big difference to the morbidity of surgery in these patients. This is discussed under Conn's syndrome in section 'Conn's syndrome (Excess aldosterone)', but minimal trauma is even more important with Cushing's patients.

Another way of improving the quality of life in patients with Cushing's syndrome is to try to control the secretion of cortisol medically with drugs that suppress cortisol production. Metyrapone and ketoconazole are reasonable choices, although they are not really long-term solutions, as they can cause nausea and are associated with abnormal liver function tests on prolonged use. However, they can be used as a stop-gap while arranging surgery.

Following adrenalectomy, patients with Cushing's syndrome are likely to be Addisonian for up to a year. The contralateral adrenal, while normal, would have become suppressed because the high cortisol level from the tumour would have inhibited pituitary ACTH release by negative feedback.

Consequently the normal adrenal often atrophies, and it can take up to a year for it to recover fully. For this reason, patients should be put on hydrocortisone replacement until recovery occurs.

Appendix: Clinical Scenarios

Clinical Scenario 10.1

Mr Leroy arrives at his GP surgery complaining of feeling continually tired, particularly at work in the building industry, and of having an infected cut on his arm which does not heal. He also remarks that he has put on weight recently. On examination, Dr Moffat notices that his face is florid and more round than he remembered from a previous visit, and that while he has clearly put on weight in the abdominal region his arms and legs seem thin. Measurement of Mr Leroy's blood pressure gives two readings that are both around 165/90 mm Hg, indicating hypertension. The wound on his arm looks bad and certainly requires attention. A finger-prick blood sample indicates a capillary glucose concentration of 12 mM.l^{-1}. Doctor Moffat confirms from Mr Leroy that he is not taking any treatments at present.

The doctor arranges for Mr Leroy to be further investigated at the local endocrine clinic.

Questions

Q1. What is the diagnosis that Dr Moffat has reached on the basis of his initial examination and findings?

Q2. How are these observations and findings explained by this diagnosis?

Q3. What further investigations would be helpful in confirming the diagnosis?

Q4. What treatments might be considered?

Answers to Clinical Scenario 10.1

A1. What is the diagnosis that Dr Moffat has reached on the basis of his initial examination and findings?

The initial diagnosis would be Cushing's syndrome. One cannot distinguish the clinical presentation of the different causes of Cushing's syndrome without doing further investigations, and it would be useful to compare the results (given here) of Mr Leroy with those of Mrs Jones in Chapter 5.

A2. How are these observations and findings explained by this diagnosis?

Being tired, particularly if doing heavy labouring, might well be due to proximal myopathy caused by increased protein catabolism resulting in wasted muscles (thin arms and legs). The weight gain, particularly in the abdominal region, could be accounted for by a characteristic lipogenesis which occurs in the presence of excessive glucocorticoids. Cortisol stimulates gluconeogenesis so the blood glucose level rises (secondary diabetes). Insulin is released in increasing amounts, and this induces the lipogenesis. The hypertension could be due to direct effects of glucocorticoids on the cardiovascular system and increased mineralocorticoid activity in the kidneys leading to fluid retention (which could contribute to the weight gain). The poor wound healing is accounted for by the anti-inflammatory response seen in the presence of excess glucocorticoids.

A3. What further investigations would be helpful in confirming the diagnosis?

Random blood cortisol and ACTH concentration measurements might indicate raised and lowered values respectively. In addition one can get Mr Leroy into hospital overnight for 24-hour measurements in the morning (e.g. 8:00 AM) and in the late evening (e.g. 11:00 PM). If both values are raised (i.e. loss of circadian rhythm), this would confirm Cushing's syndrome. An undetectable blood ACTH concentration would suggest an adrenal tumour rather than a pituitary tumour. A further investigation would be an abdominal MRI scan.

The results of the diurnal cortisol levels revealed that the 9:00 AM cortisol was 560 nM and the midnight sleeping cortisol was 540 nM. At both times, an ACTH level was undetectable. This patient also had a low-dose dexamethasone suppression test, and at the end of the test the cortisol was 550 nM. This confirms that an adrenal tumour is likely to be secreting the excess cortisol as it is doing this without the need for any ACTH. All interventions that would affect the pituitary including dexamethasone suppression have no effect on this adrenal tumour that secretes the same amount of cortisol regardless of the time of day or the presence of dexamethasone.

A4. What treatments might be considered?

Treatment for an adrenal tumour would be medical treatment initially followed by surgery. Drugs which block cortisol synthesis, such as ketoconazole or metyrapone, might be used to improve wound healing. The dose of these drugs will slowly be increased until the cortisol has been suppressed to approximately 150–300 nM. If surgery can be performed either laparoscopically or retroperitoneoscopically, then medical treatment might not be required, as the wound is extremely small. What operation is deemed suitable depends on the decision of the multidisciplinary team (MDT) and the experience of the surgeon.

Clinical Scenario 10.2

Mr Smithers has been increasingly tired over the last few months. He had been losing weight; he also noticed that he would feel dizzy when he stood up, and that he had a particularly good tan, despite not going on holiday for some time. Things became particularly bad recently, culminating in a collapse at work and urgent admission to hospital.

Examination revealed a remarkably low blood pressure of 80/40 mmHg. He appeared dehydrated and was confused.

The admitting doctor sent off blood electrolytes, and these revealed:

- Sodium 134 mM (NR 135 to 145)
- Potassium 5.9 mM (NR 3.5 to 5.5)

Questions

Q1. What is the likely diagnosis?
Q2. What investigations should be performed?
Q3. What treatment is required?

Answers to Clinical Scenario 10.2

A1. What is the likely diagnosis?

This sounds like a patient with Addison's disease who is now presenting with an Addisonian crisis.

A2. What investigations should be performed?

The 9:00 AM cortisol level should be checked, and this should be followed by a short synacthen test. The results of the synacthen test revealed a 9:00 AM cortisol level of 80 nM, and after the synacthen, the level rose only slightly to 130 nM. In a normal individual, the baseline cortisol should be at least 270 nM, with a rise to over 500 nM after synacthen. Thus the diagnosis of Addison's disease presenting with an Addisonian crisis has been confirmed.

A3. What treatment is required?

The treatment urgently required is an intramuscular injection of 100 mg of hydrocortisone, and an intravenous infusion of normal saline. Mr. Smithers required 4 litres of saline to normalise his blood pressure after which time he felt much better. He was then put onto regular oral hydrocortisone (10 mg at 6:00 AM, 5 mg at noon and 5 mg at 4:00 PM was suggested) as well as 50 mcg daily of fludrocortisone. He remained well on this treatment. He was also given a steroid card, which he now carries in his wallet at all times, so that in case he was involved in a car accident, the ambulance staff would know that he needed an injection of extra hydrocortisone. A copy of a steroid card can be found on http://www.imperialendo.com.

Further Reading

Gomez-Sanchez, E.P. (2011) Mineralocorticoid receptors in the brain and cardiovascular regulation: minority rule? *Trends in Endocrinology & Metabolism*, 22, 179–87.

Henley, D.E. & Lightman, S.L. (2011) New insights into corticosteroid binding globulin and glucocorticoid delivery. *Neuroscience*, 180, 1–8.

Wang, W.H. & Giebisch, G. (2009) Regulation of potassium (K) handling in the renal collecting duct. *Pflugers Archiv*, 458, 157–68.

Constantinides, V. & Palazzo, F. (2012) A comparison of retroperitoneal and laparoscopic adrenalectomy. In press.

CHAPTER 11

The Adrenal Glands (2): Adrenal Medulla

Introduction

As described in Chapter 10, the adrenal medulla is the central 'core' of the adrenal gland. The cells are of neural origin, developing from the fetal neural crest, and they are innervated by preganglionic sympathetic fibres reaching the gland via the splanchnic nerve (Figure 11.1). The neurotransmitter released from the myelinated presympathetic nerve endings is acetylcholine, so this is the main stimulus for the release of hormones from the adrenal medulla. The cells contain numerous secretory granules filled with molecules which stain readily with (i.e. have a high affinity for) yellow-coloured chromic acid salts, hence called chromaffin cells. These molecules are the adrenomedullary hormones which are derived from an initial precursor which is the amino acid tyrosine. The adrenomedullary cells can therefore be considered to be specialised postganglionic sympathetic nerve fibres which actually release their chemicals not across synapses as neurotransmitters, but into the bloodstream as hormones.

The preganglionic fibres innervating the chromaffin cells are part of the lesser splanchnic nerve which leaves the spinal cord from thoracic segments T10–T12 and passes through the celiac ganglion before reaching the adrenal medulla.

Synthesis, storage, release and transport of adrenal medullary hormones

Chromaffin cells, like postganglionic sympathetic fibres, synthesise noradrenaline from the precursor dopamine by the catalytic action of an enzyme called dopamine β-hydroxylase which, unusually, is located within synaptic vesicles where this reaction takes place. Dopamine itself is produced in the cytoplasm of the nerve terminal from dihydroxyphenylalanine (DOPA) by the enzyme DOPA decarboxylase and is then actively transported into the vesicles. DOPA is produced from the initial amino acid precursor tyrosine by the enzyme tyrosine hydroxylase, this being the rate-limiting step in

Integrated Endocrinology, First Edition. John Laycock and Karim Meeran.

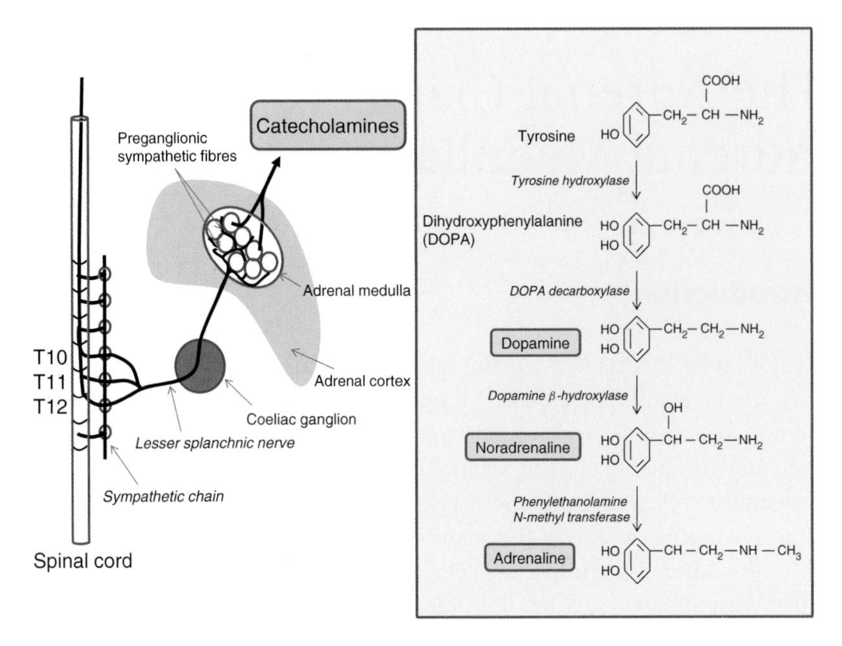

Figure 11.1 Diagram illustrating the link between the sympathetic nervous system and the adrenal medulla (left) and the synthesis pathway indicating the catechols and the relevant converting enzymes (see text).

the synthesis pathway. Unlike the postganglionic sympathetic fibres, however, the chromaffin cells also contain the enzyme phenylethanolamine N-methyl transferase (PNMT) which methylates noradrenaline, released from the vesicles into the cytoplasm, to adrenaline (see Figure 11.1).

The adrenaline is then taken up by other vesicles and stored in the nerve terminals. Adrenaline and noradrenaline are also called epinephrine and norepinephrine respectively; both terminologies are in common use.

The end products in chromaffin cells are secreted by exocytosis of the vesicle contents to the exterior when the cell is stimulated by acetycholine released from the innervating sympathetic preganglionic fibres. In humans, unlike many other species, the main catecholaminergic product from the chromaffin cells is adrenaline (approximately 80% of normal output), with noradrenaline forming most of the remaining 20%, and dopamine being released in minute quantities. In addition, chromaffin cells synthesise and release other molecules such as the enkephalins which are opioids in minute amounts, and these are released together with the catecholamines whenever the cells are stimulated by acetylcholamine from the preganglionic sympathetic nerve terminals. The vesicle contents released from the chromaffin cells enter adjacent capillaries and hence reach the bloodstream (see Figure 11.2).

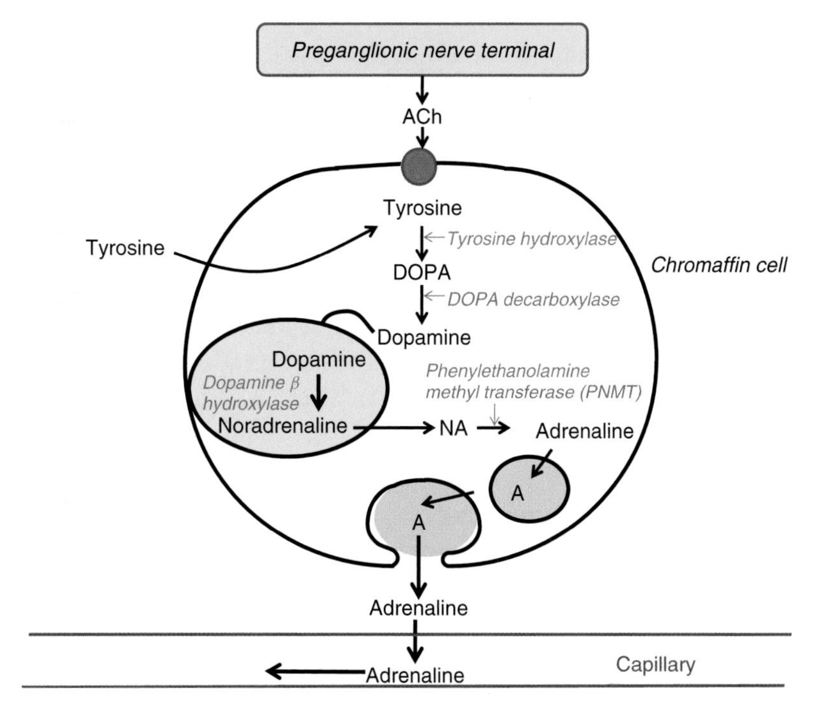

Figure 11.2 Diagram illustrating the synthesis of adrenaline in the chromaffin cell. Enzymes in blue. A = adrenalinee, ACh = acetylcholine.

The adrenal medulla receives arterial blood from the arterial network lying below the capsule from which blood passes either directly through arterioles passing through the outer cortex, or (mainly) from blood which drains between the cords of cells comprising the cortical zones through fenestrated capillaries into venules in the medulla. Blood passes through the venules into a central adrenal vein. The right adrenal vein drains directly into the inferior vena cava while the left adrenal drains into the left renal vein. The provision of arterial blood to the adrenal medulla is of particular interest because it has not only a direct, albeit limited, fully oxygenated supply but also a partly deoxygenated supply which drains through the outer cortical regions, into which corticosteroids may have been released. Indeed, glucocorticoids stimulate tyrosine hydroxylase, dopamine β-hydroxylase and PNMT activities thereby exerting a controlling influence on catecholamine synthesis.

Because the chromaffin cells can be considered as modified postganglionic nerve fibres, the adrenal medulla is essentially an extension of the sympathetic nervous system, its output differing because it reaches all parts of the body in contact with the general circulation. Once in the circulation, the catecholamines have short half-lives of less than 3 minutes, being rapidly inactivated either by uptake into sympathetic nerve terminals or by the enzymes catechol-O-methyl transferase (COMT) and monoamine oxidase (MAO), both found mainly in the liver, kidneys and brain. Noradrenaline uptake into the presynaptic sympathetic nerve terminals is by a noradrenaline transporter (NAT) mechanism called uptake 1, after which it is either recirculated back into granules or metabolised by MAO, located on the outer surface of the intracellular mitochondria. Noradrenaline can also be taken up by the postsynaptic terminals by a mechanism called uptake 2, which results in its breakdown by COMT. The final end products of metabolism, called metanephrins, are vanillyl mandelic acid (VMA) which is mostly excreted unconjugated in the urine and 3-methyl, 4-hydroxy phenylethyleneglycol (MOPEG) which is also excreted in the urine but mostly conjugated as a glucuronide or sulphate (Figure 11.3).

Catecholamine receptors

Of the three major catecholamines, dopamine is of particular importance as a neurotransmitter in the brain. It also has an endocrine role as a hypothalamic inhibitory hormone controlling the release of prolactin from the anterior pituitary (see Chapter 3), and has its own dopamine receptors (DR1 and DR2). The two other catecholamines, adrenaline and

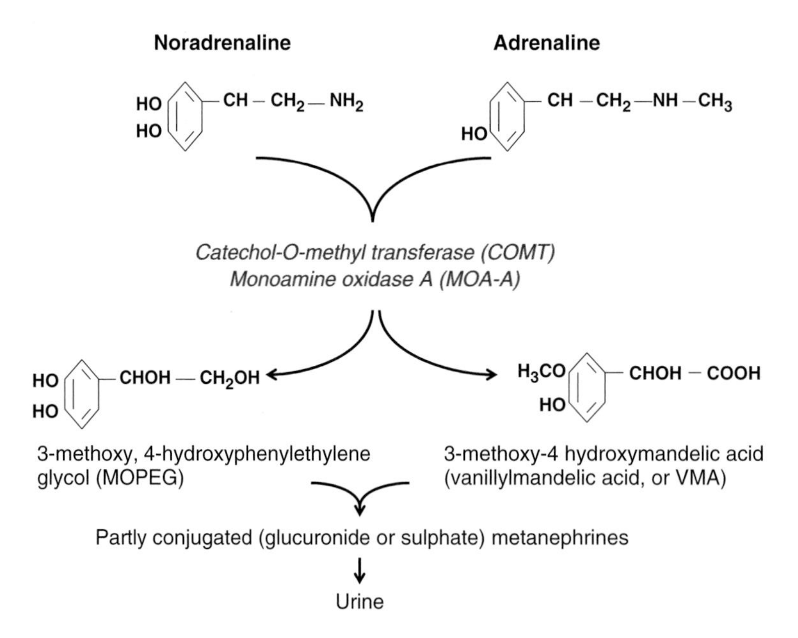

Figure 11.3 Diagram illustrating the two principal metabolic products derived from noradrenaline and adrenaline which are excreted in the urine. Enzymes in blue.

noradrenaline, exert their effects via adrenergic receptors divided into two main groups: α and β adrenoceptors. These two groups can be further subdivided on the basis of specific ligand binding into α1, α2, β1, β2 and β3 receptors.

These receptors are all members of the large seven-transmembrane domain family of receptors (see Chapter 1) and they are all linked via G proteins to intracellular second messenger systems (see Figure 11.4). In general, α1 receptors have a greater affinity for noradrenaline than adrenaline, while α2 receptors have a similar affinity for these two molecules. The α2 receptors are found within the central nervous system (CNS) where they are presynaptic and hence when stimulated inhibit further neurotransmitter release. Of the β receptors, adrenaline generally has a greater binding affinity than noradrenaline, with β1 receptors being particularly important regarding control of cardiac smooth muscle and β2 receptors being involved in mediating many of adrenaline's metabolic effects and the relaxation of bronchiolar smooth muscle. The β3 receptors are interesting in that they mediate the effect of adrenaline on brown adipose tissue which is present in new-born babies (around 5% of body weight) but is pretty well absent in adults. Brown adipose tissue has a much greater vascularity and a more pronounced sympathetic innervation than ordinary white fat. In contrast to other cells, including white

RECEPTOR	SECOND MESSENGER SYSTEM	TARGET TISSUE	RESPONSE
α1	Phospholipase C-inositol triphosphate and diacylglycerol	1. Central and peripheral nervous systems 2. Contraction of vascular and non-vascular smooth muscle (e.g. GI tract) except sphincters; other tissues (e.g. liver, heart and salivary glands)	1. Excitatory Mostly post-synaptic 2. Excitatory Intrasynaptic or on cell membranes
α2	Decreased adenyl cyclase-cAMP (resulting in decreased intracellular [Ca^{2+}])	1. Central and peripheral nervous systems; inhibition of noradrenaline release 2. Other tissues (e.g. platelets and vascular smooth muscle)	1. Inhibitory Presynaptic 2. Excitatory Post-synaptic (e.g. platelet aggregation and vascul smooth muscle contraction)
β1	Increased adenyl cyclase-cAMP	Cardiac smooth muscle and GI sphincters	Excitatory (contraction of cardiac muscle and relaxation of sphincter smooth muscle)
β2	Increased adenyl cyclase-cAMP	Smooth muscle relaxation (e.g. bronchial smooth muscle), metabolic effects (in e.g. liver)	Excitatory (e.g. hepatic lipolysis or glycogenolysis)
β3	Increased adenyl cyclase-cAMP	Adipose tissue (mainly brown adipose tissue in new-born or young)	Excitatory (e.g. lipolysis or calorigenesis)

Figure 11.4 Illustrating the various adrenoceptor subtypes, and the second messenger systems, the main target tissues and the principal responses associated with them.

adipocytes, brown adipocytes express mitochondrial uncoupling protein 1 (UCP1), which gives the cells' mitochondria the ability to uncouple oxidative phosphorylation, and thus triglycerides are metabolised to generate heat rather than ATP. This may be important in a new-born, or young, child with a relatively large ratio of surface area to body mass available for heat loss compared with an adult.

Physiological actions

It will be appreciated that since the chromaffin cells are innervated by preganglionic sympathetic nerve fibres, the adrenal medullae are extensions of the sympathetic nervous system (SNS). When the SNS is activated, catecholamines (mainly adrenaline) are released into the general circulation as part of the overall response. These hormones can then amplify the effects of synaptically released noradrenaline, and reach adrenoceptors elsewhere in the body with particular emphasis regarding β receptor–mediated actions which are particularly adrenaline sensitive.

The overall potentiating effect of circulating catecholamines is of particular importance during the SNS-mediated 'fight or flight' response first described by the American physiologist Walter Bradford Cannon in 1915. This is a generalised acute response by the body's tissues to a stressor, and includes features identified in Figure 11.5. Thus there is a generalised increase in arterial blood pressure associated with changes in the cardiovascular system (including an increased heart rate and stroke volume, and vasoconstriction in skin and the splanchnic bed); an increased airway diameter allowing for an increased air flow to the lungs; increased metabolism mainly in the liver, muscle and adipose tissue (e.g. lipolysis, gluconeogenesis and glycogenesis) resulting in a raised blood glucose concentration; an accompanying increased temperature generation and sweating; pupil dilation and an increased mental alertness.

It is worth noting that the eccrine and apocrine sweat glands are innervated by postganglionic sympathetic fibres. Eccrine sweat is mainly concerned with the loss of latent heat by evaporation of water from the skin surface while apocrine sweat, which is much more restricted to certain areas of the body such as the axillae of the arms (armpits) and the genital region, involves the secretion of an initially odourless mucoid secretion which develops a characteristic odour once it comes into contact with bacteria on the skin surface. In animals apocrine sweat plays an important sexual role as an olfactory signal, for instance to members of the opposite gender. There is evidence to suggest that it might have a similar, but much less important, role in humans with other cues (e.g. visual) being more significant. While apocrine glands are stimulated by the

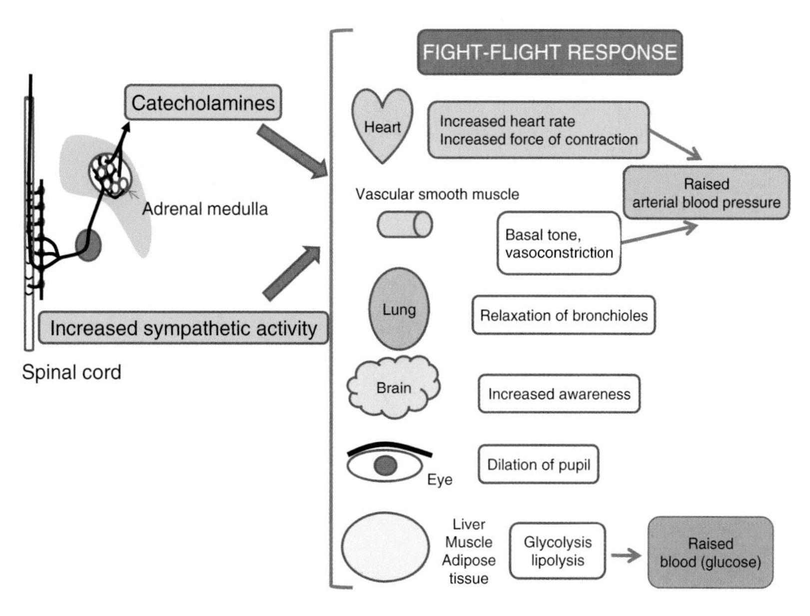

Figure 11.5 Diagram illustrating the principal actions associated with activation of the sympathetic nervous system, and the subsequent release of catecholamines from the adrenal medulla.

normal sympathetic neurotransmitter noradrenaline, the eccrine glands are actually innervated by cholinergic postganglionic sympathetic fibres, this being an exception to the general rule. Stimulation of the SNS is therefore associated with increased sweat production through direct innervations of the sweat glands, but only the apocrine glands will be stimulated by the circulating catecholamines from the adrenal medullae.

Control of release

It will be apparent that the main control over adrenal medullary catecholamine secretion is by the preganglionic sympathetic nerve fibres which innervate the chromaffin cells. Thus any stimulus which results in increased sympathetic activity will induce the release of the adrenomedullary hormones into the bloodstream. The sympathetic preganglionic nerves have their cell bodies in the intermediolateral columns of the thoracic regions (T10–T12) of the spinal cord. Sympathetic drive itself originates in the hypothalamus, pons and medulla of the CNS which receive input from a variety of cortical and limbic sources including the sensory organs and the viscera. Thus emotional or environmental stimuli, acting as stressors for instance, can elicit the mass discharge stimulation of the SNS including the release of adrenomedullary catecholamines as

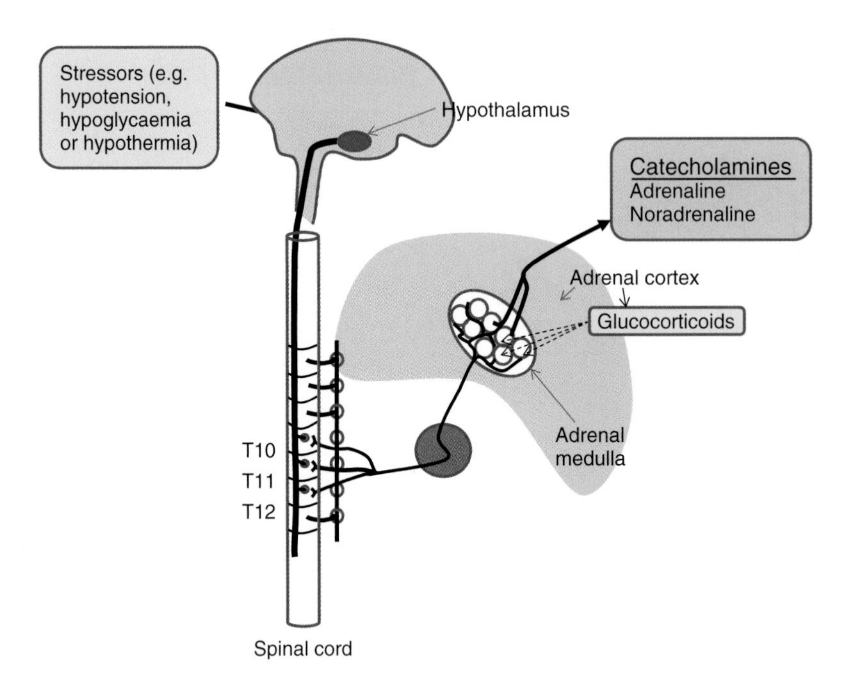

Figure 11.6 Diagram illustrating the principal mechanism of control over the adrenal medulla, involving the hypothalamus and the (preganglionic) sympathetic pathway.

the fight-or-flight response, although more specific effects can also be generated (e.g. visceral responses or pupil diameter changes). Various stimuli such as hypoglycaemia and hypotension influence the central neuronal component of the SNS (see Figure 11.6).

Catechols within the chromaffin cell cytoplasm have a (cryptocrine) negative feedback controlling influence by inhibiting the activity of the key enzyme involved in catecholamine synthesis, tyrosine hydroxylase. When catecholamines are secreted following stimulation of the chromaffin cells, the intracellular catechol levels also become depleted; consequently the negative feedback is decreased allowing for enhanced catecholamine synthesis. Also, as mentioned earlier, glucocorticoids from the cortex influence adrenomedullary catecholamine production directly, by stimulating tyrosine hydroxylase, dopamine β hydroxylase and PNMT activity.

Adrenal medullary disorders

Adrenomedullary failure

Adrenal medullary failure does not cause any specific, measurable problem in humans, because most of the effects of catecholamines are mediated by the activation of sympathetic nerve terminals all over the body. The effect of circulating adrenaline is mainly to increase β2 receptor activation, but this does not appear to be missed in patients who have adrenal failure or who have their adrenals removed.

Adrenomedullary tumours

As with tumours of the adrenal cortex, functioning tumours of the adrenal medulla can also cause hypertension. When hypertension occurs in young people, it is essential to consider secondary causes such as an adrenal tumour. There may be clues in the history to make one suspect the correct diagnosis.

Phaeochromocytoma

A secreting tumour of the adrenal medulla is known as a phaeochromocytoma. It was named thus because of its pathological appearance, being a dusky (*phaeo* in Greek) colour (*chromo*). Such tumours tend to be larger than benign adrenal cortical tumours, and when they secrete catecholamines, the effects of increased α and β adrenoceptor stimulation occur. This can be dangerous. Alpha receptors cause vasoconstriction, and beta receptors can cause myocardial irritability. The most dangerous nature of a phaeochromocytoma is its tendency to behave as a syncytium, with degranulation from a large number of these cells causing the release of large amounts of catecholamines at the same time. This can cause a dangerous surge of adrenaline, which can induce a sudden increase in blood pressure that may result in cerebral haemorrhage, myocardial infarction, arrhythmias and sudden death. The suspicion or finding of such a tumour should thus be treated as an emergency.

Symptoms in a patient include palpitations, panic attacks, episodic severe hypertension and sweating. In between attacks, patients are often completely normal and the diagnosis can thus be readily missed. Typically, functional symptomatic phaeochromocytomas are large (usually more than 2 cm by the time they cause symptoms).

The diagnosis can be confirmed by measuring catecholamines, or the metabolites of catecholamines, the metanephrines, either in the urine or in the plasma. Once confirmed, the patient needs to be made safe by urgent alpha and beta receptor blockade. Traditionally alpha blockade is carried out with phenoxybenzamine, and beta blockade using propranolol. Once these receptors are all fully blocked, the tumour can be removed surgically, ideally laparoscopically or, more recently, retroperitoneally. The latter approach has been used since 2009, results in the least morbidity and more rapid discharge of the patient from hospital and is probably now the gold standard operation for adrenal surgery. Patients who have a phaeochromocytoma removed, and who are either young or have a family history should be referred to a genetic clinic to consider genetic causes such as MEN2 (see Chapter 19) or one of the mutations of succinate dehydrogenase subunits (SDH-D or SDH-B most commonly) or neurofibromatosis.

Appendix: Clinical Case

Clinical Scenario 11.1

Mrs Newman, a 45-year-old teacher, visits her GP because she is not sleeping well and seems to get agitated a lot for no apparent reason. The GP examines her and finds that she has a tachycardia (heart rate 105 beats per minute) and a raised blood pressure (165/95 mm Hg). She informs you that she is quite stressed at work and is separating from her husband after 20 years of marriage. On enquiry, she says that she is always hot, sweats quite a lot and has noticed that she drinks more often, even at night. Her cycles are irregular.

The doctor arranges for her to have a blood sample taken and a urine analysis undertaken.

Questions

Q1. What might the GP's initial differential diagnosis be?
Q2. What could the doctor wish to have measured in the blood and urine samples?
Q3. On the basis of the laboratory results, what would the GP wish to have done next?
Q4. If the GP's diagnosis is correct, what treatment would be instigated?

Answers to Clinical Scenario 11.1

A1. What might the GP's initial differential diagnosis be?

The GP might have two, or even three, initial diagnoses in mind initially: she may simply be stressed (agitated, tachycardia and hypertension), could be having an early menopause (hot and sweaty as well as agitated) or could have excess circulating stress hormones because of the stressed state or a tumour (all of the signs and symptoms). Certainly, the increased drinking might suggest excess glucose in the urine (diabetes mellitus) as a consequence of prolonged raised circulating catecholamine (or cortisol) levels.

A2. What could the doctor wish to have measured in the blood and urine samples?

The doctor might wish to have blood gonadotrophin (luteinising hormone (LH) and follicle-stimulating hormone (FSH)) levels measured to check on the menopausal status. If she is menopausal, LH and FSH levels will be raised, and oestrogen levels low. Separating the stressed state from an adrenal tumour could be initiated by looking at urinary metanephrins, and also blood cortisol, ACTH and adrenaline levels.

A3. On the basis of the laboratory results, what would the GP wish to have done next?

In this case, LH and FSH levels are not raised, but there is a raised urinary vanillyl mandelic acid concentration, while the plasma adrenaline level is raised. The plasma cortisol concentrations lie within the normal range for a random sampling.

On the basis of the laboratory and other findings, the GP concludes that a medullary tumour is a likely diagnosis. He would wish to get an abdominal MRI scan done.

A4. If the GP's diagnosis is correct, what treatment would be instigated?

He would prescribe adrenoreceptor-blocking drugs. Removal of the tumour by surgery would be contemplated.

Reference

Cannon, W.B. (1915) *Bodily Changes in Pain, Hunger, Fear, and Rage*. New York: Appleton.

CHAPTER 12

The Endocrine Control of Salt and Water Balance

Introduction

Land animals have evolved over millions of years from an aquatic past when the surrounding medium was a saline solution, namely, the sea. The sea nowadays has a salinity of approximately 3.5% (i.e. 35 g.L^{-1}) or, given a molecular weight of 58.5 for NaCl, a molarity of approximately 600 mMol.L^{-1}. This is considerably more hypertonic than our own internal environment, the extracellular fluid, which has an approximate molarity of 150 mMol.L^{-1}. That evolutionary process has involved the development of various mechanisms which have allowed these animals, including humans, to survive in a very different, essentially dry, environment. For example, our bodies are protected by an impermeable skin allowing us to retain water, which we normally imbibe in a controlled manner, and our cells have developed mechanisms which maintain an intracellular environment suitable to allow vital enzyme-catalysed reactions to take place in order to sustain life. Central to these mechanisms is the appreciation that water moves across cell membranes up an osmotic gradient so that the regulation of osmotically active particles is essential to the maintenance of that internal environment. With sodium and chloride ions making up almost all the osmotically active particles in the extracellular fluid, giving a normal osmolality of nearly 300 mOsmol.kg^{-1} of water, the importance of salt regulation becomes apparent.

Water is essential for life, and represents approximately two-thirds of our body weight (i.e. of the order of 40 L in a 70 kg individual). Of this total volume of water, approximately two-thirds is intracellular with the remainder being extracellular. The extracellular fluid volume (ECFV) itself is comprised mostly of interstitial water (around 10 L) and plasma (around 2.5 L). The maintenance of an intracellular medium compatible with the vital activities of enzyme-driven cellular reactions can be considered to be essentially down to the activity of Na^+–K^+–ATPase pumps present in all cell membranes. These active pumps ensure that Na ions are continually pumped out of cells as fast as they enter down their own electrochemical

---

Integrated Endocrinology, First Edition. John Laycock and Karim Meeran.

gradient, in exchange for K ions, thus ensuring that the tonicity of the intracellular fluid is closely regulated. Indeed, a sizeable part of the basal metabolic rate is associated with the continuous functioning of these membrane pumps which maintain the optimal intracellular environment for our cells.

Any increase in ECFV within a restricted space will be associated with an increase in pressure. A consequence of sodium retention will be an expansion of the ECFV, since water will follow the sodium-induced osmotic gradient, and this in turn can chronically result in an increase in the mean arterial blood pressure (MABP). Thus the regulation of salt and water is also of fundamental importance with respect to the control of the arterial blood pressure, which provides the essential force driving blood to the various tissues of the body. In this chapter we shall briefly consider the whole concept of salt and water regulation by hormones, and its link to the control of the MABP and therefore, by extension, to the all-too-common clinical condition of hypertension.

Salt and water regulation and the human body

The control of salt and water balance is directed at those organs and tissues involved in the absorptive and excretory processes of these molecules which are the gastrointestinal (GI) tract, the kidneys and to a lesser extent the skin and lungs (Figure 12.1).

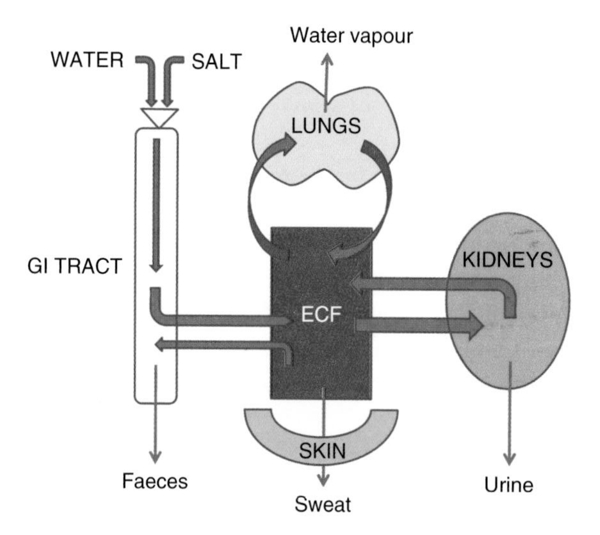

Figure 12.1 Diagram illustrating the principal organs and tissues in the body associated with the regulation of salt and water balance.

Gastrointestinal (GI) tract

Large volumes of NaCl and fluid are absorbed from the gut lumen daily. Large volumes of fluid and salt are also secreted into the GI lumen daily, not only by glands of the intestinal tract but also by the exocrine pancreatic and hepatic cells which produce the pancreatic juice and bile respectively. Thus salt and water movement along the gastrointestinal tract is bidirectional. This is the basis for the concept that absorptive and secretory processes determining Na (and Cl) ion and water movement across the intestinal epithelium are regulated in parallel in order to maintain balance between the systemic fluid volume and the necessary hydration of the luminal contents, together with the necessary release of digestive enzymes and other molecules in their fluid medium.

Salt and water are absorbed through mechanisms involving channels and pumps present in cell membranes. The active pump-mediated transport of sodium (and passively of chloride) ions out of cells at the basolateral membranes provides the necessary gradient at the apical membrane down which these ions can enter the cells through carrier mechanisms, often co-transporting monosaccharides or amino acids. This movement of Na^+ and Cl^- ions across the epithelial cells provides the osmotic gradient up which water flows, via water channels. These are protein molecules with central hydrophilic cores, called aquaporins (AQPs), a number of which (AQPs 1, 3, 4, 5, 7 and 8) are found localised in different segments of the GI tract, and in either basolateral or apical membranes.

The kidneys

Once in the general circulation, salt and water are mainly regulated by the kidneys which ultrafiltrate the blood in the glomeruli of the renal nephrons, of which there are approximately one million per kidney. Along with other small molecules, water, Na^+ and Cl^- ions pass through each glomerulus to enter a proximal convoluted tubule along which approximately two-thirds are reabsorbed. This proximal tubular reabsorption is driven by the Na–K–ATPase pumps in the basolateral membranes which maintain the low intracellular sodium ion concentration. At the apical (tubular) membrane, sodium ions move into the cells down their electrochemical concentration gradient by transporter molecules which also carry glucose into the cells. Water follows passively, through the water channels. More of the remaining water and ions are then reabsorbed along the descending limb of the loop of Henle while some dilution of the tubular fluid takes place as the fluid passes up the thick ascending limb where NaCl is pumped into the interstitium. Near the top of the ascending limb, as it becomes the distal convoluted tubule, are located the specialised sodium-sensing cells of the macula densa. These cells are adjacent to the juxtaglomerular (JG) cells of the endothelium of the afferent arterioles from which blood enters the glomerular capillaries. These JG cells are the source of the enzyme renin which is secreted into the blood, where it splits off the decapeptide angiotensin I from the circulating precursor protein angiotensinogen (see Chapter 10).

As the tubular fluid enters the distal convoluted tubule to reach the cortical collecting duct, it reaches the section of the nephron along which most of the remaining Na ions are reabsorbed. The distal convoluted tubule is relatively impermeable to water, so the final concentration of the tubular fluid takes place along the collecting duct. It is along this cortico-medullary section of the collecting duct that the final concentration of the tubular fluid occurs, following hormone-sensitive water reabsorption. Of the glomerular filtrate entering the nephrons (approximately 125 $ml.min^{-1}$), about 12 $ml.min^{-1}$ normally reaches the collecting ducts, with only 1–2 $ml.min^{-1}$ entering the ureters to reach the bladder.

The kidneys therefore play a crucial role in regulating the amounts of water and Na ions in the body.

The skin and lungs

There is another way by which water and salt can be lost from the body, and that is in the sweat. Sweat is produced by two types of sweat gland: the eccrine glands which are found on the surface of most of the skin, and the apocrine glands which are found exclusively in the axillary and pubic (armpit and groin) regions of the body. While the latter may have a role to play as the source of potential pheromones involved in the attraction of the opposite sexes to each other for example, it is the eccrine glands which produce the sweat which promotes cooling when the body temperature rises. The eccrine glands are innervated, unusually, by cholinergic sympathetic nerve fibres, and as NaCl and water are secreted onto the skin surface, their evaporation involves the loss of latent heat which decreases the body temperature as part of a normal homeostatic cooling process.

Water is also lost from the lungs on expiration, as water vapour. Thus there is a continual slight loss of body water through this essential process of respiration which needs to be taken into account whenever fluid balance regulation is under consideration. This is known as insensible water loss as it is essentially unregulated.

Regulatory mechanisms

As will now be appreciated, the processes of salt and water transport across cells (transcellular transport) require the presence of specific channels for water and ions, as well as sodium pumps and carriers, and their regulation is determined mainly by hormones. The regulation of salt and water balance occurs at the points of intake and output, namely, the gastrointestinal tract and the kidneys, with sweat production from the skin also a controllable mechanism but one which is linked primarily to thermoregulation. The central nervous system (CNS) also has a clear role in the overall regulatory process.

Salt regulation

Gastrointestinal absorption

Hormones regulate the amount of NaCl absorbed along the intestinal tract. Much of the sodium absorbed across the apical membranes of intestinal cells is by facilitated transport, involving carriers which simultaneously transfer monosaccharides or amino acids. The driving force is provided by the low intracellular Na+ concentration maintained by the activity of the basolateral Na^+–K^+–ATPase pumps.

Both angiotensin II (ATII) and aldosterone receptors are present along different sections of the GI tract. Angiotensin type 1 (ATR1) receptors are found mainly in smooth muscle (e.g. of the oesophagus and jejunum) and angiotensin II stimulates contraction through this receptor. In contrast, angiotensin type 2 receptors (ATR2) are associated with epithelial cells and may mediate actions on transport processes such as stimulating bicarbonate and K secretion into the lumen. Whether it also has effects on Na absorption is presently unclear. Mineralocorticoid receptors (MRs) are also present along much of the intestinal tract, particularly the colon, but are absent from the gastric mucosa. Aldosterone has long been known to influence absorptive processes for electrolytes along the small and large intestine, presumably through the MR. As in its principal target tissue, the kidney, it stimulates sodium absorption and potassium secretion along the intestinal tract, particularly along the colon. Presumably, glucocorticoids such as cortisol, which also bind to MR, can also have some effect on salt (and thus water) absorption when circulating concentrations are high, as in Cushing's syndrome. Progesterone has an anti-mineralocorticoid effect on sodium reabsorption in the renal distal nephron, so it is likely that this inhibitory effect on sodium absorption also occurs in the GI tract. Furthermore, iodothyronines stimulate sodium absorption which may explain why hypothyroidism is occasionally associated with a craving for salt, as a symptom.

Renal reabsorption

The kidneys are the organs which are central to the overall regulation of NaCl in the body, because they can increase either the amount reabsorbed back into the general circulation, or the amount excreted in the urine, depending on the homeostatic status of the body. All sections of the renal nephron are involved in NaCl handling and a number of factors, principally hormones, influence the reabsorptive process. However, it should also be borne in mind that the nervous system, particularly the sympathetic nervous system, also has an important controlling influence, albeit mainly indirect (see section 'Central regulation').

The renin–angiotensin–aldosterone system (RAAS)

As described in Chapter 10, the renin–angiotensin system is the principal mechanism regulating the production and release of mineralocorticoids (chiefly aldosterone) from the zona glomerulosa of the adrenal cortex. Renin is released from juxtaglomerular cells along the afferent arteriole when the renal perfusion pressure is reduced, when direct sympathetic stimulation is increased or when there is a decreased presentation of tubular Na reaching the adjacent macula densa cells (see Figure 10.9, Chapter 10). Renin is an enzyme which cleaves the circulating protein angiotensinogen to produce a decapeptide, angiotensin I (AT1). This is then enzymatically cleaved by circulating angiotensin-converting enzyme (ACE) to produce the active octapeptide molecule angiotensin II (ATII). ATII stimulates proximal tubular sodium reabsorption directly, and this is accompanied by an osmotically driven water reabsorption in this part of the nephron, involving aquaporins (mainly AQP1) which are hormone independent. ATII is also the main controlling influence for aldosterone production from the zona glomerulosa cells of the adrenal cortex. Aldosterone is the principal mineralocorticoid hormone which regulates renal Na handling in humans, by stimulating its reabsorption mainly in the distal convoluted tubule and cortical collecting ducts.

MRs have an equal affinity for the main glucocorticoid in humans, cortisol. Despite the fact that the normal circulating concentration of cortisol is of the order of 1000 times greater than the concentration of aldosterone, the MR is essentially protected from a potentially overwhelming glucocorticoid stimulation because it is immediately converted from its active form to the inactive cortisone in the kidneys. This is brought about by an enzyme called β-hydroxysteroid dehydrogenase (β HSD) which exists in two isoforms, β HSD1 which converts cortisone to cortisol (e.g. in the liver) and β HSD2 which converts cortisol to cortisone (e.g. in the kidneys).

Mineralocorticoids exert profound effects on renal salt (and water indirectly) reabsorption and hence can cause hypertension when produced in excessive quantities chronically, such as in Conn's syndrome. Furthermore, when glucocorticoids such as cortisol are produced in excessive amounts as in Cushing's syndrome, then hypertension becomes a common manifestation. When there is an excessive production of ATII as a consequence of a renin-producing tumour (e.g. Bartter's syndrome), then there is the potential for hypertension not only via increased aldosterone production but also directly from the vasoconstrictor effect of the ATII.

Natriuretic peptides (NPs)

Natriuretic peptides, as their name suggests, are peptides which initiate the loss of sodium in the urine (natriuresis). In essence, they are important in counteracting the vasoconstrictor and sodium-retaining effects of the RAAS.

There are three natriuretic peptides: atrial (ANP), brain (BNP) and c-type (CNP). They all consist of a ring of amino acids with one or two side chains; ANP and BNP have two straight side chains, while CNP has just the one. ANP is a 28–amino acid polypeptide synthesised and released by myocytes in the atria (particularly the left atrium) of the heart and hence is called atrial natriuretic peptide. BNP, a 32–amino acid peptide, was originally shown to be synthesised by neurones in the brain (and hence was named brain natriuretic peptide), but far more of it is actually produced by the ventricles of the heart. CNP is produced in the brain and also in endothelial cells. The NPs can also be synthesised in other tissues (e.g. the ovaries and testes). The two main NPs, ANP and BNP, are both produced and released from the cardiac cells when stretched, for instance by an increased venous return. Indeed, a raised circulating BNP is a good marker for heart failure. They are both synthesised initially as larger pro-hormones, and have half-lives of approximately 20 min in the blood where they are inactivated by a circulating neutral endopeptidase.

There are two membrane-spanning receptors (NPRA and NPRB) which mediate the effects of the NPs, both being associated with intracellular guanyl cyclase and cGMP generation. ANP and BNP bind mainly to NPRA and have similar effects, while CNP binds more to NPRB. They all directly dilate the vasculature, thereby decreasing central venous and arterial pressures. The NPs can also stimulate the production of calcitonin gene-related peptide (CGRP) which also causes dilation of blood vessels. In addition, ANP and BNP have important renal effects. They increase glomerular filtration rate (GFR) by dilating the afferent, relative to the efferent, arterioles, and the filtration fraction is consequently raised, increasing the sodium load in the tubular fluid entering the proximal convoluted tubules. They also decrease proximal convoluted tubule and cortical collecting duct sodium reabsorption, inducing a profound natriuresis. There is some evidence to suggest that the NPs can also influence water reabsorption in the medullary collecting duct by increasing the movement of AQP2 water channels from the apical membranes back into the cytoplasm, decreasing

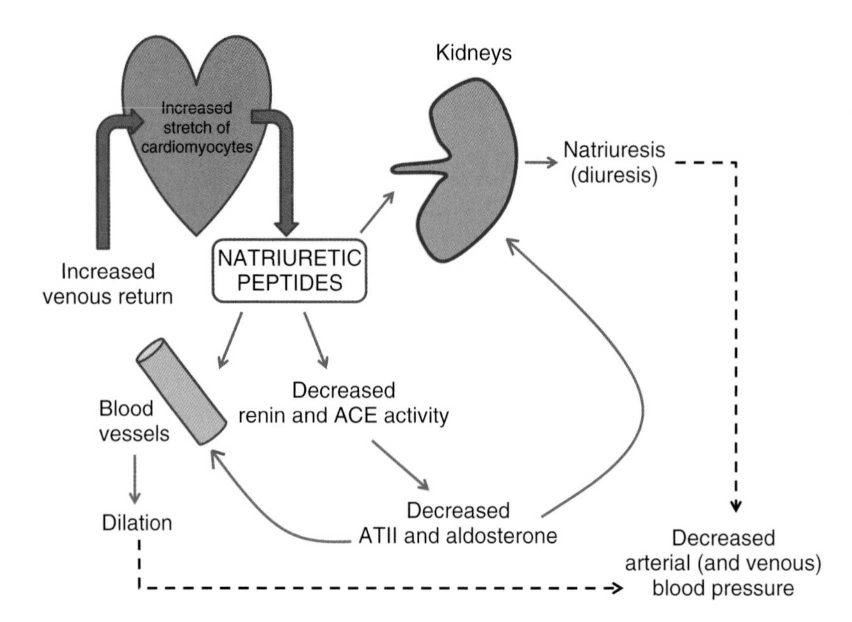

Figure 12.2 Diagram illustrating the release of natriuretic peptides from the heart, and their effects on salt and water regulation.

water reabsorption and thereby exerting a diuretic effect. In addition, they decrease renin release and ACE activity; consequently angiotensin II and aldosterone production are reduced, further enhancing the urinary loss of sodium and water (see Figure 12.2).

As a consequence of all their actions, the NPs will decrease the arterial blood pressure.

Calcitonin

Although the physiological role of calcitonin remains unclear, its principal effects are generally believed to be associated with decreasing the plasma calcium ion concentration (see Chapter 18). However, in the kidneys, it also inhibits sodium reabsorption in the ascending limb of the loop of Henle. Consequently, there is an osmotically driven loss of water along the collecting duct, the diuresis accompanying the natriuresis. These effects are likely to occur when circulating levels are relatively high, at least in humans, so here too its physiological significance is unclear.

Androgens

That men have a greater tendency for hypertension than premenopausal women of a comparable age suggests that androgens may have effects on cardiovascular and renal systems. One mechanism that may be involved is associated with the relationship which exists between the increasing natriuresis that occurs with an increased arterial blood pressure. Androgens appear to cause a shift of the pressure–natriuresis relationship to the right, so that as blood pressure increases, there is a relative decrease in sodium excretion (i.e. increased sodium retention). At least part of the effect is due to an increased activity of the RAAS. Androgens (and oestrogens) stimulate the hepatic synthesis of angiotensinogen, and circulating renin levels are also higher in normotensive men compared with age-matched women. Both these effects would increase ATII synthesis and aldosterone production. Furthermore testosterone, via its intracellular androgen receptor, can up-regulate the mineralocorticoid-sensitive apical sodium channel mRNA. Thus androgens are capable of increasing renal sodium reabsorption along the distal nephron directly, as well as indirectly via the RAAS. This, together with a consequent increase in water reabsorption, could account for the gender-linked increase in arterial blood pressure and its associated risk factors. Androgens are associated with decreased levels of circulating NPs, which may also contribute to the gender difference regarding proneness to hypertension and cardiovascular disease.

Oestrogens and progestogens

As indicated earlier, premenopausal (but not postmenopausal, women) are less prone to cardiovascular problems such as hypertension and the risk of strokes and cardiac arrest than similarly aged men. This suggests a protective effect provided by circulating oestrogens and/or progestogens.

Oestrogens, like androgens, increase angiotensinogen synthesis in the liver. However, they decrease ACE activity and also decrease AT type 1 receptor density in vascular tissue, hence reducing ATII synthesis and activity. Furthermore, there is evidence to suggest that oestrogens stimulate the synthesis and release of ANP and BNP. Thus oestrogens generally seem to oppose the effects of androgens on the RAAS and NP production.

Progestogens such as progesterone have some affinity for aldosterone receptors to which they can bind. They then block any binding by other MR ligands. Thus initially they promote renal sodium reabsorption but then lead to a block and hence increased natriuresis. Progestogens are also associated with increases in ANP.

Catecholamines

Sympathetic stimulation and circulating catecholamines clearly influence renal function. For instance, the juxtaglomerular cells of the afferent arterioles receive a direct sympathetic innervation which, when stimulated, results in their secretion of renin. Consequently, the whole RAAS is activated (see earlier section). Furthermore, catecholamines are associated with an increased GFR due to a greater vasoconstriction of the efferent over the afferent arterioles resulting in an increased presentation of sodium to the proximal tubules, and a direct stimulation of proximal tubular sodium reabsorption.

Iodothyronines

There is some evidence to suggest that the iodothyronines T3 and T4 can stimulate proximal tubular sodium reabsorption, and consequently water reabsorption along this section of the nephron.

Vasopressin

Vasopressin, and maybe to a limited extent oxytocin, stimulates NaCl reabsorption in the thick ascending limb of the loop of Henle, and also increases its reabsorption along the collecting duct, mainly by stimulating the synthesis and insertion of specific Na channels in the apical membrane. This effect is associated with the vasopressin type 2 receptor (V2).

Water handling

Water absorption is clearly associated with water intake, and this is generally related to food intake (i.e. intake of osmotically active molecules) under normal circumstances. In humans, water (or, more generally, fluid) intake accompanies food intake like in other animals may be driven by an osmotic stimulus, but it is also an in-built, socially directed action associated with established routines such as tea and coffee breaks and social drinking. Drinking behaviour is directed from the CNS, and the hypothalamus plays an important role in controlling it (see section 'Central regulation').

Water balance is directed by the presence or absence of an osmotic gradient across an epithelial lining such as the gastrointestinal wall or the renal tubule. That osmotic gradient is generally linked to the transepithelial transport of sodium ions which occurs by either active or facilitated transport mechanisms. Furthermore, water, being lipophobic, cannot readily cross cell membranes, so the presence either of gaps (i.e. absence of tight junctions) between cells resulting in intercellular (or paracellular) transport or, more likely, of water channels in cell membranes, allowing transcellular transport, is a necessary feature.

Gastrointestinal absorption

Most water absorption takes place along the small and, particularly, the large intestine. Although some water may well move between the epithelial cells in the presence of an osmotic gradient by paracellular transport, most absorption probably takes place transcellularly through specific water channels (aquaporins), again driven principally by the presence of an osmotic gradient maintained by the active absorption of Na ions. A number of aquaporins are present in the various cells comprising the different sections of the gastrointestinal tract, as indicated earlier, but none of these are believed to be regulated by hormones. Instead, hormonal control is directed at Na absorption which provides the osmotic gradient for the movement of water through the various aquaporins.

Renal reabsorption

Water reabsorption takes place mostly along the proximal tubule (approximately 60%), along the thin descending limb of the loop of Henle (approximately 28%) and finally (the last 12%) along the cortical and medullary sections of the collecting duct where final concentration of the tubular fluid takes place.

Hormones that influence sodium reabsorption in the proximal and distal tubules also, indirectly, influence the accompanying osmotically driven water reabsorption along this section.

Thus, the actions of ATII and aldosterone are both directed at raising the plasma sodium concentration, and it is this increase in the plasma osmolality that stimulates the osmoreceptors located in the circumventricular organs, those specific parts of the brain which lie outside the blood–brain barrier. These osmoreceptors, when activated, stimulate the release of vasopressin and the hypothalamic 'thirst centre'.

Vasopressin (VP)

The vasopressin which is released into the general circulation from the nerve terminals in the posterior pituitary is actually synthesised in the hypothalamus, specifically in the cell bodies of the magnocellular neurones located in the supraoptic and paraventricular nuclei (see Chapter 4). Vasopressin plays a particularly important role in the regulation of water reabsorption because it is the principal controlling influence over the synthesis and intracellular direction of a specific aquaporin (AQP2). This AQP2 is located in the principal cells of the final concentrating segment of the nephron, the collecting duct. The AQP2 molecules are water channels which are inserted into the apical membranes of the principal cells of the collecting duct. Another aquaporin (AQP3) which is located in the basolateral membranes and is involved in the movement of water out of the principal cells into the interstitial fluid and ultimately the plasma is also believed to be at least partially under the control of vasopressin. Yet another AQP (AQP4), also located in the basolateral membranes, is not influenced by VP. Since vasopressin also stimulates sodium reabsorption along this section together with aldosterone (certainly in the cortical collecting duct), it helps provide the osmotic gradient necessary for the vasopressin–AQP2 (and vasopressin–AQP3) regulated water reabsorption.

Oxytocin

This hormone may have a minor influence on water reabsorption along the collecting ducts, and this action could involve the V2 receptors for vasopressin as well as the oxytocin receptors, although its affinity for V2 receptors is minimal.

Central regulation

The CNS clearly plays an important role in exerting an overall control over the behaviours and mechanisms involved in maintaining salt and water balance.

Regarding behaviours, there are various centrally regulated activities which ensure that salt and water, so crucial to life, are actively sought when required. For this purpose, salt craving and thirst are stimuli (drivers) which will direct any animal including humans to seek sources of these necessities. Salt appetite is believed to involve inhibitory serotoninergic and oxytocinergic neurones, at least in rodents. The hypothalamus has been implicated as an important neural centre involved in salt regulation. Indeed the hypothalamus is also connected to the sympathetic nervous system, with fibres innervating the renin-producing juxtaglomerular cells of the renal afferent arterioles. Hence detection of a low sodium ion concentration (e.g. by a decreased osmotic stimulation of osmoreceptors) could, via the hypothalamus, lead to a stimulated RAAS which in turn would bring about an increase in intestinal sodium absorption and renal reabsorption.

There is also evidence for a hypothalamic thirst centre which responds to changes in plasma osmolality via the central osmoreceptors. Thus an increased osmolality would not only turn off the salt-regulating system but also stimulate the thirst centre and vasopressin release. The behavioural response in seeking fluid to drink would be associated with an increased release of vasopressin which, via its antidiuretic action in the kidneys, would maintain or restore the plasma osmolality back to normal levels. Interestingly, the central renin–angiotensin system is involved in stimulating the thirst centre. Angiotensinergic neurones certainly innervate the hypothalamus, and intracerebroventricularly administered ATII is associated with increased vasopressin secretion from the magnocellular neurones as well as increased thirst.

Clinical correlates

Hyponatraemia

Hyponatraemia is the condition which arises when the plasma sodium concentration is lower than normal (i.e. less than approximately 145 mM).

It is quite common in hospitalised patients, and can become a serious problem if the plasma Na+ concentration falls to below 120 mM, when it is associated with an increased morbidity and mortality.

Hyponatraemia can arise when the total body sodium concentration is generally reduced (solute depletion state), or when there is an excess of water in the ECFV compartment. There are two main clinical endocrine causes for this hyponatraemic state. The first occurs when there is a decreased reabsorption of Na^+ from the distal nephron because of a decreased mineralocorticoid activity, due to a lack of aldosterone (and cortisol). The second is when there is an excessive reabsorption of water in the collecting duct diluting the plasma Na^+ concentration because of an abnormal, increased production of vasopressin. The former condition is called Addison's disease (see Chapter 10), and the latter condition is known as the syndrome of inappropriate antidiuretic hormone hypersecretion (SIADH; see Chapter 5). SIADH is often associated with an indirect, vasopressin-independent natriuresis brought about because the expected expansion of the ECFV induces a compensatory increased GFR resulting in a pressure diuresis, and/or because of the increased release of natriuretic peptides.

Hypothyroidism is occasionally associated with a hyponatraemia, possibly as a consequence of water retention. This might be due to an impaired water excretion following a reduced GFR and a subsequently decreased distal tubular flow rate. It could also be related to a decreased sodium reabsorption resulting in a slight natriuresis.

Hypernatraemia

This situation will arise clinically when the Na+ concentration rises above the normal range, and again can be associated mainly with adrenal and posterior pituitary disorders. An excessive mineralocorticoid activity, as occurs in Conn's and Cushing's syndromes, will be associated with an excessive reabsorption of Na^+ along the distal nephron. This in turn would be expected to result in an expansion of the ECFV following the release of vasopressin due to enhanced stimulation of the central osmoreceptors. As it happens, the expansion of the ECFVs limited to no more than 10% because of a compensatory 'escape phenomenon' whereby the increased GFR results in increased loss of sodium and water in the urine (the pressure diuresis described in section 'Hyponatraemia'). Thus, while a hypernatraemia can occur, the main problems arise because of the concomitant hypokalaemia and metabolic alkalosis, as well as the chronic pressor effect of the ECFV expansion.

The principal endocrine cause of a hypernatraemia is the lack or absence of the antidiuretic action of vasopressin. As a consequence of the decreased reabsorption of water in the collecting ducts, the plasma Na^+ concentration increases. The clinical condition is called diabetes insipidus, and it can be either central or nephrogenic (see Chapter 5).

Hormones and blood pressure regulation

As indicated earlier, hormones which can increase renal sodium reabsorption potentially raise the arterial blood pressure, chronically resulting in hypertension. In most cases of hypertension the precise cause is unclear (essential hypertension), but some cases (approximately 2%) are clearly identified as disorders associated with raised circulating levels of specific hormones, namely aldosterone and cortisol. Thus a very high percentage of patients with Conn's and Cushing's syndromes will manifest hypertension as a presenting sign. Another endocrine cause of hypertension is the excessive production of catecholamines by a tumour of chromaffin tissue (a phaeochromocytoma, usually in the adrenal medulla). While endocrine causes of hypertension may represent only a small fraction of patients with chronically raised blood pressure, they do nevertheless represent a readily treatable group, in general.

Perhaps surprisingly, conditions associated with raised circulating levels of either vasopressin or ATII are not commonly associated with hypertension despite the fact that both of the polypeptides have extremely powerful vasoconstrictor activities. The generally accepted explanation

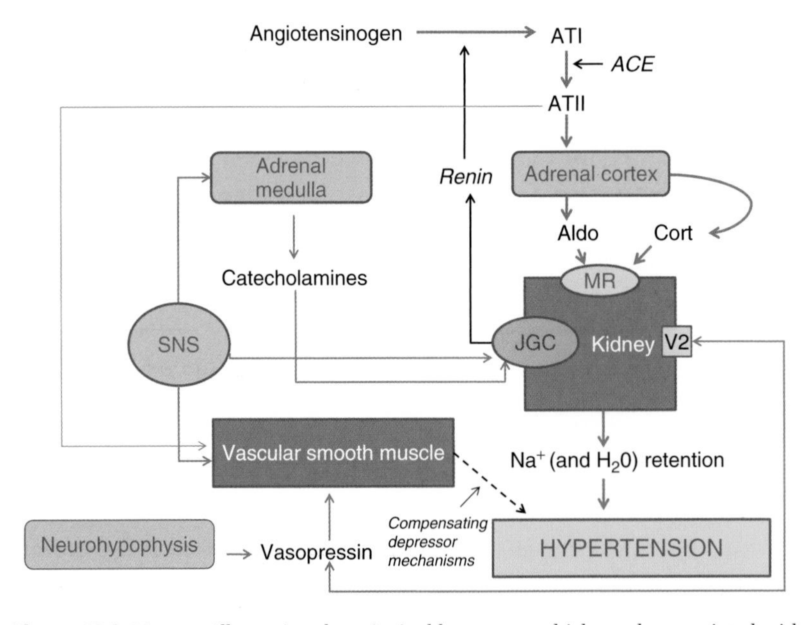

Figure 12.3 Diagram illustrating the principal hormones which can be associated with the development of hypertension. AI and AII are angiotensin I and II respectively. For further details, see text.

for this apparent paradox is that both hormones, by raising the arterial blood pressure, normally induce secondary compensating mechanisms which prevent the development of a chronic hypertensive state. These compensatory mechanisms are likely to include a reflex decrease in sympathetic activity consequent upon an expansion of the ECFV. Thus natriuretic peptides (ANP and BNP) can be released from the cardiomyocytes in order to promote a natriuresis (and subsequent diuresis). An increase in arterial blood pressure also results in a decrease in renal nerve sympathetic activity which in turn results in an increased GFR and a decrease in RAAS activity, both of which would also lead to a natriuresis and diuresis and the restoration of a normal arterial blood pressure (Figure 12.3).

One interesting endocrine condition which can also be associated with hypertension is the increasingly common type II (non-insulin-dependent) diabetes mellitus, which is often also linked to obesity. Various possible explanations have been advanced including effects on ion transport (e.g. calcium) in vascular smooth muscle, increased renal sodium retention and increased sympathetic activity.

Conclusion

Various hormones influence renal salt and water handling. As with other important physiological regulatory systems of homeostasis, control is not left to a single mechanism. Both humoral and neural mechanisms are involved, with the hypothalamus likely to be an important central regulator. By their role in regulating salt and water balance, these hormones can, when produced in excess, be associated with endocrine hypertension.

CHAPTER 13
The Thyroid Gland and Its Iodothyronine Hormones

Introduction

The thyroid gland is a relatively large endocrine gland which has an interesting historic background, not least because when it is enlarged (called a goitre) it is usually very obvious as a swelling in the neck, as depicted in various classical paintings such as in *The Crucifixion of St Andrew* by Caravaggio. Various functions were ascribed to it in the past, including for instance its role as a filter of the blood, as a mechanical protection of the underlying blood vessels and nerves from compression effects or merely as a cosmetic enhancement of beauty in women by 'preserving the contour of the neck' (see the review article in the *British Medical Journal* by V. Horsley, 1892). Interestingly, the summary of this fascinating review concludes with what we generally associate the thyroid gland with nowadays: namely, an important metabolic influence.

The hormones associated with the thyroid are mainly the iodothyronines, tri-iodothyronine (T3) and tetra-iodothyronine (T4, also known as thyroxine). These hormones do not naturally fall within either of the two main categories of hormones, the polypeptides/proteins or the steroids. Indeed, being amines, they have properties which allow them to span between these two groups, as will become clear in this chapter. They are also important clinically since thyroid disorders are relatively common endocrine disorders, second only to diabetes mellitus.

In addition to T3 and T4, a polypeptide hormone called calcitonin is synthesised by certain specific cells within the thyroid tissue. This hormone is concerned with calcium metabolism and will be discussed in Chapter 18.

Embryological derivation, general structure and histology

The thyroid gland develops embryologically as an outgrowth of the developing floor of the pharynx around the fourth week of pregnancy. By the

Integrated Endocrinology, First Edition. John Laycock and Karim Meeran.

12th week of pregnancy, growth of the thyroid becomes more marked, coinciding with the pituitary gland beginning to secrete its various hormones including thyrotrophin (i.e. thyroid-stimulating hormone (TSH)). The outgrowth, later known as the thyroglossal duct, as it develops caudally forms two lobes as it comes into contact with the fourth pharyngeal pouch to which it fuses. Normally, the anterior part of the thyroglossal duct is lost but its site of origin can be identified as the foramen caecum, a small dimple medially located on the base of the tongue. The caudal part becomes the small pyramidal lobe which may or may not be discernible, and which joins a narrow band of tissue called the isthmus linking the two main well-developed lobes of the thyroid. In an adult the thyroid normally weighs approximately 15–20 g, but when enlarged (goitre) it can be many times heavier, weighing up to a few hundred grams. It is also highly vascular, typical of any endocrine gland. Indeed, blood flow to the gland is higher than to the kidneys at 4 $ml.min^{-1}.g^{-1}$, compared with approximately 3 $ml.min^{-1}.g^{-1}$. Blood reaches the gland via two main sources: (i) the superior thyroid arteries which branch off the external carotid arteries, terminating in the upper parts of each lobe, and (ii) the inferior thyroidal arteries which come off the subclavian artery and provide blood to the lower parts of each lobe. Venous blood from the superior and middle thyroid veins drains into the internal jugular, while venous blood from the inferior venous vein enters the left brachiocephalic vein. The nerve supply to the thyroid gland is scanty, consisting mainly of sympathetic nerve fibres, from the superior cervical ganglion which appear to innervate the vasculature rather than the thyroid tissue itself, although some terminals are on follicular cells. There is also evidence for some parasympathetic innervation of the thyroid from the vagus, but its role is unclear.

The general structure of thyroid tissue is very characteristic, consisting of single-cell layers, or 'balls', of follicular cells surrounding a central region filled with a yellowish protein rich gel called colloid. These 'balls' of cells surrounding central colloid are called follicles (Figure 13.1). Between 20 and 40 follicles are grouped together by surrounding connective tissue to form lobules, each lobule receiving its own blood supply and having the potential to function independently of other lobules. The follicles vary in shape and size depending on their degree of stimulation, with highly active follicles consisting of thick columnar follicular cells while relatively inactive follicles would have thinner more cuboidal follicular cells. Interspersed between the follicles are small clumps of very different cells called parafollicular cells. These cells operate completely independently of the main thyroidal tissue and produce a polypeptide hormone which is involved in blood calcium regulation called calcitonin. This hormone is considered in Chapter 18.

The thyroidal follicular cells, also occasionally called thyrocytes, produce hormones called iodothyronines. These cells have a characteristic

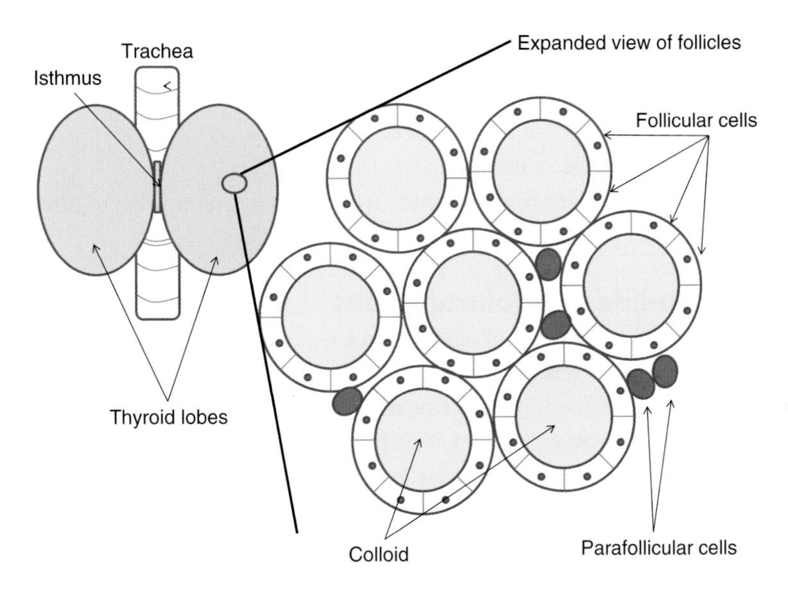

Figure 13.1 The general structure of the thyroid and an exploded view of a small section of thyroid tissue in cross-section illustrating the follicles containing colloid, surrounded by the follicular cells. Also seen are parafollicular cells which produce the calcium-regulating hormone calcitonin.

appearance under the electron microscope: the inner (apical) membrane facing the colloid is a site of much activity, being associated with the hormone synthesis process itself and also with the movement of synthesised hormone into, and out of, the colloid. This latter activity is associated with numerous microvilli generally seen projecting from the apical membrane into the colloid, capturing clumps of colloid and drawing them into the cell. There is a well-developed endoplasmic reticulum within the follicular cell and many mitochondria, as well as an abundance of lysosomes filled with enzymes.

Synthesis, storage and release of iodothyronines

The iodothyronines, as their name implies, are iodinated molecules, and this is specific to the thyroid hormones produced by the follicular cells. There are various iodothyronines synthesised in the thyroid and in peripheral cells, depending on the presence of different deiodinases. However, the two main, physiologically active, thyroid hormones which are considered in this chapter are T3 and T4. There are seven distinct stages in the synthesis of these iodothyronines:

1. Uptake of iodide into follicular cells.
2. Synthesis of thyroglobulin protein.

3. Iodination and organification reactions forming mono- and di-iodotyrosines.
4. Coupling reaction, with the formation of tri- and tetra-iodothyronines (T3 and T4).
5. Storage in the follicle (colloid).
6. Uptake from colloid and release into circulation.
7. Peripheral conversions.

Uptake of iodide into follicular cells

The concentration of inorganic iodide (I) in the general circulation is of the order of 10–15μg.L^{-1} blood, while its concentration in the follicular cells is some 25–50 times higher, under normal conditions (and much greater when the thyroid is hyperactive). Furthermore, this anion would have to enter the cell against an electrical gradient, the cell interior normally being negative with respect to the exterior, as in all living cells. Iodide therefore has to move against an impressive electrochemical gradient in order to enter the follicular cells. It does this by means of a sodium–iodide symporter (NIS) in the basolateral (plasma) membrane which transports two sodium and one iodide ions into the cell, down the sodium gradient. This membrane glycoprotein is also found on the basolateral membranes of salivary glands, gastric mucosa and mammary glands during lactation all being capable of taking up iodide. The energy necessary to drive the symporter is provided by the sodium–potassium (Na–K) ATPase which pumps three sodium ions out of the cell in exchange for two potassium ions. The Na–K ATPase is ouabain sensitive, and can be blocked not only by ouabain but also by drugs used to block thyroid function such as perchlorate and thiocyanate. The NA–I symporter itself is under the control of the anterior pituitary hormone thyrotrophin which binds to its specific membrane receptor and stimulates the second messenger adenyl cyclase–cyclic AMP system which mediates its actions. Iodide is also transported out of the follicular cell at the apical membrane, and into the follicular lumen, a process also stimulated by thyrotrophin. This transport process is at least in part mediated by an iodide transporter called pendrin. An exchange with chloride ions may be part of the precise mechanism (Figure 13.2).

Thyroglobulin synthesis

The follicular cell synthesises a large glycoprotein homodimer called thyroglobulin which consists of two similar chains each of 2748 amino acids of which 67 are tyrosyls (a tyrosyl is a tyrosine amino acid residue as found within a peptide chain). Its synthesis is also stimulated by thyrotrophin. The glycoprotein is then exported (probably by a secretion process) from the follicular cell into the follicle lumen through the apical membrane (Figure 13.2).

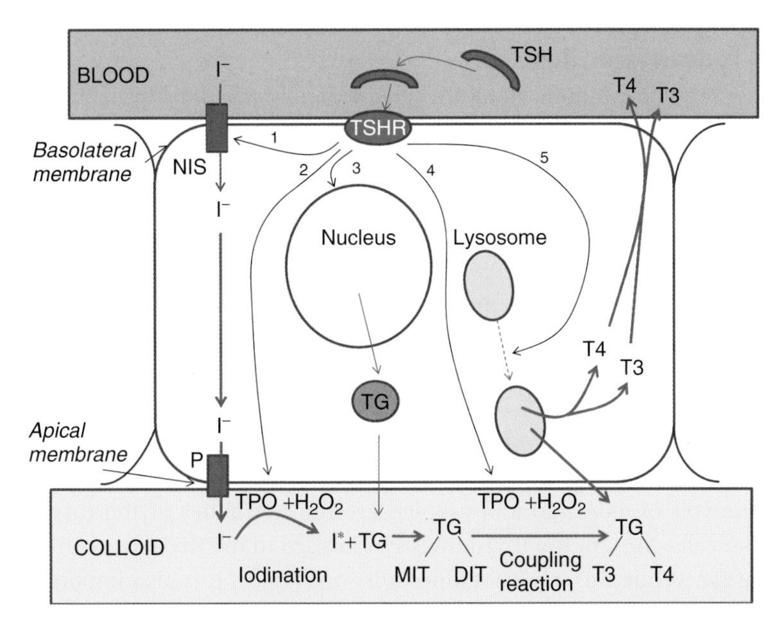

Figure 13.2 Diagram illustrating the follicular cell separating the external fluid (extracellular fluid: blood) from the internal follicular colloid. Steps stimulated by TSH (thyrotrophin) are (1) stimulation of NIS, (2) activation of TPO resulting in iodination reaction, (3) nuclear stimulation of TG synthesis, (4) stimulation of the coupling reaction and (5) stimulation of lysosome movement towards apical membrane and fusion with endocytosed colloid containing the TG with its T3 and T4. $2I^-$ = iodide; H_2O_2 = hydrogen peroxide; NIS = sodium iodide symporter; P = pendrin; TG = thyroglobulin; TPO = thyroperoxidase; TSHR = thyroid-stimulating hormone (thyrotrophin) receptor; T3 = tri-iodothyronine and T4 = tetra-iodothyronine (thyroxine). See text for further details.

Iodination and organification reactions forming mono- and di-iodotyrosines

Once iodide has been brought into the follicular cell, it reaches the apical membrane through which it is transported into the follicular lumen by the pendrin transporter. Along the apical membrane–colloid border, the inorganic iodide gets oxidised to a highly active, short-lived form of iodide known as 'reactive iodine'. This iodination reaction is catalysed by thyroidal peroxidise (TPO) in the presence of hydrogen peroxide. In this form it immediately gets 'organified' by binding to positions 1 and 2 of certain tyrosyls which are incorporated into the thyroglobulin, forming mono- and di-iodotyrosyls (MIT and DIT). Of the total 134 tyrosyls in each thyroglobulin dimer, only 25–30 of these are actually iodinated. The iodination and organification reactions both involve TPO which is one of the many targets stimulated by thyrotrophin.

Coupling reaction, with the formation of tri- and tetra-iodothyronines

Still in the follicle lumen along the apical membrane, the thyroglobulin undergoes a configurational change, known as the coupling reaction, in which the molecule undergoes a structural realignment such that specific di-iodotyrosyls link up with either mono- or di-iodotyrosyls, to form tri- and tetra-iodothyronines (a thyronine being the result of linkage between two tyrosyls; see Figure 13.3). Approximately 3–4 thyronines are formed on each thyroglobulin, under normal conditions. The coupling reaction is also under the control of thyrotrophin.

Storage in the colloid

The follicle lumen is filled with iodinated thyroglobulin protein which forms a yellowish, thick, gel-like substance known as colloid. This acts as a reservoir of iodothyronines which are the hormones of the thyroid follicular cells. Most of the thyronines synthesised in the thyroid are in the form of T4. While this is a hormone in its own right, it is also important because it can be deiodinated to T3, which is by far the more potent of

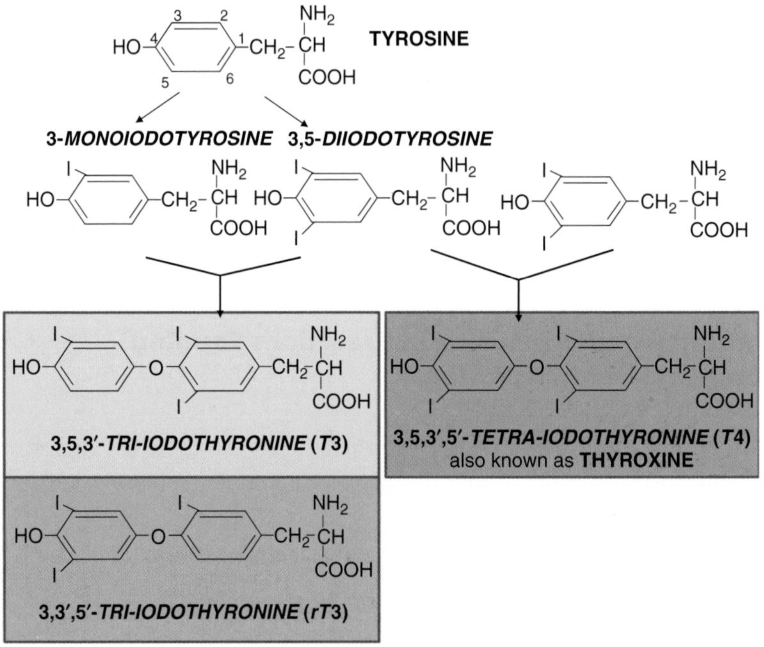

Figure 13.3 The chemical structures of tyrosine which is first iodinated to 3-monotyrosine and 3,5 di-iodotyrosine, which are then coupled to form either 3,5, 3′ tri-iodothyronine (T3) or 3,3′,5′tri-iodothyronine (reverse T3, rT3) and 3,5,3′, 5′ tetra-iodothyronine (T4, thyroxine).

the two molecules, mostly in peripheral tissues. Up to 3 months' worth of iodothyronines are stored in the follicular colloid of the thyroid gland.

Uptake from colloid and release into circulation

When the follicular cells are stimulated by thyrotrophin, in addition to its various stimulatory activities as described earlier, it is also associated with the pinocytosis of colloid, drawing bits of it back into the follicular cells. Once the iodinated thyroglobulin is back in the follicular cell cytoplasm, it is taken up by lysosomes directed towards the apical membrane, in response yet again to thyrotrophin. These lysosomes contain various enzymes, including proteases and deiodinases, which break down the thyroglobulin, releasing iodine, tyrosine and the thyronines. The iodine and tyrosine residues of MIT and DIT can be recirculated within the cell for reuse. Far more T4 is synthesised in the thyroid than T3. The ratio of thyroidal T4 to T3 is normally of the order of 15:1. The presence of deiodinases in the follicular cell (as well as in other target tissues) promotes the formation of T3 from T4, the deiodinases involved being types D1 and D2.

The iodothyronines T3 and T4 are secreted through the basolateral membranes into the general circulation by means of a transporter mechanism.

Peripheral conversions

Most of the iodothyronines released from the thyroid gland are T4 molecules. However, thyroxine plays a major role in acting as a prohormone in addition to its direct role as a hormone itself. Numerous peripheral tissues contain deiodinases which can deiodinate T4 to the far more active T3 by the removal of the iodine atom at position 5′ (see Figure 13.3). However, a different deiodinase can act on the prohormone T4 to remove a different iodine atom, this time at the 5′ position; this T3 is called reverse-T3 (rT3). While T3 is more biologically active than T4, rT3 is metabolically inactive. Hence peripheral tissues under specific conditions have the ability to direct the nature of the deiodination pathway depending on their environmental circumstances, either increasing or decreasing the biological activities associated with these thyroid hormones.

The biological half-life of T3 is approximately 15 hours compared with the 6.5 days for T4. Likewise, the latent period for T3 is much shorter, being about 6 hours compared with 72 hours for T4.

Transport of iodothyronines in the blood

Once released into the bloodstream, the iodothyronines preferentially bind to plasma proteins, the main one being thyronine-binding globulin (TBG) which is specific for T3 and T4. Approximately 80% of T3 and 70% of T4 are bound to this globulin, while the remainder is mostly bound to a pre-albumin (called transthyretin) and albumin itself. The prealbumin binds approximately 10% of the two hormones, while albumin (which has a low affinity, but a high capacity, for hormone binding) is associated with approximately 10% T3 and 20% T4. The binding of hormone to plasma protein is in dynamic equilibrium (see Chapter 1) so that at any time a small amount of the hormone is present in the free, unbound state. For T3 and T4 respectively, these free amounts represent 0.3% and 0.02% of the totals respectively. While the total amounts of T3 and T4 in the serum are approximately 2 and 100 nmol.L^{-1}, the free hormone concentrations are normally of the order of 5 and 20 pmol.L^{-1}. This is the component in the circulation which is active (i.e. can bind to its receptor).

One example of how the plasma concentration of binding protein in the blood influences the total (and free) circulating iodothyronine levels is pregnancy. During pregnancy, the increasing circulating levels of oestrogen stimulate the hepatic production of TBG which increases the amount of bound iodothyronine in the circulation. The (transient) accompanying decrease in free iodothyronines, via the negative feedback they exert on the pituitary and hypothalamus, results in more iodothyronine being produced and released from the thyroid gland. While this would restore the equilibrium between free and bound iodothyronine states, it would also result in an overall increase in total (as well as free and bound) thyroid hormone levels in the circulation. Thus pregnancy is associated with an increase in circulating iodothyronine concentrations.

The iodothyronine receptors and mechanism(s) of action

The free, unbound iodothyronines are able to cross peripheral cell membranes by means of specific transporters. These include mono-carboxylate transporters 8 and 10 (MCT8 and MCT10) and organic anion transporting protein1c1 (OATP1c1). In their target cells the iodothyronines cross the nuclear membrane, also probably via a transporter system, and subsequently bind to specific thyronine receptors (TRs, also known as thyroid hormone receptors) of which there are at least three: TRα1, TRβ1 and TRβ2. The first two of these receptors are located widely in peripheral target tissues, their relative expressions varying in different tissues and at different stages in development, while TRβ3 is restricted to the hypothalamus and pituitary where it is associated with the negative feedback actions of the iodothyronines. All these receptors have a far greater (e.g. 15-fold) affinity for T3 than for T4, and intracellular deiodinases (mainly D2)

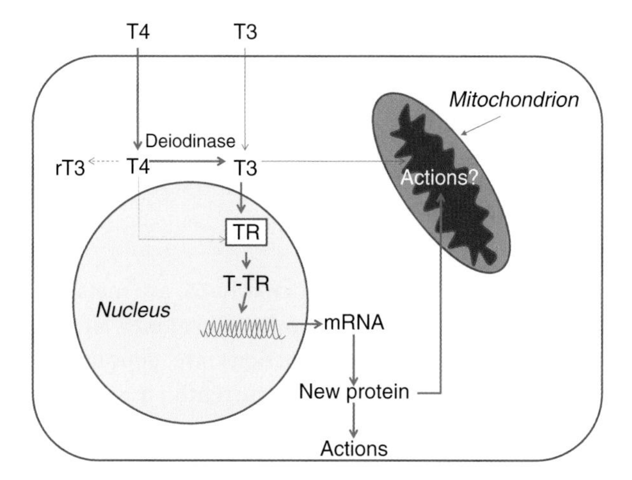

Figure 13.4 The intracellular deiodination of T4 to the more active T3 (or to the inactive rT3) within the target cell, the preferential binding of T3 to the nuclear thyroid receptor (TR) and the transcription of new protein which can have actions within the cytoplasm or on intracellular components such as the mitochondria. Also shown is the possible direct action of T3 on a mitochondrion. Non-genomic effects are also a possibility.

convert much of the T4 to the more active T3 molecule. In peripheral tissues the deiodinase D3, if activated, can convert T4 to the inactive rT3 molecule (see Figure 13.4).

The T3 (or T4) molecule, once bound to its nuclear receptor, will then locate a specific domain on a target chromosome and activate (or inhibit) a specific gene transcription process resulting in the synthesis (or inhibition) of new protein.

In addition to this generally accepted nuclear genomic mechanism of action, two other potential mechanisms could exist: a mitochondrial genomic action, and a direct non-genomic mitochondrial action. The mitochondria are organelles within the cell which are associated with most of the energy provision of the cell in the form of ATP, and they also have a small amount of DNA (of purely maternal origin). There is evidence to support the likelihood that (i) high-affinity binding sites for the iodothyronines are present in mitochondrial membranes, and (ii) they influence mitochondrial metabolism directly by acting on mitochondrial DNA. In addition, there is evidence for a non-genomic action, and that the iodothyronines can influence mitochondrial membrane transport processes directly, once bound to membrane-located binding sites. One such suggested mitochondrial action has been the uncoupling of energy production from heat synthesis, and another concerns the exchange of intramitochondrial ATP for cytosolic DTP. Certainly, non-genomic mechanisms of action could account for those effects which appear to have a much shorter latency than those normally associated with T3 (and T4).

Furthermore, there may be plasma membrane TRs which mediate rapid, direct, non-genomic actions of iodothyronines on ion channels and pumps.

Physiological actions

The principal actions of the iodothyronines, particularly T3, are metabolic, but there are also important interactions with other hormones and regulatory systems. As mentioned earlier, most actions are genomic but non-genomic effects (e.g. on membranes) are currently a source of much interest.

Basal metabolic rate (BMR)

One important action of T3 is the regulation of the basal metabolic rate in most tissues of the body, exceptions being principally the brain, spleen and testis (Figure 13.5). The basal metabolic rate is that rate of metabolism (e.g. measured as oxygen consumption) which exists when the fasted body is at complete rest in an environmentally neutral medium (e.g. during sleep). T3 increases ATP turnover and utilisation. A considerable

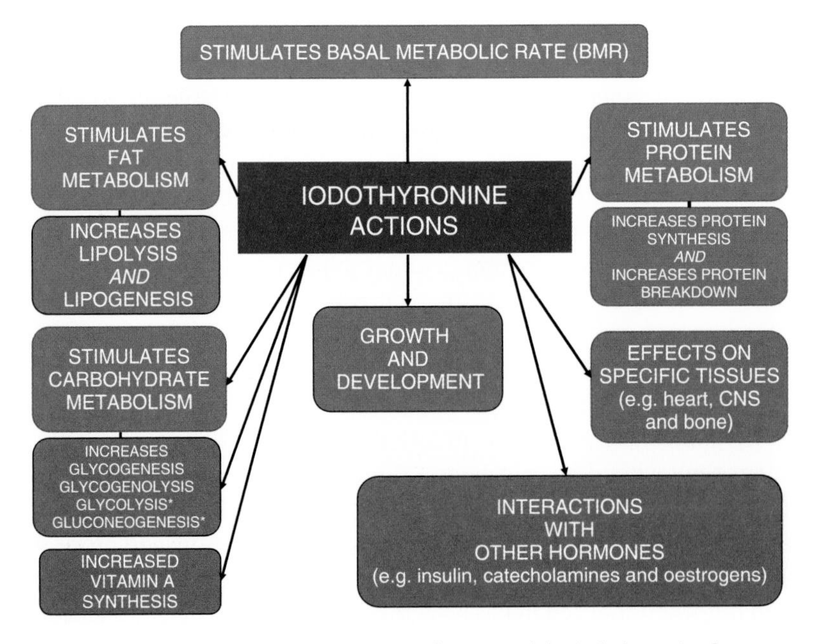

Figure 13.5 Diagram illustrating the principal actions of the iodothyronine hormones T3 and T4.

proportion of the BMR is utilised simply for the maintenance of the cellular membrane pumps such as the Na–K–ATPase which, apart from anything else, maintain cell volume. One major consequence of T3-stimulated BMR is increased heat generation (thermogenesis). Thermogenesis not only is an obligatory component (by-product) of energy expenditure, but also can be an adaptive mechanism against cold temperatures. A key tissue involved in this adaptive thermogenesis in babies is brown adipose tissue (BAT), and the mechanism involved in its regulation is the sympathetic nervous system in the presence of T3. At least part of the effect of T3 on BAT is due to an increased expression of the mitochondrial uncoupling protein (UCP1). Furthermore, T3 increases tissue responsiveness to catecholamines by increasing cAMP generation as well as interacting with cAMP-dependent transcription factors to increase the expression of specific genes associated with metabolism.

The clinical correlates are that hypothyroidism is associated with decreased heat production and the patient feels the cold, while the hyperthyroid patient generates a greater quantity of heat and becomes heat intolerant.

Protein metabolism

Many aspects of protein metabolism are influenced by the iodothyronines, and these relate to both anabolic and catabolic actions. These effects vary according to the tissue, the stage of development and the level of the circulating hormones in the blood. Thus during childhood growth and development, the iodothyronines increase the synthesis of proteins to a greater extent than the breakdown of protein, while in adulthood this balance tends to be the reverse. Hypothyroidism is associated with decreased protein catabolism, while hyperthyroidism is characterised by protein loss (which contributes to the accompanying loss in body weight). The iodothyronines influence protein metabolism mainly by acting via the TR as nuclear transcription factors, either stimulating or repressing gene transcription processes influencing protein synthesis. However, a non-genomic action on cell membranes associated with the stimulation of amino acid transport into the target cell has also been suggested.

As an example of the temporal difference in T3 activity, hyperthyroidism in children is associated with enhanced linear growth resulting in earlier fusion of the bone epiphyses, the end result being reduced height. At least part of this effect is due to the effect of T3 on specific TRs (mainly TRα1 but also TRβ1) in bone end-plate chondrocytes, bone marrow stromal cells and osteoblasts resulting in overall stimulation of bone synthesis, for example in bone matrix (protein) production. These receptors are also present in osteoclasts, but the function of T3 in these cells is unclear. In contrast, however, hyperthyroidism in adults is associated with a bone resorption which is promoted to a greater extent than bone synthesis, so that there is an increased loss of bone matrix (increased risk of osteoporosis). This may be a reflection of greater effects of T3 on osteoclasts in adults.

Carbohydrate metabolism

As with protein metabolism, both anabolic and catabolic components of carbohydrate metabolism are influenced by the iodothyronines. When normal circulating levels of iodothyronines are present, the overall effect is glucose storage (glycogenesis), while higher, more pharmacological amounts (such as in hyperthyroidism) are associated overall with glycogenolysis resulting in moderate hyperglycaemia. For example, glycogen synthase activity is enhanced increasing glycogen synthesis, while phosphorylase kinase, one enzyme which stimulates glycogenolysis, is also stimulated; the balance depends on the overall thyroid hormone level in the circulation. Furthermore, in addition to direct effects on relevant enzymes, at least some of the T3 effects on carbohydrate metabolism are mediated by its interactions with other hormones. Thus euthyroid subjects are sensitive to the effects of insulin, indicated by the enhanced peripheral uptake of glucose into cells, increased glycolysis and increased hepatic glycogen synthesis, while the catabolic effects of hormones such as the catecholamines (adrenaline and noradrenaline) are also enhanced when T3 levels are raised, resulting in increased glycogenolysis. The iodothyronines also promote gluconeogenesis indirectly by increasing the amounts of substrates such as glycerol and lactate.

Lipid metabolism

The iodothyronines stimulate both anabolic and catabolic lipid metabolism, again depending on factors such as the stage of development and the levels of circulating hormones in the blood. Overall, the higher the level of circulating iodothyronines the greater the breakdown of body fat (lipolysis) relative to fat synthesis (lipogenesis). Increased lipolysis is therefore associated with decreased levels of triglycerides and increased levels of fatty acids and glycerol. Cholesterol levels are also decreased with higher thyroid hormone levels in the circulation. Thus in hypothyroidism, there is an increase in fat storage (and body weight) while in hyperthyroidism there is a decrease in body fat (and hence body weight).

Vitamin A synthesis

Because the thyroid hormones stimulate metabolic pathways, there is an increased demand for other molecules which play an essential role to the various metabolic processes, such as vitamins acting as co-enzymes. T3 is of particular importance for one such vitamin, vitamin A, because it stimulates its synthesis from precursor carotenes as well as its conversion to retinene. In hypothyroidism the reduced synthesis of vitamin A can result in a build-up of carotenes in the blood which manifests itself as a yellowish coloration which can be observed in the skin. This is similar to the yellowish tinge commonly associated with the build-up of bilirubin in the blood which is a sign of hepatitis. The two differ in that the sclera of the eye is spared this coloration in hypothyroidism but not in hepatitis.

Brain energy balance regulation

Because of the metabolic effects associated with the iodothyronines, body weight is clearly influenced, as seen when there is thyroid dysfunction. The increasingly common problem of obesity has focused attention on the possible additional roles that T3 may play in the brain regarding the regulation of energy balance. T4 and T3 can cross the blood–brain and blood–cerebrospinal fluid (CSF) barriers by means of two transporter proteins, OATP1c1 and MCT1 and hence reach specific target neurones within the neural tissue. Of particular interest is the widespread distribution of TRs within the hypothalamus, a region of the brain known to play a critical role with regard to the control of appetite and satiety. One crucial hypothalamic region involved in appetite regulation is the arcuate nucleus and there are many TRs here, implicating iodothyronines in the overall control process. The control of appetite is discussed in more detail in Chapter 16.

Growth and development

It is clear that T3 and T4 play an important part in growth and development *in utero* as well as in the growing child, and indeed the adult. The best way to appreciate the involvement of these iodothyronines in these processes is to consider the clinical conditions associated with low and high circulating levels of these hormones in the blood. The most important growth phase occurs *in utero* and also in the growing child. Lack or absence of these thyroid hormones is manifested by the abnormal, delayed growth and development seen following birth, a condition known as cretinism. The lack of growth and development, which is not only physical but also mental, becomes irreversible if treatment is not initiated as soon after birth as possible. Indeed, one of the first checks done on a newborn baby is the measurement of circulating T4 and TSH levels, because the condition is eminently treatable nowadays (see the section on neonatal hypothyroidism). In the adult the lack or absence of T3 and T4 is known as hypothyroidism, and it too is characterised by an impaired metabolism and decreased mentation (a slowing down of thought processes), which are characteristic of the disorder. The overactive thyroid (hyperthyroid) state is also associated in changes in metabolism with aspects of increased metabolism becoming apparent, as well as changes in cerebral processes, with the patient become hyperactive and anxious. These various clinical conditions are considered in detail later, and in the clinical scenario in this chapter's Appendix.

Interactions with the sympathetic nervous system

As mentioned earlier, there are various interactions between the thyroid (and its hormones) and the sympathetic nervous system (SNS). For instance, increased sympathetic stimulation is associated with a decreased blood flow through the thyroid. This could influence follicular cell activity, for instance by decreasing the endocrine regulatory signal provided by thyrotrophin and the provision of necessary metabolic products (e.g. iodide) for their normal functioning.

T3 acts as a transcription factor by stimulating the transcription of the β1-adrenergic receptor gene thereby playing an important role in amplifying catecholaminergic effects in specific tissues such as the heart (myocardial cells). Thus in hyperthyroidism, the commonly observed tachycardia can be effectively controlled using a β-blocker.

CNS effects

The widespread distribution of thyroid hormone receptors (TRs) in the brain indicates that T3 and T4 play a significant role in regulating neuronal activity. They access the brain by the transporter systems OATP1c1 and MCT8 located in the endothelial cells of the blood–brain barrier capillaries, and to a lesser extent at the choroid plexus–CSF interface. For example, the TRs are prominent in the hypothalamus, particularly in the paraventricular and arcuate nuclei, and here they may well be involved in the regulation of appetite and satiety (see Chapter 16). The hypothalamus is also involved in the control of thyroid hormone production, so TRs will be part of the negative feedback influence of T3 and T4 here. Elsewhere, the effects of T3 are less clearly defined. Since many neurones contain intracellular deiodinases, T4 can be converted to either active (T3) or inactive forms. T3 certainly acts as a transcription factor on various genes and is associated with the expression of various proteins, including mitochondrial and cytoskeletal proteins, and myelin. Indeed, during neonatal development hypothyroidism is associated with a delayed, impaired myelination of nerve fibres. Furthermore, as mentioned earlier, hypothyroidism is associated with slow mental processes (a classical symptom in adults), while hyperthyroidism is linked to increased nervousness (anxiety) and irritation. Exactly how T3 and T4 are involved, however, is still unclear, particularly since mutant mice lacking TRs (Null-TR) fail to show much in terms of central nervous defects.

Effects on bone

Iodothyronine transporters are present in chondrocytes, osteoblasts and osteoclasts, as are TRs (mainly the TRα1 isoform). Hypothyroidism in children is associated with a delayed bone age and failure to grow, resulting in short stature if untreated. In contrast, hyperthyroid children have an advanced bone age, accelerated growth and premature fusion of the end plates of the long bones also resulting ultimately in short stature if untreated. Thus T4 and T3 are clearly involved in skeletal development. Similarly, in adults hypothyroidism and hyperthyroidism are associated with a decreased and increased bone turnover, respectively. In hypothyroidism, both resorption (by osteoclasts) and remodelling (by osteoblasts) phases are lengthened, but the remodelling phase is affected more, so the condition is associated with an increased bone mass. In hyperthyroidism, both phases of bone resorption and remodelling are shortened since both processes are accelerated. Thus in hyperthyroidism there is a decreased bone mass and an increased risk of osteoporosis, and consequently of bone fractures.

Regulation of thyroid iodothyronine hormones

Control of thyroid iodothyronine synthesis, storage and release is essentially due to the hormone thyrotrophin (TSH) from the anterior pituitary thyrotroph cells. It binds to its membrane receptor and activates its G protein which in turn results in increased adenyl cyclase activity. The second messenger cAMP then initiates various intracellular actions associated with the thyrotrophin. These include the stimulation of the sodium–iodide pump in the cell membrane, synthesis and stimulation of the intracellular proteins thyroid peroxidise and thyroglobulin, stimulation of the iodination and coupling reactions, the process of colloid endocytosis and fusion with follicular cell lysosomes and finally the release of T4 and T3.

Thyrotrophin synthesis is primarily under the control of the hypothalamic thyrotrophin-releasing hormone (TRH), so the hypothalamus exerts an important, indirect regulatory influence over thyroid hormone production and release (see Chapter 2). The circulating levels of free T3 and T4 have direct and indirect negative feedback effects on the adenohypophysis and hypothalamus, respectively (see Figure 13.6).

An additional action of thyrotrophin appears to be the stimulation of α1 adrenoceptor synthesis on follicular cell membranes. These are receptors for ligands such as noradrenaline and adrenaline, and mediate effects on iodide uptake and iodination in the follicular cells. In addition, other factors influence the thyroidal production of iodothyronines. These include various other hormones including glucocorticoids and the oestrogens.

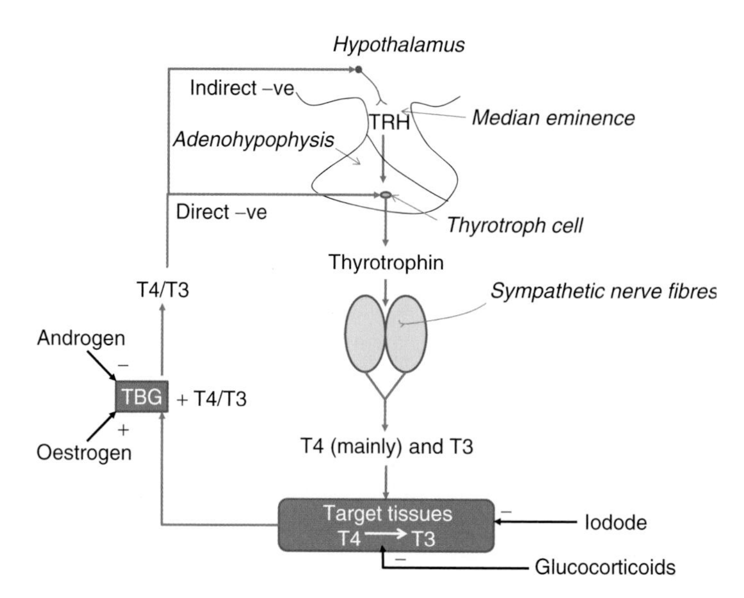

Figure 13.6 Diagram illustrating the main factors regulating T3 production. Glucocorticoids can inhibit the target cell deiodination of T4 to T3, while oestrogens stimulate, and androgens inhibit, hepatic TBG synthesis thereby altering total circulating T3 and T4. TBG = thyronine-binding globulin and TRH = thyrotrophin = releasing hormone.

High concentrations of glucocorticoids decrease the deiodination of T4 to the more potent T3 in target tissues. This is another way in which biologically active thyroid hormone levels are controlled at the target cell level.

Gonadal steroids also influence thyroid status. Certainly, clinical conditions associated with the thyroid gland are far more common in women than men (of the order of 8:1), suggesting a gonadal hormone influence. One such influence is that oestrogens increase, and androgens decrease, the hepatic synthesis of plasma proteins including thyronine-binding globulin. As the concentration of TBG in the blood changes, so does the amount of total T3 and T4, because of the negative feedback influence of the thyroid hormones on the anterior pituitary and hypothalamus. For example, oestrogens stimulate TBG production, and consequently there is a greater binding of free thyroid hormones to the plasma protein fraction with the consequences described earlier. Androgens (and glucocorticoids) have the opposite effect, being associated with a decreased production of hepatic TBG. The gender difference is also likely to involve the hepatic deiodination of T4 to T3 which is certainly greater in female than in male rats. Whether this is the same in humans is currently unknown.

Finally, large amounts of iodide (taken in or administered as iodine) also block follicular cell function leading to a decreased production of iodothyronines (the Wolff–Chaikoff effect) and this is used clinically.

Diseases of the thyroid gland

Thyroid failure, overactivity of the thyroid and lumpiness of the thyroid gland (with normal thyroid function) are the most common disorders affecting the thyroid gland. Change in the function of the thyroid presents with systemic symptoms such as tiredness or palpitations, whereas tumours or nodules simply cause a lumpy neck. Often a lumpy neck is completely benign, but it is important that the correct examination is carried out in such patients because occasionally a lump in the thyroid gland turns out to be a thyroid cancer.

Thyroid failure

Thyroid failure affects about 2% of the population and occurs most commonly because of autoimmune destruction of the thyroid gland. Other causes include removal of the thyroid gland, and pituitary failure where the secretion of TSH is affected.

Primary hypothyroidism

When the cause is due to the thyroid gland itself failing, this is known as primary hypothyroidism. The condition is also called myxoedema. As the circulating concentration of thyroxine falls, the basal metabolic rate decreases and the patient slows down, physically and mentally. Utilization of energy therefore falls, and patients lose their appetite and eat less. However, as the BMR falls, as explained earlier, patients put on weight despite losing their appetite. The consequences of thyroid failure are entirely predictable: (i) tiredness and lethargy, (ii) lack of mental agility and (iii) feeling cold.

In primary hypothyroidism, it is possible for the thyroid gland to be enlarged (goitre) because the low circulating T3 and T4 levels exert a reduced negative feedback on the hypothalamo-adenohypophysial axis, with a subsequent increase in TSH production by the anterior pituitary. This TSH excessively stimulates the follicular cells which can hypertrophy. The diagnosis of primary hypothyroidism can be made by finding a low level of thyroxine (T4) in the circulation at the same time as a raised level of TSH.

Treatment of thyroid failure simply involves replacing the missing hormone, thyroxine. This can be easily administered orally. The dose of thyroxine should be slowly increased until the TSH is normal.

Myxoedema coma

This is a special case and needs particular mention. If the diagnosis of hypothyroidism is not considered for several years in a patient with primary hypothyroidism, the condition slowly worsens until the patient loses consciousness. At this point the patient is said to have myxoedema coma, a condition with a high mortality if untreated. Because thyroid function tests, and in particular the TSH assay, are now widely available, myxoedema coma is unusual but it still occurs in elderly patients who have no relatives. Such patients who are in residential care may not be regularly reviewed, and when they become confused they may be thought to be suffering from senile dementia. Thus it is essential to check the TSH before suggesting any patient has dementia. In the unhappy situation in which such patients are not referred for endocrine investigation, several months of slowly worsening hypothyroidism occur until the patient develops coma. Because they are unconscious, this usually results in referral to casualty, where the diagnosis will be made on admission both because of the clinical appearance of the patient and because any such confused patient would have a TSH measured.

If a patient is admitted with Myxoedema coma, it is essential to start the patient on very small doses of T3. A usual starting dose is 5 mcg, and the reason for using such a small dose is that since the patient will have had hypothyroidism for many years, a sudden normalization of the free T3 (FT3) may cause a sudden tachycardia, which could precipitate cardiac disease and ischaemia, which in turn can be fatal. It is obviously important to make such changes gradually. If there are no adverse events for 8 hours after the first dose of T3, the dose can safely be repeated every 8 hours until the patient awakens. It is also important to exclude other causes of coma, and since hypothyroidism is an autoimmune disease, it is possible that the patient will have more than one such autoimmune disorder. Therefore patients are often treated for autoimmune Addison's disease, just in case the patient has both myxoedema and Addison's disease. The treatment of Addison's disease is covered in Chapter 10. The combination of hypothyroidism with Addison's disease is known as Schmidt's syndrome.

Neonatal hypothyroidism

In the United Kingdom, 1 in 4000 live births has congenital hypothyroidism. In 90% of cases, this occurs because of an anatomical defect, such as an absent thyroid (thyroid agenesis) or ectopic thyroid. In the other 10% of cases, the condition is due to a biochemical defect due to an enzyme deficiency in thyroxine synthesis, such as a congenital deficiency of a deiodinase, an enzyme important in the recycling of iodine within the follicular cells.

Thyroxine is essential for normal brain development. Cretinism is the condition caused by untreated neonatal hypothyroidism, where there is irreversible brain damage. In some parts of the world, iodine deficiency used to cause a condition called endemic cretinism, where iodine deficiency in mothers and neonates causes thyroxine deficiency either *in utero* or shortly after birth, which in turn leads to irreversible brain damage. Because of the general availability of iodine replacement (it may be added to salt, and to various foods such as bread), endemic cretinism is generally a disease of the past. However, sporadic cretinism still can occur.

Even in the absence of a thyroid gland in the fetus, thyroxine is detectable in umbilical cord blood, indicating that thyroxine crosses the placenta, so that at birth neonates can have normal brain development due to the mother's thyroxine. Such individuals have a rapid fall in circulating thyroxine days after birth, coupled with a rapid rise in TSH. These neonates are at risk of cretinism if thyroxine is not commenced within a few days. Because of this, there is a national screening service to prevent neonatal cretinism. All newborns have a heel prick blood sample, which is screened for a number of metabolic disorders, and also hypothyroidism. If a raised TSH is discovered, neonates are rapidly commenced on thyroxine.

Overactivity of the thyroid gland

An overactive thyroid gland occurs when either part of the gland starts to oversecrete thyroxine, or the whole gland does so.

Graves' disease

This condition was first described in 1796 by Robert Graves in Dublin who noticed that patients would present with weight loss, tachycardia and palpitations that would slowly get worse. In those who perished, the post mortem revealed an enlarged vascular thyroid gland (goitre).

Graves' disease is an autoimmune disease with a number of different antibodies. One of these binds to, and stimulates, the TSH receptor and causes hyperthyroidism. A slightly different antibody binds to the same receptor but causes hypertrophy of the gland rather than hyperthyroidism. These commonly occur together, so that patients with Graves' disease often present with both a thyroid goitre and hyperthyroidism, although they *can* present with one or the other. A third antibody binds to and stimulates growth factor receptors behind the eye, which cause muscle hypertrophy. These are responsible for the appearance of exophthalmos (the forward protrusion of the eyeballs), sometimes called proptosis (Figure 13.7).

A fourth antibody binds to and stimulates growth factor receptors at the front of the shin, causing pretibial myxoedema (not to be confused with the condition of myxoedema which is primary hypothyroidism; see section 'Primary hypothyroidism'). Any of these four antibodies *can* appear in any order, although *commonly* the first two appear first, causing hyperthyroidism. The third and fourth usually appear about 1 year later, so that patients often (wrongly) blame their anti-thyroid therapy for their worsening eye disease. Sometimes the third antibody can appear by itself. These patients present with exopththalmos (proptosis) first, and see the ophthalmologists, who check their thyroid function tests and find them to be normal. Note that proptosis and exopthalmos represent the same condition, the latter term usually being used when associated with an endocrinopathy. Patients with Graves' eye disease can become thyrotoxic up to 5 years later, but some never do.

Graves' Disease Autoantibodies

- Antibodies bind to and stimulate the TSH receptor in the thyroid.
  - Cause smooth goitre (1) and hyperthyroidism (2).
- Other antibodies bind to muscles behind the eye.
  - Cause exophthalmos (3).
- Other antibodies bind to the front of the shin.
  - Cause pretibial myxoedema (hypertrophy). (4)

Lid lag occurs due to hyperthyroidism of any cause. The muscle that opens the eye (levator palpabrae superioris) has two nerve supplies, one from the third cranial nerve (voluntary) and the other sympathetic (autonomic). When a patient has hyperthyroidism, the β receptors are sensitised to systemic adrenaline levels. This causes the eye to open or close more slowly than normal. It is the opposite of the ptosis that occurs when the sympathetic nerve is severed. Lid lag thus occurs in patients with any cause of hyperthyroidism, including Graves' disease, a toxic nodular goitre and viral thyroiditis, and also in patients who take a large overdose of thyroxine.

Eye Signs in Graves' Disease

1. Lid lag caused by high T4 which sensitises levator palpabrae superioris to catecholamines.
2. Periorbital oedema and chemosis (swelling of conjunctiva).
3. Proptosis (exophthalmos) (Figure 13.7).
4. Diplopia (double vision).

Note: Graves' is the commonest cause of unilateral exophthalmos.

The diagnosis of Graves' disease can be confirmed by detecting anti-TSH receptor antibodies, in the presence of hyperthyroidism. Because the whole thyroid gland is overactive, an uptake scan will show increased uptake in all areas of the thyroid (Figure 13.8).

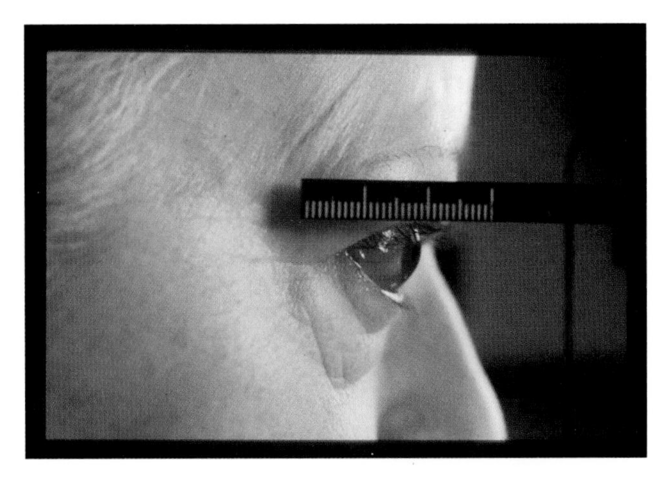

Figure 13.7 Lateral view of a Graves' patient's eye.

A hot thyroid nodule

This occurs when a clone of cells in the thyroid gland becomes autonomous. There may be one nodule or several. A toxic nodular goitre can occur with only very minor derangement of thyroid function, or occasionally with frank hyperthyroidism. A radioactive iodine uptake (RIU) test is performed for diagnosis. This involves a scan of the thyroid, following oral administration of radioiodine, and it shows an increased uptake in the hot nodules, with suppression of the rest of the thyroid gland. Antibodies are negative (Figure 13.9).

de Quervain's (viral) thyroiditis

The natural history of viral (de Quervain's) thyroiditis starts with infection and inflammation of the thyroid gland. The patient complains of painful swallowing (dysphagia), fever and features of hyperthyroidism. The thyroid gland then stops synthesising any new thyroxine, and makes more viruses instead. Thus the uptake of iodine on a scan falls to zero. Damage occurs to the thyroid follicles, which release their stored thyroxine in large amounts, making the patient truly hyperthyroid clinically and biochemically, with a raised FT3 and FT4, and a suppressed TSH. The outcome is that the patient develops a combination of clinical hyperthyroidism together with a painful neck. Because the follicular cells are not synthesizing more thyroxine, the stored thyroxine runs out a month or so following infection, when the circulating thyroxine concentration starts to fall. The half-life of thyroxine is about 10 days, so the patient will move slowly from being thyrotoxic to being hypothyroid, about a month later. After a further month, with the viral infection under control, the patient returns to normal.

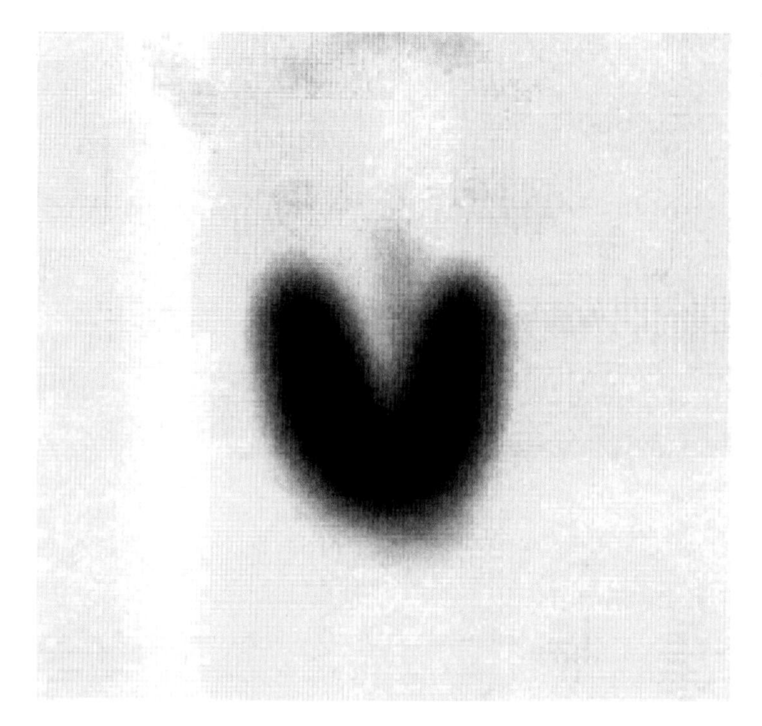

Figure 13.8 Scan of a thyroid gland in Graves' disease.

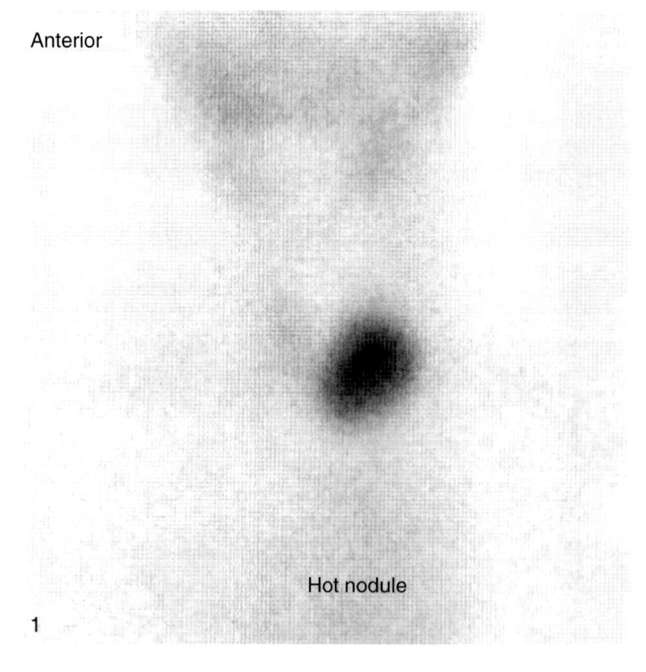

Figure 13.9 Thyroid scan of a single hot nodule.

The natural history of viral (de Quervain's) thyroiditis is:

1. The virus attacks the thyroid gland and causes pain and inflammation, a fever and a high erythrocyte sedimentation rate (ESR).
2. The thyroid gland starts to make new viruses and stops making new thyroxine. Iodine uptake therefore stops.
3. The damaged thyroid gland releases any stored thyroxine that is within the thyroid gland, causing hyperthyroidism.
4. The stored thyroxine becomes depleted, and once the released thyroxine has been metabolised, the patient will move from being hyperthyroid to hypothyroid. This occurs after about a month, as the half-life of circulating thyroxine is about 10 days.
5. The patient becomes hypothyroid and stays so for about a month.
6. There is still no iodine uptake into the thyroid.
7. Once the virus is cleared, thyroxine synthesis is restored.
8. The best way to confirm the diagnosis of viral thyroiditis is to show that there is no iodine (or technicium) uptake when the patient is toxic.
9. The antibody status may mislead you, as after thyroid gland damage, any thyroid antigens get into the circulation and "immunise" the patient. Thus you may detect anti-thyroid antibodies for about a year after an attack of viral thyroiditis.
10. The best treatment for viral de Quervain's thyroiditis is to wait for the patient to recover. The treatment below refers to Graves disease or a toxic nodule.

Treatment of hyperthyroidism

Because thyroxine sensitises the beta adrenoceptor, hyperthyroidism responds quickly to beta blockade. In addition, there are three other important treatment options for patients with hyperthyroidism.

1. Anti-thyroid drugs such as carbimazole or propylthiouracil (PTU). These block components of the iodothyronine synthesis pathway.
2. Thyroidectomy. This was the original treatment used by Robert Graves.
3. Radioiodine, which has been available since about 1960.

Hyperthyroidism is usually straightforward to treat. However, occasionally patients become extremely thyrotoxic and this can be life threatening. Severe hyperthyroidism is also known as thyroid storm which is defined in any patient who is thyrotoxic together with any two of:

- Hyperpyrexia $> 41^\circ$ C
- Accelerated tachycardia ($>$140 beats.min^{-1})
- Any arrhythmia
- Cardiac failure
- Delirium or frank psychosis
- Hepatocellular dysfunction

In a patient with thyroid storm, treatment starts with high-dose beta blockade (60 mg propranolol every 4 hours). In an asthmatic, guanethidine or reserpine would be used if beta blockade is contra-indicated. Propylthiouracil (PTU; 250 mg, four hourly) can then be given and this will take about an hour to block iodine organification. The combination of beta blockade and PTU is usually sufficient. Rarely, full control is not achieved, and the next step depends on what treatment is necessary in the longer term. Corticosteroids block T4 to T3 conversion. Iodine, either potassium iodide (KI) 60 mg three times daily (tds) or Lugols iodine (a mixture of iodine and KI in distilled water) can be used if surgery is planned. Iodine has the advantage that it reduces the thyroid vascularity, making surgery easier, provided this occurs within 10 days of the starting of the iodine. Aspirin is contra-indicated, as it competes with thyroxine, worsening the storm.

Management of a non-functioning thyroid lump

Patients who find a lump in the neck seek medical attention because they are worried that they have a thyroid cancer. This is indeed an important diagnosis to exclude. Patients who detect a thyroid lump need to have a careful clinical examination. A solitary nodule needs careful cytological examination. Patients who have a multinodular goitre do not need any further intervention. If, however, a dominant nodule is found, then that lesion also needs cytological examination. This can be performed either using ultrasound or, if the nodule is easily palpable, then by inserting a needle directly into the nodule.

Guidelines suggest that such a fine needle aspiration (FNA) should be undertaken twice (6 months apart), and if both aspirates confirm a benign cytology then no further intervention is required.

If thyroid cancer is suggested, then the patient should be reviewed by a thyroid multidisciplinary team. A total thyroidectomy should be performed, and careful histological examination of the removed thyroid carried out. If papillary or follicular thyroid cancer is confirmed, patients are given ablative (high-dose) radioiodine therapy to try to remove any TSH-responsive cells. Patients should then be followed up in a specialist thyroid cancer clinic, where they are usually given a generous dose of thyroxine to suppress the TSH. Annual iodine uptake scans are often used to ensure that patients remain in remission, and if any recurrence occurs, further doses of radioiodine can be administered. In general the prognosis for papillary and follicular thyroid cancer is excellent.

Medullary thyroid cancer is a more aggressive cancer, and is a malignancy of the C-cells. These cells do not respond to TSH or take up iodine, so if medullary thyroid cancer is suspected, blood calcitonin levels should be checked, and if raised, a total thyroidectomy is indicated. Patients with medullary thyroid cancer might in fact have multiple endocrine neoplasia 2 (MEN2). For further details, see Chapter 19.

Appendix: Clinical Scenario

Clinical Scenario 13.1

Mrs Andrews, a 45-year-old mother of two children aged 18 and 13, phones her GP clinic and requests a doctor's appointment. The receptionist notes that the woman's voice is rather husky and she seems to have difficulty in following the conversation, as she seems very lethargic. At the time of the appointment, her GP Dr Watson notes that she has put on weight, and calculates her BMI as 29. She certainly seems slow in answering his questions and seems somewhat depressed. Her heart rate is 52 beats per minute; her skin is dry, feels cold and has a slight yellowish tinge to it (although the sclera of each eye seems normal) and she has lost part of her eyebrows. Mrs Andrew reports that she always feels tired, she feels the cold more than she used to and she has infrequent, but heavy, periods, and she also complains of being constipated. Dr Watson also notes a slight swelling in the neck, on both sides of the trachea. From Mrs Andrew's past history, he recalls that her mother had a thyroid disorder.

Dr Watson organises a phlebotomy appointment for Mrs Andrews at the local hospital pathology department, and arranges for her to return to see him 2 weeks later, when he should have received the results of certain thyroid tests.

(*Note:* Phlebotomy is the taking of a blood sample.)

Questions

Q1. What condition does Dr Watson initially suspect that Mrs Andrews might be suffering from, and why?
Q2. How could you explain the various signs observed in Mrs Andrews, and the symptoms that she is experiencing?
Q3. What would be the likely results of her blood thyroid test?
Q4. What treatments could be considered for Mrs Andrews?

Answers to Clinical Scenario 13.1

A1. What condition does Dr Watson initially suspect that Mrs Andrews might be suffering from, and why?

Dr Watson might well suspect that Mrs Andrews has hypothyroidism. She has a goitre in the neck, there is a family history of thyroid problems, her voice has noticeably deepened, she is at the top of the overweight range (BMI between 25 and 30) and her skin is dry, cold and has a yellowish tinge to it. She is also experiencing menstrual problems. These are all classic signs of hypothyroidism.

A2. How could you explain the various signs observed in Mrs Andrews, and the symptoms that she is experiencing?

The deepening (husky) voice is due to a thickening of the vocal cords due to the retention of mucopolysaccharides and other water-retaining molecules (which would also account for the puffiness of the skin that can also be observed). The general tiredness (lethargy) and the cold intolerance are associated with decreased metabolic activity. The increase in body weight is also a feature of the decreased metabolic rate, although it is not necessarily diagnostic. The yellowish tinge is due to the decreased synthesis of vitamin A from its precursor carotene which consequently can build up in the circulation giving the skin a characteristic yellowish colour. The yellowish colouration of the skin can be confused with jaundice but in this case the sclera of the eye becomes equally tinged, unlike in hypothyroidism.

The loss of hair (alopecia) is also common in hypothyroidism and can sometimes be manifest in the loss of the outer third of the eyebrows. Constipation is a common symptom of hypothyroidism and is probably due to the slowing down of bowel movements consequent to the decreased general metabolism. The decreased heart rate (bradycardia) is also characteristic of hypothyroidism, as are the slow mentation and the increased incidence of depression.

The disruption to menstrual cycles in premenopausal women is likely to be due to a lack of inhibition by thyroidal iodothyronines on the hypothalamo-pituitary axis. This would result in an increased thyrotrophin-releasing hormone (TRH) production by the hypothalamus which stimulates prolactin release from the pituitary. Consequently, prolactin inhibits pituitary gonadotrophin production resulting in a decreased gonadal hormone production by the ovaries.

A3. What would be the likely results of her blood thyroid test?

The concentrations of iodothyronines T3 and T4 in the blood would be low, and the thyrotrophin (TSH) concentration would be raised, in primary hypothyroidism (Hashimoto's disease). Thyroidal antibodies also might well be raised. If by any chance T3, T4 *and* TSH levels are low, then secondary hypothyroidism (due to decreased pituitary, or even hypothalamic, function) would be suspected.

A4. What treatments could be considered for Mrs Andrews?

The most likely treatment for primary hypothyroidism is L-thyroxine replacement therapy.

Reference

Horsley, V. (1892) Remarks on the function of the thyroid gland: a critical and historical review. *British Medical Journal,* 1 (1622), 215–19.

Further Reading

Bizhanova, A. & Kopp, P. (2009) Minireview: the sodium-iodide symporter NIS and pendrin in iodide homeostasis of the thyroid. *Endocrinology,* 150, 1084–90.

Gogakos, A.I., Bassett, J.H.D. & Williams, G.R. (2010) Thyroid and bone. *Archives of Biochemistry and Biophysics,* 503, 129–36.

Goglia, F., Moreno, M. & Lanni, A. (1999) Action of thyroid hormones at the cellular level: the mitochondrial target. *Federation of European Biochemical Societies (FEBS) Letters,* 452, 115–20.

van de Graaf, S.A.R., Ris-Stalpers, C., Pauws, E., Mendive, F.M., Targovnik, H.M. & de Vijlder, J.J.M. (2001) Up to date with human thyroglobulin. *Journal of Endocrinology,* 170, 307–21.

CHAPTER 14

The Islets of Langerhans and Their Hormones

Introduction

Hormones play a crucial role in regulating the metabolic requirements of the body. They control both anabolic and catabolic reactions by stimulating transport processes and enzymes involved in the synthesis of proteins, fats and complex carbohydrates, as well as their breakdown to the constituent building blocks, the amino acids, fatty acids and monosaccharides respectively. The involvement of many of the hormones in these metabolic processes has already been considered in previous chapters covering the iodothyronines, growth hormone, adrenaline, the glucocorticoids and the gonadal steroids, for instance.

One key energy substrate utilized by all tissues in the body, and of special importance for neurones in the central nervous system (CNS), is glucose. Indeed, its importance is emphasised by the fact that many, if not all, hormones released by stressors stimulate an increase in the blood glucose concentration in order to provide this essential substrate to all tissues in times of emergency. For example, the acute fight–flight response induced by the activation of the sympathetic nervous system, and involving the release of adrenaline from the adrenal medulla, is associated with the breakdown of glycogen to glucose (glycogenolysis). Likewise, the longer term release of the glucocorticoid cortisol from the adrenal cortex, as an essential component of the stress response, is also associated with an increase in blood glucose concentration. In addition, other hormones such as growth hormone (e.g. released during exercise), the iodothyronines (e.g. released during exposure to cold temperatures) and even vasopressin (e.g. released during haemorrhage) all have effects which will tend to increase the blood glucose concentration. Thus, all stress hormones increase the blood glucose concentration to a greater or lesser extent.

Another endocrine gland, so far mentioned only in passing, plays an essential role in regulating general metabolism, with particular relevance to the control of the blood glucose concentration. This gland is actually

---

Integrated Endocrinology, First Edition. John Laycock and Karim Meeran.

comprised of groups of cells concentrated into islets distributed throughout much of the pancreas, called islets of Langerhans after the doctor who first described them. These islets act as an endocrine gland, and are the subject of this chapter. Different molecules are produced by the different cell types of the islets of Langerhans. These include yet another hormone which generally acts to increase the blood glucose concentration called glucagon, and, most importantly, another hormone called insulin. Insulin has various actions but it is of particular importance because it is the one hormone which actively decreases the blood glucose concentration (Figure 14.1).

Glucose is a simple monosaccharide which can either be utilised directly as a source of energy (glycolysis) or be stored as a complex carbohydrate called glycogen (glycogenesis) in most cells, but particularly in the hepatocytes of the liver. Glycogen can be metabolised back to glucose by a process called glycogenolysis, and the liver is an important source of glucose which can be readily released into the general circulation when required by the cells in other parts of the body.

Glucose is a hydrophilic molecule which naturally exists as a D-isomer (i.e. it rotates polarised light to the right) and is sometimes called dextrose. It does not readily diffuse across cell membranes but requires specific glucose transporter (GLUT) molecules. These molecules form a large family of proteins, found not only in outer (plasma) cell membranes (class I) but also in the membranes of intracellular structures such as in nuclear and Golgi membranes (class II). There are four well-defined class-1 glucose transporters.

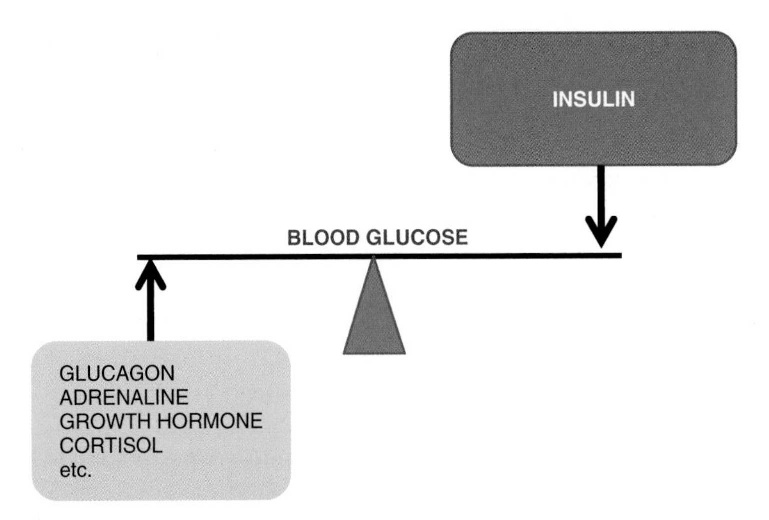

Figure 14.1 Diagram illustrating the principal hormones involved in regulating the blood glucose concentration, with those hormones actively increasing it on the left of the balance, and insulin the only hormone actively decreasing it on the right.

GLUT1 is found in all plasma cell membranes to some extent, and is associated with the basal uptake of glucose. This transporter is particularly important for glucose transport in erythrocytes, and endothelial cells of the blood–brain barrier. It is also expressed in fetal tissues.

The GLUT 2 transporter is located in the renal tubules, the hepatocytes of the liver and the β-cells of the islets of Langerhans.

The GLUT 3 transporter is found in neuronal membranes, and also in the placenta.

GLUT 4 transporters are expressed in skeletal and cardiac muscle cells and in adipose tissue. This is the only glucose transporter system which is influenced by insulin.

Generally, this form of transporter-mediated transport depends on the presence of a concentration gradient across the cell membrane, but does not require energy in the form of adenosine-5′-triphosphate (ATP) directly, and is described as a facilitated transport mechanism. Once glucose has entered the cell, it is immediately phosphorylated to glucose-6-phosphate, which keeps the intracellular concentration of glucose very low, maintaining the concentration gradient across the cell membrane.

The pancreatic islets of Langerhans: Anatomy and structure

The pancreas is an organ which lies in the abdomen, and is closely associated with the small intestine. It is derived from the developing endoderm of the foregut, initially as ventral and dorsal buds which ultimately fuse to form the primitive pancreas. The ventral bud becomes the posterior part of the head of the pancreas, while the rest develops from the dorsal bud. Most of the pancreas consists of acinar cells grouped into lobules which secrete a proteinaceous fluid directly into common ducts which join up to form the main pancreatic duct. The pancreatic duct joins the bile duct from the liver, and the two together release their secretions into the duodenum which is the first section of the small intestine (Figure 14.2). The sphincter of Oddi controls the flow of these digestive juices into the duodenum. Relaxation of the sphincter smooth muscle is induced by the gut hormone cholecystokinin which stimulates the local release of vasoactive intestinal peptide (VIP). Scattered throughout the pancreatic lobules are small, easily distinguishable collections of cells called islets of Langerhans. There are approximately one million islets in a normal human pancreas, representing about 2% of the pancreatic mass. The islets start to become distinguishable in the human embryo from about the 12th week developing from the pancreatic duct, with their own independent blood supply by the 16th week.

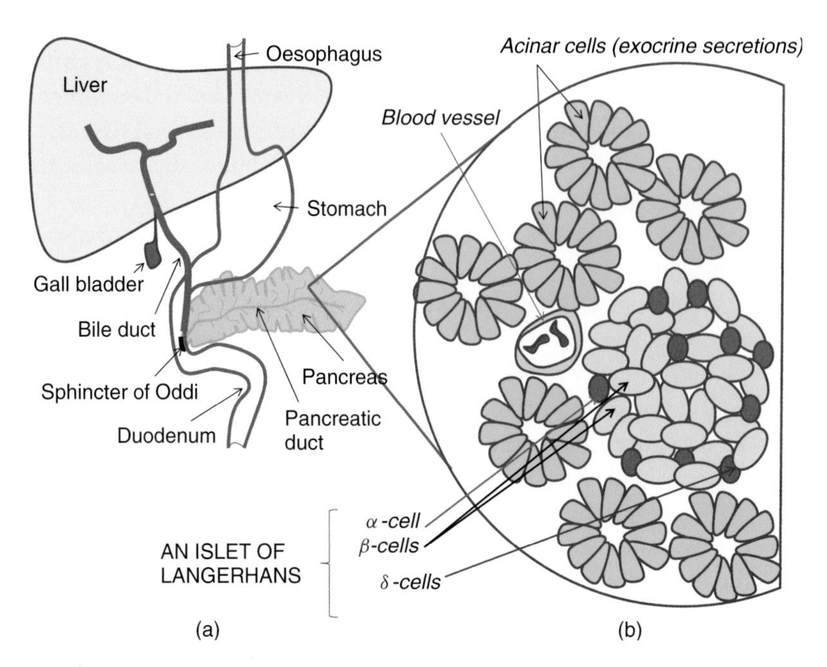

Figure 14.2 Diagram illustrating (a) the position of the pancreas in relation to the main visceral organs (liver and intestinal tract), and (b) a section through the pancreas indicating the many exocrine ducts surrounded by acinar cells, and the principal cells comprising an islet of Langerhans (alpha (α), beta (β) and delta (δ)).

Blood reaches the pancreas via the superior and inferior pancreatico-duodenal arteries. The superior branch arises from the gastroduodenal artery and supplies blood to the body of the pancreas, while the inferior branch comes off the superior mesenteric artery and supplies the head of the pancreas. Both the body and the tail of the pancreas also receive blood from branches of the splenic artery. The veins draining the pancreas join the portal vessels passing to the liver.

The innervation of the pancreas is autonomic with sympathetic and parasympathetic supplies to both exocrine and endocrine tissues.

Each islet comprises at least four cell types which can be differentiated by appropriate immunostaining. These different cells are called alpha (α), beta (β), delta (δ) and gamma (γ) cells and they are associated principally with the hormones glucagon, insulin, somatostatin and pancreatic polypeptide respectively. Another polypeptide called polypeptide YY (PYY) is found co-localised with each of the four other hormones, but its role is unclear. In human islets the α and β-cells represent approximately 15% and 80% of the total cell count respectively, the β-cells predominantly forming the core, with most of the α-cells towards the periphery. Most of the remaining 5% of the cells are δ-cells, and these are scattered throughout the islets. Many of the β-cells are linked to their neighbours by gap junctions in the

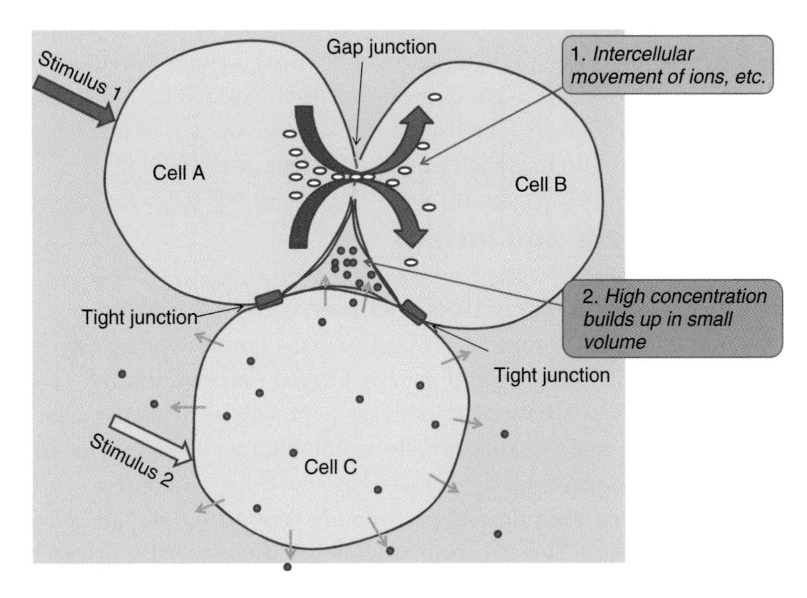

Figure 14.3 Diagram illustrating the close relationships which can exist between islet cells, such as the interconnecting cytoplasms of two adjacent cells via gap junctions, and the enclosure of tiny volumes of extracellular fluid by the presence of dynamic tight junctions.

abutting cell membranes, allowing for cross-talk between them by means of small molecules moving between the cells for instance. Furthermore, there are dynamic tight junctions which can link nearby cells to each other, trapping small volumes of extracellular fluid between the cells. This may also have a bearing on intercellular communication, since if one cell type is stimulated and releases a small quantity of hormone into the general circulation, that same small quantity released into the trapped volume of extracellular fluid could reach much higher concentrations which might then affect those nearby cells (see Figure 14.3). These are examples of paracrine activity between different cells.

Arterial blood flow to each islet reaches the outer cells of the cluster first, and then passes through the cluster towards the centre where it ultimately joins the venous outflow of the pancreas.

The pancreas has two functions:

1. The production of an exocrine secretion; fluid containing enzymes and other molecules necessary for normal digestion. This secretion passes into the pancreatic duct and then into a common duct, which also collects bile produced in the liver and stored in the gall bladder via the bile duct. The secretion can then enter the first part of the small intestine (the duodenum) leading to the exterior of the body.
2. The production of an endocrine secretion; the polypeptide hormones produced by the cells of the islets of Langerhans are released directly into

the bloodstream. These islet hormones are the focus of this chapter. By far the most important of them, particularly from a clinical perspective, is insulin. Approximately 2% of the population worldwide suffers from a metabolic disorder associated with the absence or lack of insulin, or tissue insensitivity to it: diabetes mellitus (Chapter 15).

The β (beta) cells and insulin

Synthesis, storage, secretion and transport of insulin

The predominant β-cells of the islets of Langerhans synthesise the polypeptide hormone insulin. The insulin gene is located on chromosome 11 in humans. Insulin is synthesised initially as a precursor molecule which consists of a single chain of amino acids containing two disulphide bonds. This molecule is cleaved by a peptidase in the Golgi complex to form the insulin molecule itself, and the remaining connecting peptide, or C-peptide (Figure 14.4). The two components are then stored together in secretory granules which are abundant in the cell cytoplasm. The principal stimulus for the synthesis and secretion of insulin is an increase in the blood concentration of glucose perfusing the β-cells. Glucose enters the β-cells via GLUT2 transporters and is immediately phosphorylated to glucose-6-phosphate.

The key regulatory step in stimulating the synthesis, storage and release 'machinery' in the β-cell is probably the rate of production of glucose-6-phosphate, which will be in direct proportion to the rate of entry of glucose into the cells via GLUT2. The enzyme involved in stimulating the

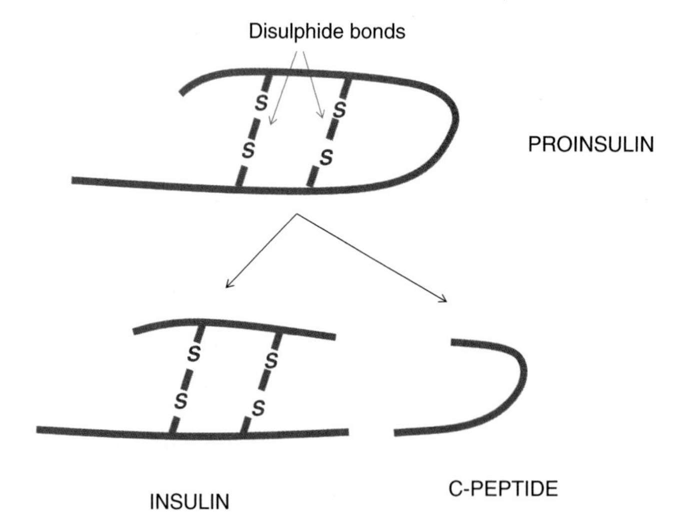

Figure 14.4 Diagram illustrating the proinsulin molecule and the insulin and C-peptide breakdown products.

production of glucose-6-phosphate is called glucokinase; the activity of this enzyme probably functions as the glucose 'sensor'. As glucose is further metabolised along the glycolysis pathway to form pyruvate, two ATP molecules are generated providing energy to the cell. Insulin synthesis is initiated, the insulin then being packaged and stored in secretory granules in the cell cytoplasm. Further ATP molecules are produced when the pyruvate enters the Krebs cycle. When the cell is stimulated, pre-synthesised packaged insulin is released from the granules. The release mechanism involves ATP which closes potassium channels in the cell membrane. The transient build-up of K^+ in the cell causes the membrane to depolarise, and this activates voltage-dependent calcium channels which open. The influx of calcium ions down its concentration gradient is associated with the movement of granules towards the cell membrane, the fusion of the granule membrane with the plasma membrane and the release of granule contents into the bloodstream by exocytosis (Figure 14.5).

There is a continual basal release of insulin, with a stimulus-directed additional release superimposed upon it. When the β-cells are stimulated and the stimulus is maintained, there is a biphasic pattern of insulin release. First there is an initial sharp rise in insulin release over a period of 10 minutes, followed by a second release of insulin which reaches a plateau approximately 2–3 hours later. It is likely that the initial phase is due to the release of pre-synthesised rapidly available insulin in granules close to

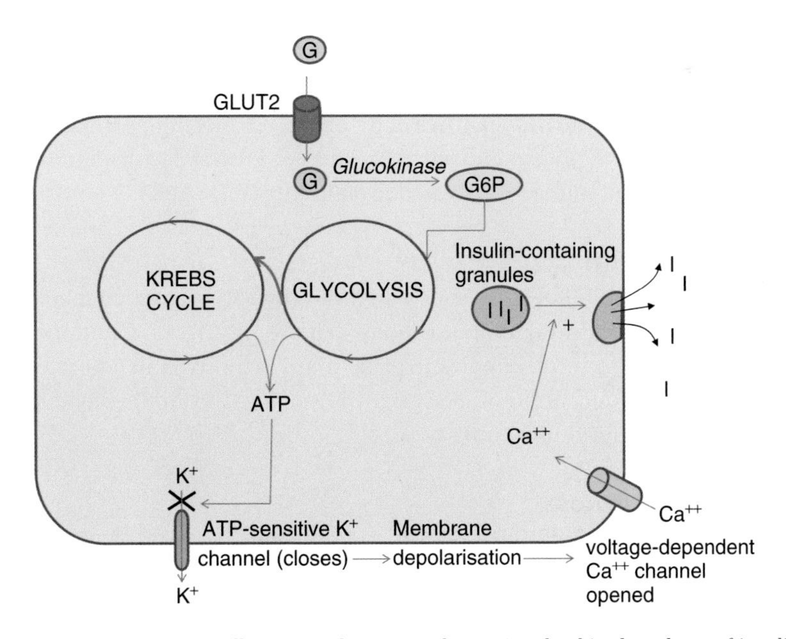

Figure 14.5 Diagram illustrating the principal steps involved in the release of insulin from a β-cell in response to an increased movement of glucose into the cell.

the membrane, and the second phase is due to the movement of granules to the cell membranes followed by the release of insulin, together with the synthesis (and incorporation into granules) of newly synthesised insulin. Once in the bloodstream, insulin circulates generally unbound to plasma proteins.

Insulin receptors and mechanism of action

Insulin receptors belong to the receptor tyrosine kinase family, each one having an extracellular domain to which the ligand binds, a transmembrane domain and an intracellular domain that includes the tyrosine kinase catalytic site. This family includes the receptors for insulin-like growth factors type 1 and type 2 (IGF1 and IGF2 respectively) with which they share certain properties including similarities in the binding sites, allowing them to bind to each other's receptors with varying affinities.

Unlike many of the tyrosine kinase receptors which exist as monomers until the binding of a ligand induces dimerisation, the insulin receptor exists as a dimer even in the unstimulated state. It is believed that activation of the tyrosine kinases on the dimerised receptor is due to the auto- and cross-phosphorylation of tyrosines in the intracellular domain. The subsequent conformational change in the receptor occurs only when the ligand links to its binding site. The activated tyrosine kinases then phosphorylate intracellular and nearby plasma membrane proteins which then mediate the actions of insulin within the target cell (Figure 14.6). Because insulin has numerous effects in different tissues, it is not surprising that the intracellular second messenger pathways are equally diverse. Intracellular molecules called insulin receptor substrates (IRS, of which there are at least four), mediate some of these pathways by acting as docking proteins for other proteins such as phosphoinositol (PI) kinases, growth factor–binding proteins (GFBP) and mitogen-activated protein kinase (MAPK).

Actions of insulin

The actions of insulin are diverse, and it influences many metabolic processes in tissues throughout the body (Figure 14.7). As mentioned earlier, probably the most important of its many actions is its effect on the blood glucose concentration, essentially removing it from the blood by stimulating its storage and/or utilisation.

Effects on blood glucose

Insulin has a number of effects which promote the removal of glucose from the blood.

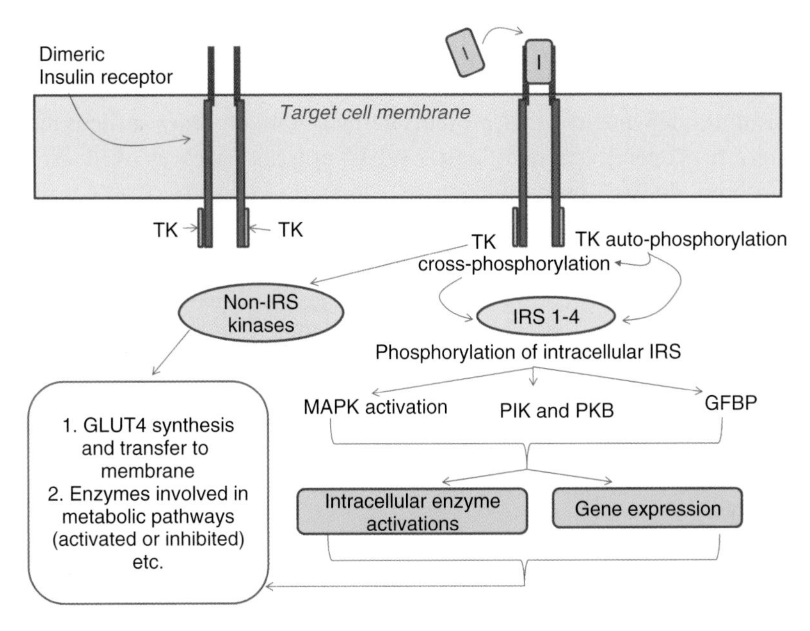

Figure 14.6 Diagram illustrating the intracellular pathways activated by insulin when it binds to its dimeric receptor. IRS = insulin receptor substrates, GFBP = growth factor binding proteins, MAPK = mitogen activated protein kinase, PKB = protein kinase B, PIK = phosphoinositol kinase.

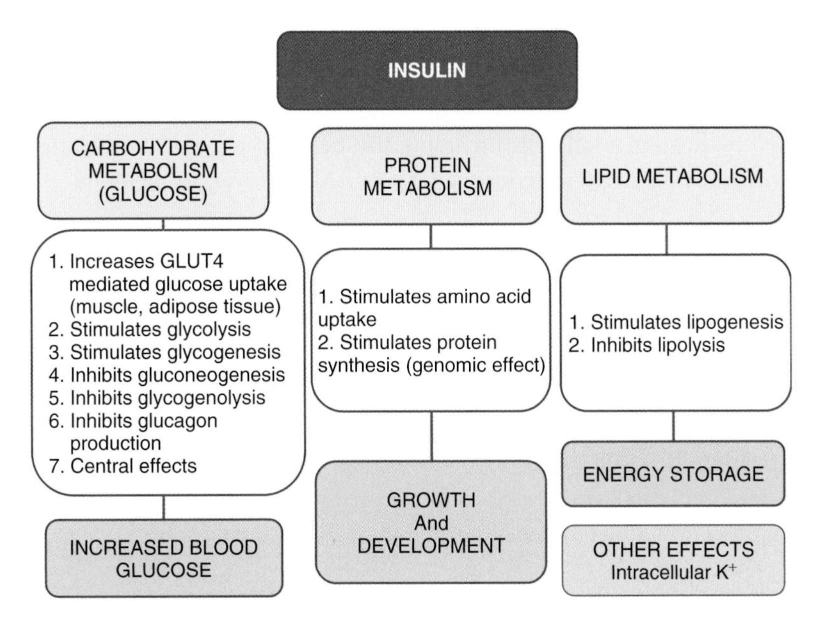

Figure 14.7 Diagram illustrating the key effects of insulin on metabolism.

Glucose transport into skeletal and cardiac muscle and adipose tissue

In many tissues throughout the body, glucose enters the cells by means of insulin-independent GLUT proteins in the cell membranes, as described earlier. In skeletal (striated) and cardiac muscle fibres as well as in adipocytes, glucose enters the cells to a limited extent via GLUT1 transporters, but this is insufficient as a means of providing sufficient glucose for anything more than basal activity. Provision of additional glucose (e.g. as required by actively contracting muscle) occurs by means of the insulin-dependent GLUT4 transporters. Insulin not only stimulates the recycling of GLUT4 transporters from the intracellular cytoplasm to the plasma membranes, but also induces GLUT4 synthesis. This means that insulin regulates this important entry pathway for glucose.

Glycolysis

Insulin stimulates intracellular glycolysis by stimulating the transcription of the three enzymes regulating key stages in glucose metabolism by this pathway: glucokinase, phosphofructokinase and pyruvate kinase. All cells are stimulated thus by insulin, not just those of muscle and adipose tissue.

Glycogenesis

Once glucose has entered a cell, it is immediately phosphorylated to glucose-6-phosphate and then either utilised immediately by the glycolytic pathway, or stored in the form of glycogen. The conversion of glucose-6-phosphate to the complex branched polysaccharide glycogen (via the process called glycogenesis) is promoted by insulin. Glycogenesis is stimulated by insulin in all cells including the hepatocytes, the liver being a major store of glycogen (hence readily available glucose) in the body.

Glycogenolysis

The intracellular store of glucose, in the form of glycogen, becomes available for glycolysis following the breakdown of glycogen by the process called glycogenolysis, whenever it is stimulated appropriately (e.g. by hormones such as adrenaline or glucagon). The hepatic store of glucose in the form of glycogen has a greater role to play, however, because when glycogenolysis is stimulated in the liver, the glucose is released into the general circulation and becomes available for use by other cells in the body. Insulin inhibits the enzymes involved in glycogenolysis.

Gluconeogenesis

Insulin inhibits the synthesis of glucose from non-carbohydrate precursors such as certain amino acids or glycerol (gluconeogenesis), which normally takes place particularly in the liver. Hence insulin prevents any increase in the blood glucose concentration due to this process.

Inhibition of glucagon production

Insulin also has an indirect effect on glucose production by inhibiting the release of glucagon. This is likely to be due mainly to a paracrine effect. When insulin levels are low or absent, as in diabetes mellitus, one might expect the raised glucose levels to have a major inhibitory effect on glucagon levels, but surprisingly levels are higher than might be expected; one explanation could be that this is due to the loss of the inhibitory effect normally exerted by insulin on the α-cells.

Protein metabolism

Insulin can be described as an anabolic hormone because it promotes the synthesis of protein. It has effects on both amino acid uptake and protein synthesis.

Amino acid uptake

Various specific transporters capable of assisting the passage of amino acids across cell membranes have been described, and some of them are stimulated by insulin. For example, the cationic amino acid transporter (CAT1) for the amino acid arginine, and the transporter for alanine, are both directly stimulated by insulin.

Protein synthesis

In addition to the likely stimulation of protein synthesis through its action on amino acid transporters in the cell membrane, insulin induces new protein synthesis by an indirect genomic effect. Thus insulin contributes to normal body growth and development. Indeed, the relevance of insulin in stimulating protein anabolism can be seen in the increased growth present in new-born babies born to untreated diabetic mothers. The explanation is at least partly because the mother's raised blood glucose concentration is associated with an increased placental transfer of glucose to the fetus. This increase is detected by the fetal islets of Langerhans which respond by increasing insulin release, the hormone then exerting its normal effects resulting in the stimulation of glucose storage – and protein synthesis (i.e. growth).

Lipid metabolism

Another metabolic pathway through which glucose can be stored is by its conversion to lipid in adipose tissue, a process called lipogenesis. When glucose enters the glycolysis pathway, the end product (pyruvate) can be converted to acetyl co-enzyme A (acetyl CoA) in the cytoplasm which is then converted to fatty acids by a complex of enzymes collectively known as fatty acid synthase. Three free fatty acids are esterified with one molecule of glycerol to form a triglyceride which is then transported, and stored, as fat. Insulin stimulates the synthesis of fatty acids and the formation of triglycerides which are deposited to some extent in a variety of tissues but particularly in adipose. Insulin also inhibits the opposite process of lipolysis which is the breakdown of triglyceride back to its constituent fatty acids and glycerol. Thus insulin once again acts as a storage hormone for glucose.

Ketogenesis

The molecule acetyl co-A can also be converted to acetoacetyl co-A in adipocytes, particularly when lipolysis is encouraged in the absence of insulin. The acetoacetyl co-A is then converted to hydroxyl-β-methylglutaryl co-A which in turn is converted to acetoacetic acid. This molecule and its two metabolites, the gaseous acetone and β hydroxybutyric acid, are called the ketone bodies. Normally, in the presence of glucose, insulin depresses the formation of ketone bodies, but when the hormone is minimally present, or absent, as in type 1 diabetes mellitus, their concentration in the blood rises. These molecules are acidic, and cell function is impaired as the pH falls.

Acetone has a characteristic smell, being described as that of pear drops or nail varnish. The excretion of acetone on the breath of a comatose person brought in to an accident and emergency department is an indicator of type 1 diabetes mellitus, and not of excessive alcohol intake!

Central actions

Various neurones in the brain, particularly in the hypothalamus, are very sensitive to changes in glucose concentration in their immediate environment. Discrete populations of these neurones are inhibited, and others stimulated, by increases in the glucose concentration. Many of these neurones express neuropeptides such as orexin, melanin-concentrating hormone (MCH) and neuropeptide Y, which are believed to be part of the central circuitry involved in the regulation of energy balance by influencing food intake. Insulin receptors are also present throughout the CNS, particularly in the hypothalamus where they are located for example in neurones of the arcuate nucleus. This part of the brain is associated with the regulation of eating and satiety. It is likely that insulin, the circulating level of which increases after a meal, is linked to the neural circuitry which indicates satiety and a decreased feeding behaviour under normal circumstances. The central regulation of food intake is discussed in more detail in Chapter 16.

Intracellular potassium

Insulin maintains the intracellular K^+ concentration, an action which becomes important in insulin deficiency states (i.e. diabetes mellitus). In the absence of insulin, K^+ moves out of cells into the extracellular fluid. Because of the glucose-induced osmotic diuresis, there will also be an increased K^+ loss in the urine which results in the plasma K^+ concentration appearing normal if measured. However, a potentially dangerous hypokalaemia can develop when insulin replacement is given, since there is a rapid increased movement of K^+ back into the cells when the body's total K^+ content is reduced.

Control of insulin release

There are two secretion levels at which insulin is released: (i) a basal secretion, which is necessary since some cells are dependent on the insulin-dependent GLUT4 transporters in order to receive sufficient intracellular glucose, and (ii) an enhanced release when blood glucose levels rise (e.g. after a meal; this is called post-prandial), which itself is biphasic with an initial increase lasting about 10 minutes followed by a second more prolonged rise which reaches a plateau 2–3 hours later. Mechanisms involved in the release of insulin are believed to include intracellular second messengers cyclic adenosine monophosphate (cAMP) and phospholipase C.

There are various factors which influence insulin release, many of them linked to the principal actions of the hormone (Figure 14.8).

Blood glucose concentration

The principal stimulus for insulin release is the blood glucose concentration. As the blood glucose level rises (e.g. after a meal), insulin output increases. This is the physiological basis for the glucose tolerance test, used for testing the function of the islet β-cells: a standard quantity of glucose (usually 75 g in 100 ml water) is taken orally, and blood samples are taken before drinking the liquid and again 2 hours later. The normal response would be for the initial sharp rise in blood glucose to be reversed, with normal levels being attained within 2 hours. If blood insulin concentrations were also measured, then there would be an increase initially in line with the raised blood glucose and then the level in subsequent blood samples would return to a lower (basal) value (see Figure 14.9).

Amino acids

Since insulin is a protein synthesis stimulator, it is not too surprising that certain amino acids (glycine, alanine and arginine) directly stimulate its release. The mechanism by which these amino acids are thought to act is similar to the one induced by glucose: namely, the cell membrane

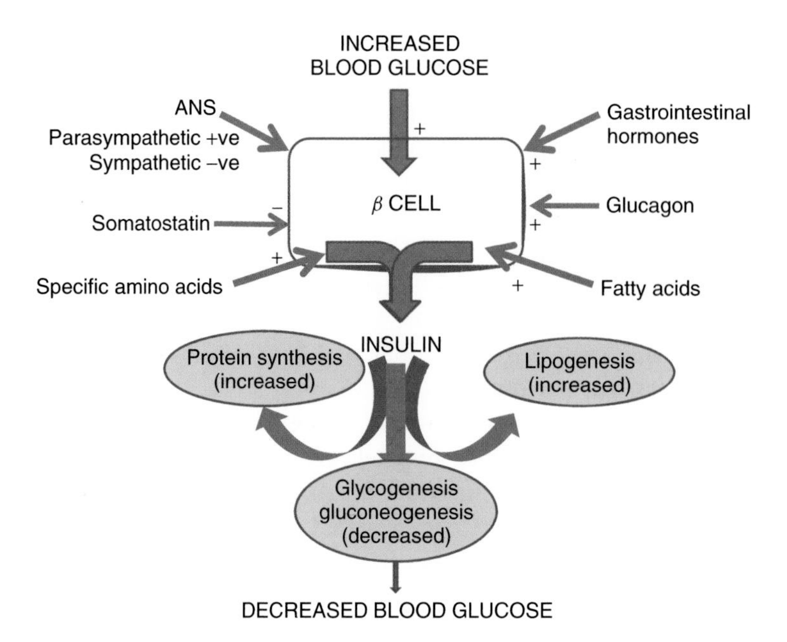

Figure 14.8 Diagram illustrating the principal factors controlling the synthesis and release of insulin from the β-cell.

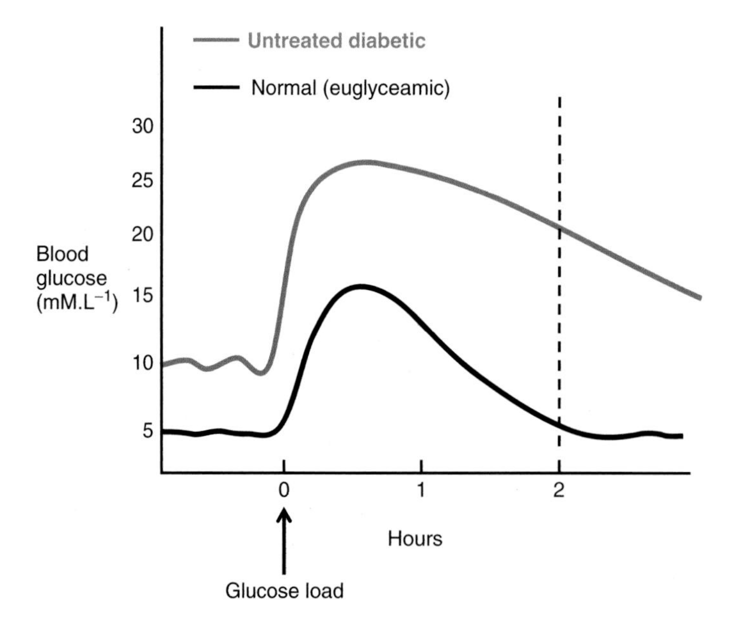

Figure 14.9 Diagram illustrating the blood glucose response to a glucose load ($75\,g.100\,ml^{-1}$) in a normal euglycaemic person and an untreated (hyperglycaemic) diabetic patient.

depolarises causing influx of Ca^{++} and subsequent exocytosis of granule contents. Arginine has its own transporter and is slightly cationic, and so causes depolarisation directly, while alanine and glycine enter the β-cells by a Na^{+}-associated transport system, the co-transported Na^{+} inducing the depolarisation.

Fatty acids

Free fatty acids (FFA) are generally potent enhancers of glucose-stimulated insulin secretion. A free fatty acid receptor (FFAR1, also known as the G protein–coupled receptor GPR40) directly mediates an increase in the intracellular Ca^{++} concentration resulting in granule content exocytosis. In addition, other second messenger systems are induced as a consequence of intracellular FFA metabolism. These include the synthesis of the second messenger molecule diacylglycerol (DAG) from membrane lipids, and long-chain acyl-CoA (LC-CoA). Since the β-cells contain lipoprotein lipase, they are capable of utilising plasma triglycerides in their immediate vicinity, as well as any intracellular stores they may contain.

Gastrointestinal hormones

A given bolus of glucose taken orally will have a greater stimulatory effect on insulin release than the same amount administered intravenously. This observation, together with other findings, points to an amplification of the normal stimulatory effect of orally administered glucose mediated by the gastrointestinal tract, known as the incretin effect. It also acts as an early warning system, informing the pancreas in advance of the nutrient absorption influx. Indeed, various molecules originally described as gastrointestinal hormones are now known to play a crucial role in regulating metabolism, by actions not only on the gastrointestinal tract influencing absorption processes, but also on the pancreas and the liver, as well as on the central (hypothalamic) regulation of food intake, hunger and satiety.

Gastrointestinal hormones which stimulate insulin production and β-cell mass (e.g. by cell proliferation) include the well-established hormones gastrin and cholecystokinin which are primarily involved in the regulation of digestion processes (acid secretion in the stomach and enzyme secretion from the exocrine pancreas respectively), as well as glucagon-like peptide-1 (GLP1) and gastric inhibitory polypeptide (GIP, also known as glucose-dependent insulinotropic polypeptide). These hormones are known as incretins, these being gastrointestinal hormones which stimulate insulin secretion. Incretins are more fully discussed on page 339 (Chapter 16).

Autonomic nervous system

The autonomic nervous system has an important role to play in the direct and indirect regulation of metabolism. For instance, sympathetic activation is associated with a catecholaminergic stimulation of glycogenolysis resulting in an increase in the circulating glucose concentration (see Chapter 11) as part of the acute stress ('fight-or-flight') response. One indirect mechanism is to influence the release of hormones from the islets of Langerhans, particularly insulin and glucagon. Animal studies demonstrate that the islets receive a dual innervation from the autonomic nervous system, with individual islet cells (α, β and γ) receiving direct innervations. However, it now appears that in humans most of the axons terminate on the smooth muscle cells of the pancreatic vasculature, with the sympathetic neurones being predominant. Nevertheless, sympathetic stimulation is associated with an inhibition of insulin secretion while parasympathetic (vagal) activity stimulates its release, but it would now appear that these effects are indirect and involve the islet vasculature. Thus, control over local blood flow by vasoconstrictor or vasodilator effects may indirectly influence islet cell activity. What is clear is that increased sympathetic and parasympathetic activities are associated with decreased and increased insulin release, respectively.

Islet cell interactions

As indicated earlier, the β-cells in each islet are generally more central than the α-cells which tend to be more peripherally located, while the δ-cells are scarce and more randomly distributed. Adjacent cells can be seen to make contact with each other, so it is generally believed that paracrine influences between them exist. Furthermore, since blood flow through each islet is from periphery to core, stimulation of the outer α-cells results in the release of glucagon and GLP1, both of which can have a more localised influence on the more central β-cells and insulin release. Indeed, glucagon appears to have a stimulatory effect on insulin release, while somatostatin from the δ-cells is inhibitory. The effect of glucagon on insulin release may at first appear to be paradoxical since glucagon actions normally oppose those of insulin on glucose, by stimulating its production and raising its concentration in the blood (see section 'Actions of glucagon'). However, if the purpose of raising the blood glucose concentration is to get more of it to the cells, then of course many cells (muscle and adipose) require insulin in order to promote GLUT4-mediated glucose uptake into those cells. The role of the inhibitor somatostatin is less clear, but it has a similar effect on glucagon release, and so acts as a generalised break on islet cell hormone production. It is relevant to note that somatostatin is also a gastrointestinal hormone, so its release from this source indicates a wider role in the overall conversation which clearly takes place between the gastrointestinal tract and the endocrine pancreas.

Recent evidence suggests that interleukin-6 (IL6), which is produced particularly by insulin-sensitive adipocytes, stimulates the release of GLP1 from the pancreatic α-cells (and intestinal L-cells). This would then stimulate insulin release. Obesity is associated with raised IL6 production which may increase tissue resistance to insulin. It is likely that the enhanced GLP1 leads to an increased insulin production and an increased sensitivity to insulin, providing a compensatory mechanism in response to the obesity-associated decrease in tissue sensitivity to insulin. Exercise also is associated with increased circulating IL6 levels, and the enhanced insulin production in this situation is also likely to be at least partly due to the enhanced GLP1 production.

The α (alpha) cells and glucagon

Synthesis, storage, release and transport

Glucagon is synthesised in the α-cells of the islets of Langerhans, initially as a pre-prohormone (pre-proglucagon) which becomes the proglucagon precursor molecule once it has lost its signal peptide upon entering the Golgi complex. Here it is further processed to the 29–amino acid glucagon molecule. The human glucagon gene is located on chromosome 2. Proglucagon can be differentially processed in other cells in the body such as in specific intestinal cells (L-cells), where the processing produces alternative molecules such as glicentin, oxyntomodulin, GLP1 and glucagon-like peptide-2 (GLP2).

The glucagon molecule once synthesised is stored within granules in the α-cell cytoplasm. It is released following stimulation of the α-cells by the process of Ca^{++}-mediated exocytosis. Once in the circulation, it remains free (i.e. it does not bind to plasma proteins) and has a half-life of approximately 5 minutes. It is metabolised mainly in the liver.

Glucagon receptors and mechanism of action

Glucagon binds to specific glucagon receptors, members of the G protein–coupled family of receptors, found on the plasma membranes of its target cells (chiefly hepatocytes and adipocytes). Its mechanism of action is to activate adenyl cyclase resulting in the formation of the second messenger cyclic AMP and protein kinase A phosphorylation. Subsequent intracellular pathway activation results in genomic and post-translational effects.

Actions of glucagon

Glucagon has effects on glucose metabolism which culminate in the raising of the blood glucose concentration; it is therefore described as having an opposing effect to that of insulin. Its main target tissue is the liver, where it stimulates glycogenolysis and gluconeogenesis. As mentioned elsewhere, glucagon also stimulates β-cells by a paracrine action.

Glucagon stimulates the breakdown of triglycerides to fatty acids in adipose tissue. The accompanying increase in glycerol production then feeds into the gluconeogenic pathway in the liver, providing a readily available source of glucose for other cells in the body.

Control of glucagon release

Glucagon release, like that of insulin, is very much related to changes in the blood glucose concentration. The stimulus for glucagon release is a fall in the blood glucose level, such as occurs during fasting. Interestingly, as with insulin, factors such as certain gastrointestinal hormones also have effects on glucagon release. Indeed, the same gastrointestinal hormones that enhance insulin release (e.g. cholecystokinin) also stimulate glucagon release (see Figure 14.10). The autonomic nervous system influences its release too, with increased sympathetic activity being associated with stimulated glucagon release (opposing the inhibitory influence on insulin release). Insulin itself has an inhibitory effect on glucagon release probably at least partly by a paracrine effect, and this probably explains why the glucagon levels in untreated diabetic patients tend to be higher than would be expected for the given raised blood glucose concentration; the inhibitory effect of insulin on glucagon release is reduced or lost. Somatostatin from the δ-cells also has an inhibitory effect on glucagon release, similar to its action on insulin release, as mentioned earlier.

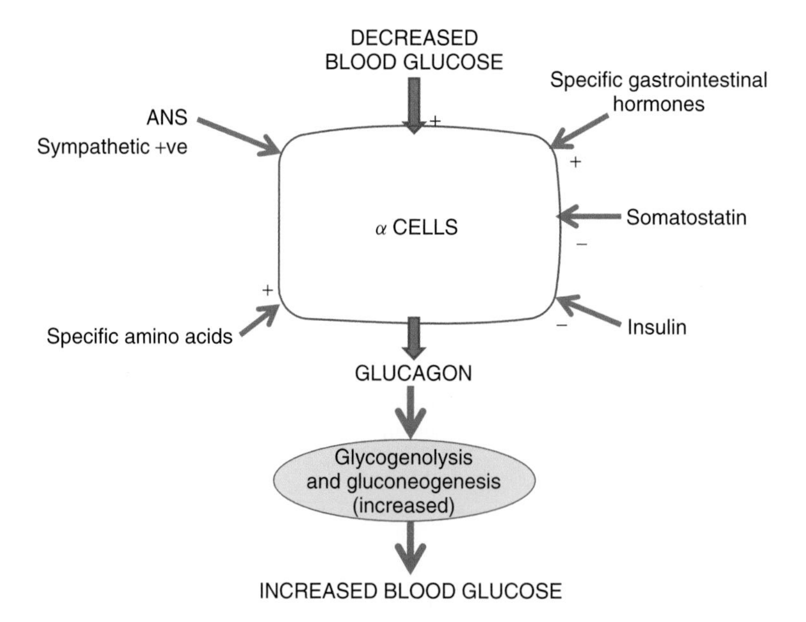

Figure 14.10 Diagram illustrating the principal factors controlling the synthesis and release of glucagon from the α-cells.

The δ (delta) cells and somatostatin

The δ-cells synthesise the 14–amino acid polypeptide somatostatin from its precursor prosomatostatin molecule. Somatostatin is quite ubiquitous, as it is found in the hypothalamus, the intestinal tract as well as the pancreatic islets, among others. Wherever it is found, it appears to be an inhibitory molecule. In the pancreas, it inhibits both insulin and glucagon production and thus appears to act as a general brake on islet cell activity. It is probable that these activities are exerted mainly as paracrine effects.

Little is known about the precise control of pancreatic somatostatin production.

The γ (gamma) cells and pancreatic polypeptide (PP)

Pancreatic polypeptide is a member of a family of hormones which includes neuropeptide Y (NPY) and polypeptide YY (PYY), all sharing a similar structure. It is a 36–amino acid polypeptide synthesised in gamma (γ) cells of the gastrointestinal tract and hypothalamus as well as the pancreatic islets. Its physiological importance is unclear, but in addition to effects on inhibiting pancreatic exocrine secretions, gall bladder contraction and intestinal motility, it stimulates glycogenolysis, and decreases fatty acid levels. It may also have a role in regulating food intake and energy metabolism by exerting effects in the hypothalamus.

Clinical conditions associated with glucose metabolism disturbances

The plasma glucose concentration is normally regulated by the various hormones mentioned in this chapter to be 4–6 mM. Glucose, being an essential energy substrate, will be associated with various detrimental effects including particularly a loss in mental function, if a decrease in its blood concentration (a hypoglycaemia) occurs. If the concentration falls below 2 mM, then a loss of consciousness, coma and ultimately death can ensue if the decrease is prolonged. This can be caused by a tumour that secretes insulin (an insulinoma) more fully considered on page 343 (Chapter 17). Likewise, an excess of glucose in the blood is also associated with detrimental effects which give rise to the disorder diabetes mellitus. Some of these effects are acutely associated with the raised blood glucose concentration (hyperglycaemia), while others are associated with the long-term complications which develop if the blood glucose concentration is not adequately regulated, for instance as a consequence of the glycation of proteins in tissues such as neurones, renal nephrons and the retina.

Diabetes mellitus (DM) is the most common of all endocrine diseases, and its prevalence is such that approximately 2% of the population worldwide will develop one of the two forms of the disease, called type 1 and type 2 diabetes mellitus. Type 1 DM is due to the absence of insulin, and is by far the less common of the two types. Type 2 DM accounts for the majority of all other cases, and most of these will be associated with the loss of sensitivity by the responsive tissues to insulin. It is also possible to develop DM as a secondary consequence of excessive glucagon, adrenaline, glucocorticoids and growth hormone, all these situations usually resulting from adenomas of the appropriate endocrine organs producing uncontrolled quantities of these glucogenic hormones. Diabetes mellitus is considered in Chapter 15.

The name diabetes mellitus is derived from ancient Greek, and it actually means large volumes (*diabetes,* meaning a syphon) of sweet-tasting urine (*mellitus* meaning honey), which is precisely what occurs with the disease. The plasma is normally filtered at the glomeruli of the renal nephrons, and all the glucose enters the luminal fluid, to be reabsorbed along the proximal tubules by a carrier-mediated transport system which co-transports the glucose with sodium ions. The sodium ions move down their concentration gradient across the luminal membranes into the cells, providing the driving force for the inward movement of glucose. The sodium ions are then pumped out of the cells into the blood in exchange for potassium ions by an active Na^+–K^+–ATPase in the serosal membranes, maintaining the low intracellular Na^+ concentration which provides the necessary gradient across the luminal membranes. The glucose is transported out of the proximal tubular cells into the blood via GLUT (1 and 2) transporters. In DM, because there is so much glucose in the blood, when it reaches the proximal tubules the transport maximum of the transfer system is surpassed. The excess glucose passes along the renal nephron taking water with it because of the osmotic effect it exerts (see Figure 14.11). Since it cannot be further reabsorbed anywhere else along the nephron, glucose and the accompanying water are excreted in the urine. Consequently, the urine volume is larger than normal, and it tastes of glucose (i.e. sweet).

The excretion of excess water in the urine (polyuria) results in the increased concentration of osmotically active particles in the blood (essentially sodium and chloride ions). The consequent increase in plasma osmolality stimulates osmoreceptors in the brain which in turn stimulate vasopressin neurones and the thirst centre. While the released vasopressin will stimulate water reabsorption along the collecting ducts in a futile attempt to counteract the osmotically driven loss of water in the luminal fluid, stimulation of the thirst centre will normally result in an increased fluid intake (polydipsia). Interestingly, the main presenting symptom of diabetes mellitus is actually the awareness of thirst, rather than the increased excretion of large volumes of urine.

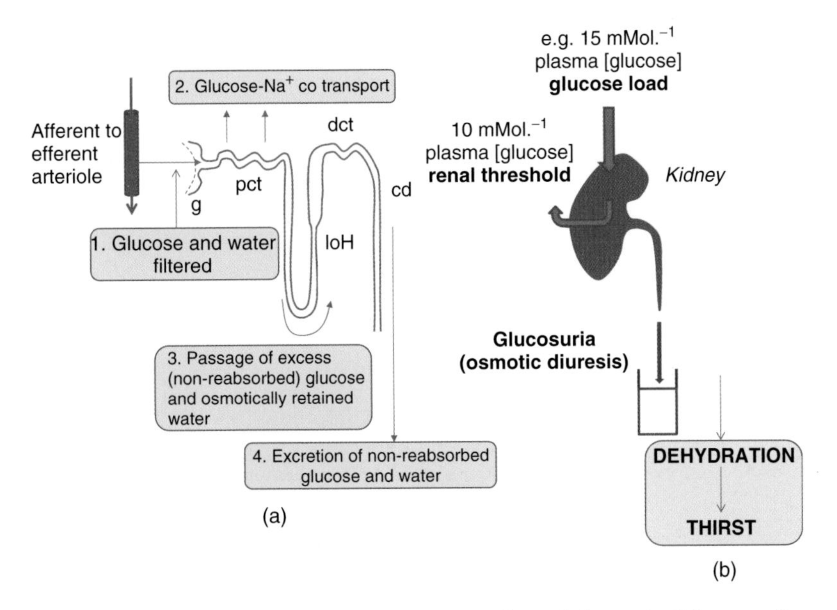

Figure 14.11 Diagram illustrating (a) the renal handling of glucose in diabetes mellitus, and (b) the consequences of glucose excretion on fluid balance once the glucose load exceeds the renal threshold for glucose. g = glomerulus; pct = proximal convoluted tubule; loH = loop of Henle; dct = distal convoluted tubule and cd = collecting duct.

Further Reading

Marston, O.J., Hurst, P., Evans, M.L., Burdakov, D.I. & Heisler, L.K. (2011) Neuropeptide Y cells represent a distinct glucose-sensing population in the lateral hypothalamus. *Endocrinology*, 152, 1307–14.

Rodriguez-Diaz, R., Abdulreda, M.H., Formoso, A.L., Gans, I., Ricordi, C., Berggren, P.F. & Caicedo, A. (2011) Innervation patterns of autonomic axons in the human endocrine pancreas. *Cell Metabolism*, 14, 45–54.

Sinclair, E.M. & Drucker, D.J. (2005) Proglucagon-derived peptides: mechanisms of action and therapeutic potential. *Physiology*, 20, 357–65.

CHAPTER 15
Diabetes Mellitus

Introduction

Diabetes mellitus literally means to pass too much sweet urine. The urine is large in volume, contains raised amounts of glucose and, as a result, will be sweet. Diabetes is caused by the lack of insulin action. This can be either because the secretion of insulin is reduced or completely absent (type 1 diabetes) or because of a combination of severe resistance to the action of insulin coupled with a lack of beta cell reserve (type 2 diabetes). Thus the two conditions are very different in their aetiology, pathogenesis and treatment (Table 15.1). Diabetes mellitus, particularly type 2, is by far the most common of all endocrine disorders, and is present in approximately two million people in the United Kingdom alone.

Type 1 diabetes (insulin-dependent diabetes)

This results from the autoimmune destruction of the beta cells in the islets of Langerhans. Insulin signals the fed state, so lack of insulin suggests starvation. Thus, in the absence of insulin, gluconeogenesis is switched on, glycogenesis is turned off and glycolysis also increases. Ketogenesis is also suppressed by insulin, so when insulin levels are low, ketogenesis occurs in large amounts. Ketones provide a small alternative energy source for the brain (and kidneys) when glucose provision is low as it is during starvation. The peak incidence of type 1 diabetes is between 12 and 14 years of age. The precise cause of type 1 diabetes is still unclear, but there is likely to be an interaction of genetic and environmental causes, and a viral infection as an activator of the immune system directed at the beta cell and the insulin molecule has not been ruled out.

As insulin levels decline, the body moves into the 'starving' state, so that plasma levels of glucose start to increase. As GLUT 4 is not activated, glucose is not transported into the relevant cells, so that although there is a lot of glucose in the circulation, it is not available for cellular metabolism. The increase in plasma glucose overcomes the renal threshold for glucose reabsorption, and glycosuria (glucose in the urine) results. The glucose draws water with it by osmosis, so that patients complain that they pass

Integrated Endocrinology, First Edition. John Laycock and Karim Meeran.

Table 15.1 Differences between type 1 and type 2 diabetes.

	Type 1 diabetes	Type 2 diabetes
Cause	Insulin deficient	Insulin resistant
Insulin levels and C-peptide levels	Very low	Normal or high
Peak incidence	Age 12–14	Age 60
Body Mass Index (BMI, $kg.m^{-2}$)	Usually normal (19–23)	Usually high (>25)
Onset	Sudden and acute	Gradual and insidious
Family history	Unusual	Very common
Ketones	Ketosis and acidosis prone	Ketosis does not occur
HLA association	DR3 and DR4	None
Islet cell antibodies	Present	Absent

large amounts of urine day and night. This in turn makes them very thirsty, so they can complain of polydipsia (increased drinking) as well as polyuria (passing lots of urine).

A more sinister aspect of type 1 diabetes is the enormous increase in ketogenesis. While ketones are a useful fuel when needed, if produced in excess they cause a fall in the pH, as they are weak ketoacids. The danger is that the pH can fall dangerously low, which in turn has a negative impact on the activity of a number of brain and other enzymes. This is known as diabetic ketoacidosis (DKA).

Type 2 diabetes (non-insulin-dependent diabetes)

This results from a combination of insulin resistance and a failure of the beta cells to be able to increase insulin secretion sufficiently in order to overcome the resistance. Obesity worsens the insulin resistance. Initially as the insulin resistance worsens, the beta cells overcome this and produce more insulin, so that the patient is hyperinsulinaemic and euglycaemic (normal blood glucose levels). Eventually the pancreas cannot keep up the large insulin production rate and the blood glucose concentration then begins to rise. There is, however, enough insulin to prevent ketogenesis, but not enough to prevent the hyperglycaemia. There is a gradual and insidious increase in plasma glucose levels over time, and patients present with polyuria and polydipsia without ketoacidosis. The glucose in the urine puts the patient at risk of urinary tract infections. There are many patients with type 2 diabetes who do not know that they have the condition because the onset of disease can be quite insidious. The cause of type 2 diabetes is probably a combination of genetics and the environment; surprisingly, there is a stronger heritable component than for type 1 diabetes.

Clinical presentation of patients with diabetes mellitus

Although patients with type 1 and type 2 diabetes all have a high plasma glucose level, as the diseases are so different they present in different ways. In type 1 diabetes, the complete and often sudden lack of insulin causes hyperglycaemia and the acidosis described above. Patients with type 1 diabetes who become unwell for the first time will have a short history of passing excess urine (polyuria) and experiencing constant thirst (polydipsia) and tend to lose weight rapidly. However, they become extremely unwell because of the systemic acidosis. Enzymes have an optimum pH at which they are most active, and in the presence of an acidosis many of the cellular enzymes become less active. In particular, the brain and liver increasingly function abnormally.

Patients may therefore present to a hospital Casualty department with confusion and apparent breathlessness. This is because, in response to the ketoacidosis, patients find themselves breathing deeper and faster because of the fall in pH which stimulates respiration. This is known as Kussmaul respiration and is deep and sighing in nature. If successful, the fall in the plasma arterial CO_2 level ($PACO_2$) compensates for the metabolic acidosis by inducing a respiratory alkalosis. The diagnosis can be confirmed with a combination of a high plasma glucose concentration (over 7 mM) and a systemic acidosis.

Treatment of diabetic ketoacidosis (DKA) involves rehydrating the patient who is likely to be extremely dehydrated due to the increased urine loss, as well as replacing insulin to (i) turn off the ketogenesis, and (ii) transport glucose into the cells by activation of GLUT4. Insulin is initially administered intravenously, but once the patients recover they can be taught to inject themselves with insulin regularly, which they will need to continue for life. If patients with type 1 diabetes ever stop their insulin treatment, they are at risk of episodes of diabetic ketoacidosis which may be fatal.

Clinical presentation of type 2 diabetes mellitus

The symptoms of type 2 diabetes can be difficult to ascertain. There are probably a million people with type 2 diabetes in the United Kingdom who do not know they have it, in addition to the million who do. The problem is that the onset of type 2 diabetes (unlike type 1) is quite insidious. Thus as the glucose begins to rise, patients may not notice anything at all. It is only when the plasma glucose concentration exceeds about 10 mM that the glucose rises above the renal threshold for its reabsorption, and glucose starts to be excreted in the urine. Once this occurs, patients start to notice some polyuria and nocturia (increased urinary excretion at night), but because they do not have a ketoacidosis they often do not seek out medical care, assuming that this is part of the normal aging process. As things get worse, they may develop urinary tract infections, because the high level of glucose in the urine encourages organisms to grow there.

Whenever patients drink glucose-containing drinks, they often develop a worsening hyperglycaemia. Severe hyperglycaemia can sometimes cause a hyperosmolar state (of the blood) that can be life threatening, occasionally causing coma. This is known as a hyperosmolar non-ketotic coma (HONKC). Not only is the blood glucose level high, but also because the glycosuria takes water with it, the plasma sodium concentration can rise, causing hypernatraemia which contributes to the hyperosmolar state. The very high osmolality can contribute to confusion and loss of consciousness.

The normal fasting plasma glucose (FPG) concentration is between 3.5 and 5.5 mM. The diagnosis of diabetes is based on a FPG of over 7.0 mM. When the glucose values are close to 7.0 mM it is advised to repeat the level, so that for a firm diagnosis one needs both symptoms (e.g. some polyuria) and two FPG levels of over 7.0 mM. The World Health Organisation (WHO, 2006) guidelines require the blood sample to be centrifuged and the plasma separated. Whole blood will have a slightly lower measured blood glucose because the red cells have lower levels of glucose and occupy just under half the volume of the blood.

Patients who have a FPG level of over 7.0 mM need no further investigations, as the diagnosis of diabetes is confirmed. However, if the fasting glucose is close to 7.0 mM or there are symptoms of diabetes, then a further test is required. This is because some patients with diabetes have a reasonable fasting glucose level, but when given an oral load of glucose (the glucose tolerance test (GTT)), they are unable to normalise the plasma glucose in a reasonable time. Normal individuals will clear a glucose load within 2 hours of administration. Patients who have a plasma glucose value of over 11.1 mM 2 hours after a glucose load also have diabetes by definition, regardless of the fasting glucose. Thus there are two different diagnostic features of diabetes: either a FPG of more than 7.0 mM, *or* a 2-hour value in a GTT of over 11.1 mM.

Patients who have a fasting glucose of between 6.0 and 7.0 mM are now defined as having impaired fasting glucose (in the United States, the American Diabetes Association have different guidelines and say that anyone with a FPG of between 5.5 and 7.0 mM has impaired fasting glucose).

Another measure that can be used is the degree of glycation of haemoglobin (HbA1) in the blood that is present. The WHO have more recently suggested that a single HbA1c value, in new units of 48 mmol of

Table 15.2 Values for fasting glucose, glucose 2 hours after a glucose load and HbA1c levels in normal subjects, in patients with impaired fasting glucose (IFG), impaired glucose tolerance (IGT) or frank diabetes mellitus.

Normal	IFG	IGT	Diabetes mellitus
Fasting glucose <5.5 mM	Fasting glucose 6.0–7.0		Fasting glucose >7.0 mM
2-hour glucose <7.8 mM		7.8–11.1 mM	2-hour glucose >11.1 mM
HbA1c	Normal	Normal	HbA1c >48 mmol/mol

HbA1c per mol of haemoglobin, could also be used, as when making the diagnosis of diabetes. Table 15.2 summarises these tests.

Treatment of type 1 diabetes

Once the patient has been treated for diabetic ketoacidosis and has survived, the longer term management needs to be considered. Essentially this involves regular administration of insulin by subcutaneous injection. Getting the dose correct can take some time, and depends on how much exercise the individual takes and what his or her diet involves. In fact, careful control of diet and an increase in exercise make the long-term control of diabetes more reliable.

Treatment of type 2 diabetes

Treatment for type 2 diabetes patients starts with weight reduction, improved diet and exercise. Monitoring of blood glucose can help with the management of the patient. If glucose levels stay high, then drug treatments can be useful. Drugs to increase insulin sensitivity include metformin, which is the mainstay of treatment. Other classes of drug include the sulphonylureas which stimulate insulin release, and the thiozolidinediones which increase insulin sensitivity. More recently, the incretins have been invoked to treat type 2 diabetes. Glucagon-like peptide-1 (GLP-1) and its analogues stimulate insulin release. Preventing the degradation of GLP-1 with dipeptidyl-peptidase 4 inhibitors is another possible treatment of type 2 diabetes. Finally, the use of insulin itself, which is not essential, certainly works. Thus, although this form of the disease is 'not dependent on insulin', insulin treatment can be used to good effect.

One of the most difficult long-term problems for patients with diabetes of either type is to control their blood glucose. The normal individual takes glucose control for granted. One can eat anything at any time, and as long as one has a normal functioning pancreas, the blood glucose concentration will hardly vary. The patient with diabetes, in contrast, has to be very careful not to eat excessively, and also has to adjust his or her insulin therapy (in type 1 diabetes) depending on what is about to be eaten. Getting this slightly wrong can result in an unpleasant hypoglycaemia if too large a dose of insulin is taken, and if the patient then does not have his or her normal meal the blood glucose level can fall to low levels. When the glucose falls to between 2 and 4 mM (hypoglycaemic stress), most patients develop symptoms of sympathetic overactivity. This means that they feel sweaty, and also might feel their heart beating more than usual. If they quickly eat something sweet, they can usually terminate the hypoglycaemic episode. However if the glucose continues to fall below 1.5 mM, neuroglycopaenia may occur. This means that there will not be enough glucose provision for the brain, and the patient will become unconscious. Patients who present with established hypoglycaemia can be treated with intravenous glucose or subcutaneous glucagon.

Some patients avoid unpleasant hypoglycaemia by deliberately running a higher glucose than normal. Unfortunately chronically this goes on to cause many of the long-term complications of diabetes. Even the best controlled patients with diabetes are more at risk of such complications than a normal individual.

Long-term complications of diabetes mellitus

High levels of circulating glucose result in the very slow rise in glycation of proteins. One can get an idea of how much glycation has occurred by measuring the proportion of haemoglobin that is glycated, as mentioned in section 'Clinical presentation of type 2 diabetes mellitus'. The more haemoglobin is glycated, the higher the mean glucose level in the plasma. Glycated haemoglobin is now measured in mmol of glycated haemoglobin per mole of haemoglobin A. There has recently been a move to make a single measurement of HbA1c the method of making a firm diagnosis of diabetes.

Glycation of endothelial proteins results in large- and small-vessel vascular disease, with the arteries being slowly occluded. This causes both microvascular and macrovascular complications.

Microvascular complications

The kidneys, retina and nerves depend crucially on very small vessels.

In the glomeruli of the kidneys, damage causes nephropathy. This can result in progressive renal impairment, which in many cases after several years results in patients requiring dialysis.

The nerves have very narrow blood vessels known collectively as the *vasa nervorum*. These can become occluded, and a neuropathy can then occur. Many nerves can be affected, and hence many different clinical presentations can occur. When the long sensory nerves affecting the feet are affected, a peripheral neuropathy is said to occur. Often patients cannot feel their feet, and thus may injure them but not be aware of this until later. When the autonomic nervous system is involved, an autonomic neuropathy is the result. This can present with gastric stasis, constipation and a loss of variation in the pulse rate. Sometimes random nerves can be affected, and this is known as mononeuritis multiplex.

In the eyes, the very small retinal arterioles may become ischaemic causing a retinopathy. Patients thus need regular retinal screening, to look for retinopathy, which can be seen at the back of the eye.

Background diabetic retinopathy

This presents with tiny microaneurysms at the back of the eye (red dots), blot haemorrhages and hard exudates. A typical appearance of this is given in Figure 15.1. Patients have normal vision, and will have no idea that they have a problem. The appearance of background diabetic retinopathy is a warning sign that blood glucose control overall is poor, and that if not improved more complications will occur. Patients should thus be informed that they have background retinopathy, and a real effort to improve glucose control must be made.

Pre-proliferative retinopathy

If control is not improved for some time (months to years), then areas of ischaemia will occur in the retina. This is associated with the release of

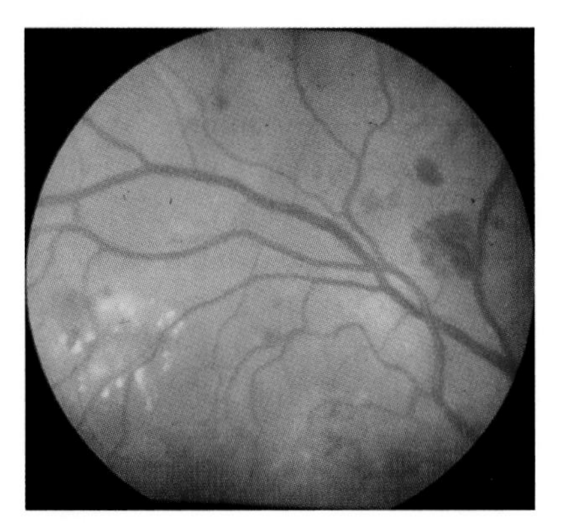

Figure 15.1 Background diabetic retinopathy with hard exudates, microaneurysms (dot haemorrhages) and blot haemorrhages. The hard exudates are due to exudates of lipid.

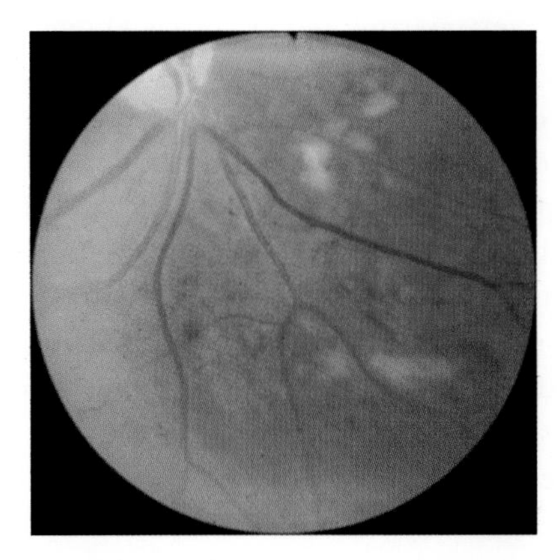

Figure 15.2 Ischaemia causes cotton wool spots. This patient also has some hypertensive retinopathy with silver wiring.

angiogenic factors which will stimulate the formation of new blood vessels to improve blood flow to the ischaemic retina. The ischaemic areas appear pale with a 'cotton wool' appearance known as cotton wool spots. A typical appearance of this is given in Figure 15.2. Unfortunately the new vessels if they appear cause even more trouble, so the aim of treatment is to reduce ischaemia by reducing the demand for oxygen. This is effectively carried out by sacrificing some peripheral retina by destroying it with a laser. Such 'pan-retinal photocoagulation' needs to be very carefully arranged, to destroy only a proportion of the peripheral retina that is not essential for normal vision.

Proliferative retinopathy

If the ischaemia is not treated, or not noticed, then new blood vessels may grow. If new vessels are observed, this is known as proliferative retinopathy. Although theoretically these new vessels may well deliver more blood to ischaemic areas of the retina, they unfortunately are extremely friable, so that minimal trauma causes bleeding into the vitreous humour of the eye. This results in instantaneous blindness when this occurs, because light cannot travel through the haemoglobin-stained vitreous humour. A typical appearance of this is given in Figure 15.3. Thus as soon as new vessels are seen, laser treatment is indicated in order to try to prevent a bleed. Clearly it would be better to prevent their development, which is why laser treatment is advocated as soon as cotton wool spots are detected. Figures 15.4 and 15.5 show the appearance of multiples laser burns.

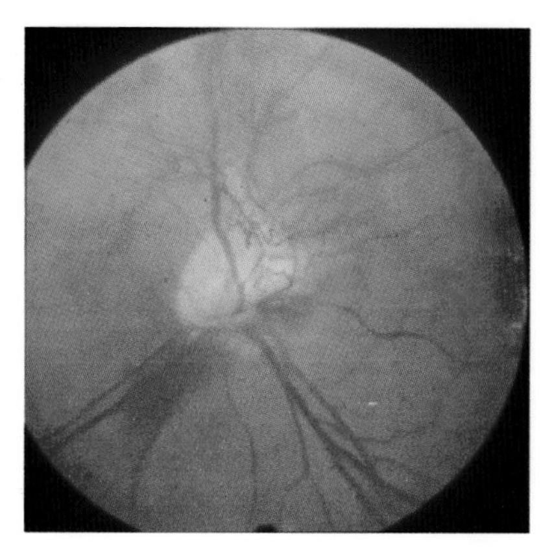

Figure 15.3 There are a large number of very fine new vessels around the disc. The patient may be unaware that these vessels are present. The danger is that these very fine vessels can easily bleed with minor trauma, and once blood has got into the vitreous humour, light cannot reach the retina and the patient becomes suddenly blind. Scarring of the vitreous will then occur, so that vision is never again normal. This is the reason why screening for retinopathy is so important. Occasionally this occurs in patients who have had undiagnosed type 2 diabetes for many years. These changes take some time to develop, so they should be prevented, but patients who do not know that they have diabetes remain at risk as they do not seek attention. It is estimated that there are one million undiagnosed patients in the United Kingdom with type 2 diabetes, in addition to the million who have had the diagnosis made.

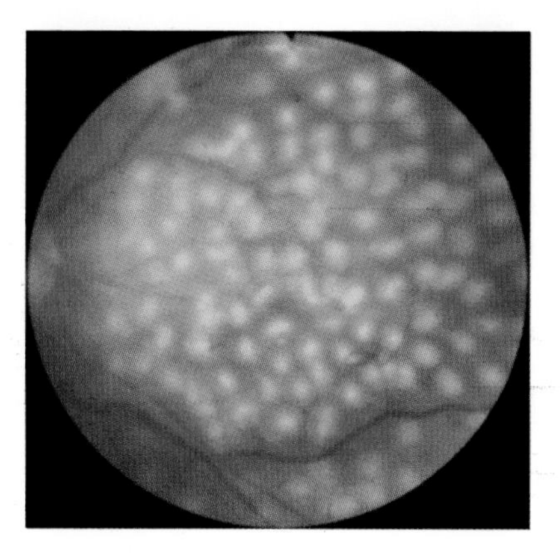

Figure 15.4 These are the multiple scars seen in patients who have had pan-retinal photocoagulation. At one edge of the photograph, the disc can be seen. The laser burns are carefully arranged to avoid damage to any vessels that can be seen in the picture. The purpose of such laser treatment is to infarct areas of the peripheral retina, so that the ischaemia disappears. By destroying about 30% of the peripheral retina, the demand for oxygen drops by a similar amount. Thus angiogenic factors (that would eventually have caused new vessels to appear as in Figure 15.3) will be reduced. Following such laser treatment, patients tend not to notice the reduction in their peripheral vision as their central vision is maintained.

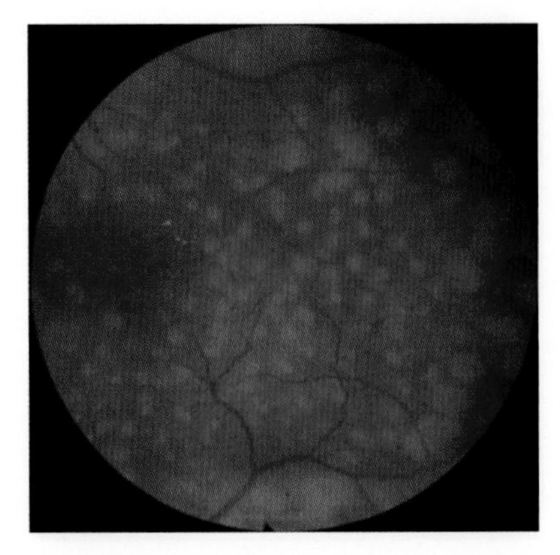

Figure 15.5 Multiple laser burns can be seen here also. In this case, the macula is in view, and the laser burns have been carefully arranged to avoid the macula, essential for central vision. There are also a few hard exudates at one edge of the macula.

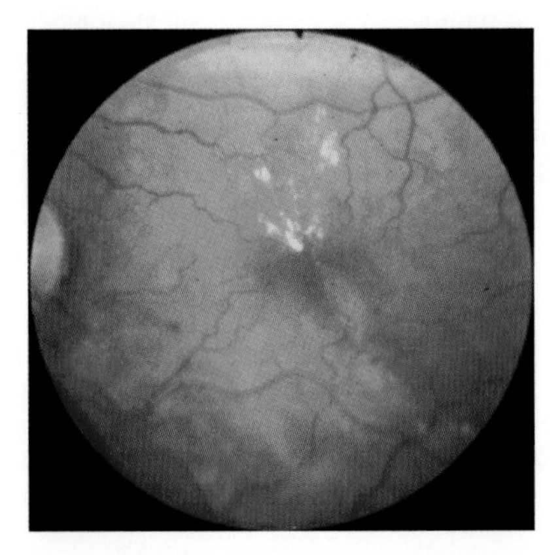

Figure 15.6 Maculopathy, as occurs in a patient who has hard exudates around the macula. These patients require a flouriscein angiogram to determine the position of any leak, followed by a small grid of laser therapy to stop the leak and preserve the macula.

Maculopathy

The appearance of hard exudates around or at the macula is known as maculopathy, and this can be seen in Figures 15.6 and 15.7. Localised laser grid therapy is indicated for patients who have maculopathy.

Macrovascular complications

Patients with diabetes are more likely to have occlusion of large arteries, and therefore are more likely to suffer from vascular disease. Thus strokes, acute myocardial infarction, and peripheral vascular disease are more

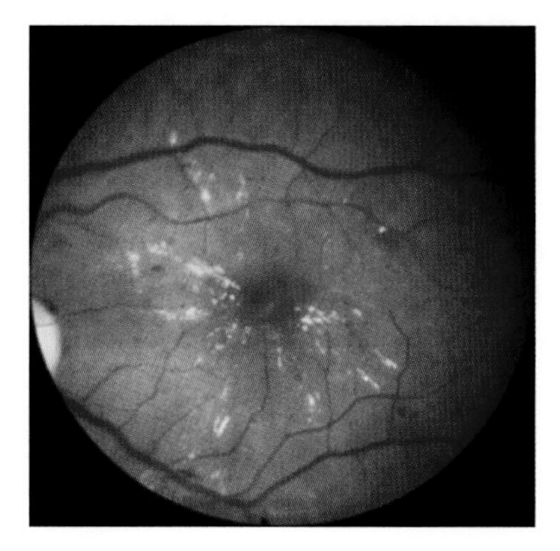

Figure 15.7 A 'macular star' where hard exudates appear around the macula, and again can threaten central vision.

common. The diabetic foot is a particular combination of reduced blood flow to the feet, often in combination with a peripheral neuropathy, making it more likely for patients to injure themselves without noticing. Damage to the skin allows organisms to enter, and infection of the skin and underlying tissue including bone can ensue. The high blood glucose concentration encourages the growth of such organisms. The poor blood flow caused by the macrovascular occlusion slows down potential healing.

Appendix: Clinical Scenario

Clinical Scenario 15.1

Mrs Al-Azzawi aged 45 visits her GP complaining of tiredness and increased weight gain. She also tells her that she is always thirsty and is drinking a lot. Her Body Mass Index (BMI) is 29.6, and a random thumb-prick blood sample gives a glucose reading of 11.6 mM. On further examination, her FPG is 7.1 mM and her glucose tolerance test (GTT) plasma glucose concentration at 3 hours is 11 mM. Her high-density lipoprotein (HDL) and low-density lipoprotein (LDL) concentrations are 4.1 and 0.9 mM respectively.

Questions

Q1. How do these various measurements compare with normal reference ranges?

Q2. What is the doctor's likely initial diagnosis?

Q3. Why is Mrs Al-Azzawi always thirsty?

Answers to Clinical Scenario 15.1

A1. How do these various measurements compare with normal reference ranges?

The BMI is at the top of the overweight range (25–30), obese being over 30 and normal weight being between 20 and 25. Her random blood glucose of 11.6 mM is just above the normal range, as is the fasting blood glucose (FBG) value of 7.1 mM. The 2-hour blood glucose after the GTT is above the normal range. The high-density lipoprotein concentration is low (normally HDL is above 1.0 mM), and the low-density lipoprotein concentration is high (normally LDL is below 3 mM).

A2. What is the doctor's likely initial diagnosis?

All these values are indicative of diabetes mellitus (DM). As she is 45, it is highly likely that it is type 2 DM.

A3. Why is Mrs Al-Azzawi always thirsty?

The thirst occurs because of the excessive urinary excretion of glucose which induces an osmotic diuresis. The plasma osmolality is raised so the thirst centre in the brain is stimulated. The increased drinking means that more water is absorbed into the body. The plasma osmolality returns to normal, at least temporarily.

A4. What treatments are recommended?

A. The obvious treatments are to reduce weight, by decreasing the intake particularly of simple carbohydrates and fats, and increasing dietary fibre which delays absorption. Increased exercise is the other obvious recommendation. If Mrs Al-Azzawi smokes or takes excessive alcohol, then a cessation of smoking and a reduction of the alcohol intake would be further recommendations.

Reference

World Health Organization (WHO) (2006) *Definition and Diagnosis of Diabetes Mellitus and Intermediate Hyperglycemia: Report of a WHO/IDF Consultation*. http://www.who.int/diabetes/publications/Definition%20and%20diagnosis%20of%20diabetes_new.pdf (accessed May 2012).

Further Reading

Misra, S., Hancock, M., Meeran, K., Dornhorst, A. & Oliver, N.S. (2011) HbA1c: an old friend in new clothes. *Lancet*, 377, 1476–77.

CHAPTER 16
The Gut–Brain Axis

Introduction

One of the most important factors in human survival is nutritional status. Evolution dictates that those animals that are well nourished are the ones that survive. Starvation has been a major cause of death over the ages, contributing to the ability to increase body weight becoming a major determinant of survival through natural selection. The drive to keep well nourished has been well balanced by a relative paucity of available energy, and a requirement to exercise in order to find or hunt for food. The sensation of hunger and resultant food-seeking behaviour are thus crucial factors in human survival.

In order to control appetite, the brain requires information about the nutritional status of an individual. After a large meal, hormonal and neural changes cause the sensation of satiety, and change behaviour from food seeking to other activities. A well-nourished individual can safely reproduce, while a starving one should not.

Such changes in behaviour are mediated by many factors, but hormones from the gastrointestinal (GI) tract are central to this role. It is likely that there remain several other such hormones, and in this chapter, some that have been discovered will be discussed.

To put the importance of the gut–brain axis in perspective, it is also necessary to appreciate that in times of plentiful food, an individual's body weight will increase, particularly if the amount of exercise does not increase in proportion. Nowadays, certainly in the Western world, this is precisely the situation: lots of affordable food and a more sedentary lifestyle have resulted in an increase in body weight. Recently quoted figures indicate that in England, for example, about 46% of men and 32% of women are overweight (BMI of 25–30 $kg.m^{-2}$), and an additional 17% of men and 21% of women are obese (a BMI of more than 30 $kg.m^{-2}$). The problem with being overweight or obese is that there is an increased morbidity and mortality associated with it, because of the increased risk of cardiovascular disease and cancer in particular. The control mechanisms which regulate behaviours such as hunger and satiety, as well as food intake and metabolism, have become key areas of research in order not

Integrated Endocrinology, First Edition. John Laycock and Karim Meeran.

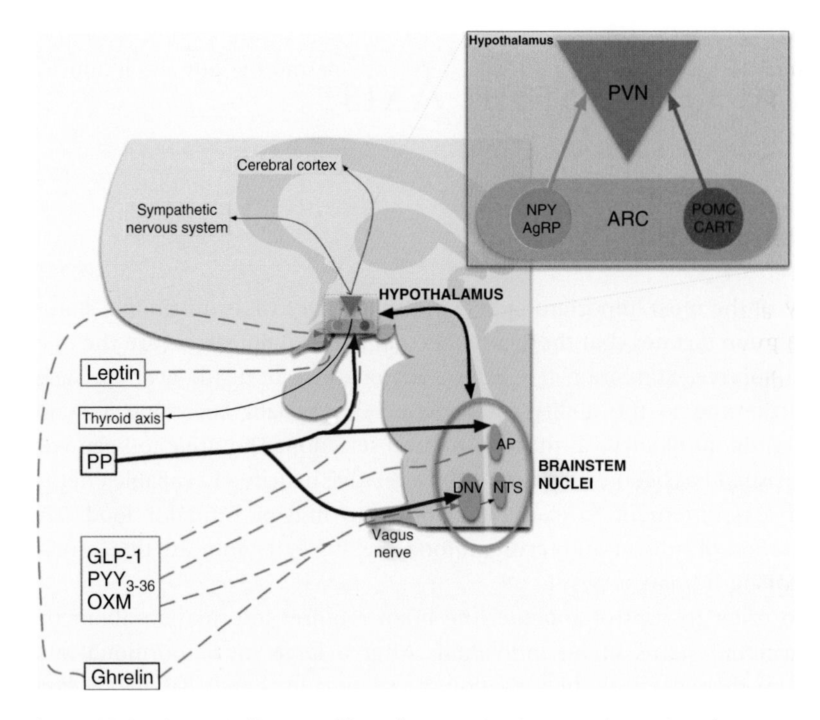

Figure 16.1 Schematic diagram of how the CNS circuits control appetite. There are reciprocal connections between the brainstem and the hypothalamus allowing both CNS centres to receive information from the periphery. Short-term satiety signals communicate meal intake, and long-term signals communicate information regarding body energy stores. This information is ultimately integrated at the level of the hypothalamus, and especially in the arcuate nucleus (ARC). AgRP = agouti-related peptide; AP = area postrema; CART = cocaine- and amphetamine-regulated transcript; DNV = dorsal nucleus of the vagus; GLP-1 = glucagon-like peptide-1; NTS = nucleus tractus solitarius; NPY = neuropeptide Y; OXM = oxyntomodulin; POMC = pro-opiomelanocorticotrophin; PP = pancreatic polypeptide and PVN = paraventricular nucleus. Note: The reader is directed to the 'References' and 'Further Reading' sections at the end of this chapter where other factors identified in the diagram are considered.

only to understand them but also to learn how to influence them in order to treat the underlying conditions of being overweight or obese (See Figure 16.1).

Neuropeptide Y (NPY)

Neuropeptide Y is a 36–amino acid peptide, so named because it is rich in the amino acid tyrosine (Y). NPY was first isolated from the GI tract of pigs by Tatemoto and colleagues in 1982. When administered into the third cerebral ventricle of a rat, it causes a remarkable stimulation of feeding. A similar peptide, peptide YY (PYY), also comes from the gut and it appears that PYY can bind to NPY receptors in the brain. PYY is released after meals and is transported across the blood–brain barrier to the arcuate nucleus in the hypothalamus. Fragments of PYY such as PYY (3–36) have been shown to bind to the NPY Y2 receptor in the brain.

NPY has a number of receptors, and determining the effects mediated by each receptor is on-going in many laboratories. Fragments of NPY and PYY are being used to try to elucidate the effects of endogenous NPY and PYY. Once determined accurately, and assuming that safe antagonists are discovered, the aim is to develop a drug that will inhibit feeding by interacting with this extremely important pathway. However, other effects of NPY in the brain, for example on behaviour, may mean that modulation of NPY needs very careful consideration.

Glucagon and glucagon-like peptide-1 (GLP-1)

Glucagon is secreted from the alpha cells of the islets of Langerhans, and has an important role in the stimulation of an increase in glucose and as a counterregulatory hormone to insulin. GLP-1 is formed by alternative splicing of the pre-proglucagon gene. It is secreted by L-cells of the small intestine, and it stimulates insulin release in response to food. It is responsible for the incretin effect which describes the influence of glucose on insulin release *by indirect means*. When glucose is given orally, there is a much bigger stimulation of insulin release than when the same amount of glucose is given intravenously. This is now known to be due to the effects of the incretin hormones, the most well described of which is GLP-1. Thus the pancreatic beta cells release insulin in response to glucose as the component of a meal, but also due to the concomitant release of GLP-1. This property of the beta cell is exploited in the treatment of type 2 diabetes by the use of GLP-1 analogues such as exenatide and liraglutide.

In addition to its role as an incretin, GLP-1 also has a role in the control of appetite. GLP-1 is a satiety ('feeling of fullness') factor. GLP-1 is thus one of the so-called gut–brain peptides, and it seems to reduce feeding. Blockade of the GLP-1 receptors in the hypothalamus increases feeding, which confirms that endogenous GLP-1 is important in the reduction in food intake, and the stimulation of satiety.

A third role of GLP-1 is that it seems to reduce gastric emptying, and thus increases transit time in the gut. Thus GLP-1 has at least three different effects when administered: (i) it increases insulin release (the incretin effect), (ii) it inhibits feeding at the level of the hypothalamus and increases satiety and (iii) it slows gastric emptying.

Early results with the use of GLP-1 analogues such as exenatide and liraglutide suggest that the use of these agents may be associated with weight loss, which would be very useful in patients with type 2 diabetes. The use of exogenous insulin in type 2 diabetes patients causes an increase in appetite and weight, so if early results are confirmed, GLP-1 analogues will form a useful long-term treatment for type 2 diabetics.

Oxyntomodulin

Oxyntomodulin inhibits acid secretion from the stomach oxyntic cells, but it also has a role in the control of appetite. It is named because of its modulating role on the activity of oxyntic cells. It is a 37–amino acid peptide, synthesised in intestinal cells by post-translational modification of proglucagon. Oxyntomodulin is secreted into the bloodstream in response to a meal, and in proportion to its calorific content. Oxyntomodulin infusion and subcutaneous injection have been shown to reduce food consumption in humans as well as other animals. Again, its potential in treating obesity and its related clinical conditions remains to be fully determined.

Leptin

Leptin is produced by adipocytes, and plasma concentrations correlate with fat mass. Leptin deficiency was first discovered in the ob/ob mouse. This mouse has been bred for many years and exhibits Mendelian inheritance. It seemed to express autosomal recessive obesity, so that when heterozygote mice are crossed, the ratio of normal to obese mice is 3:1. It was thus clear that a single gene was the cause of obesity in these animals. Sequencing of the genome in these mice resulted in the discovery of a circulating hormone which was named leptin in 1994 by Friedman's research group (Zhang *et al.*, 1994). A human homologue was discovered, and there are a very small number of humans who have a similar inherited condition with very extreme morbid obesity in children, and in whom leptin replacement has been effective. These children have a mutation in the leptin gene. However, in the majority of patients, leptin therapy does not cause such weight loss, and unsurprisingly, obesity is associated with high levels of leptin.

Leptin signals to the hypothalamus which has specific receptors for leptin. Leptin also has specific transporters to get it across the blood–brain barrier. It is a marker of the fat stores of an individual. Thus very thin people will have low levels of leptin, and this may be a marker of starvation. It appears that people with low levels of leptin may have a reduced gonadotrophin production, which in turn is associated with low fertility. Replacement of leptin results in a return to normal circulating levels of gonadotrophins, which in turn stimulate the production of sex steroids and the restoration of fertility. Therefore it appears that leptin may be a marker of nutrition, with adequate levels of leptin signalling adequate levels of nutrition and enabling reproduction. Patients who are undernourished, for example those with anorexia nervosa, are known to develop amenorrhoea due to low levels of circulating gonadotrophins. Furthermore, leptin may be one of the factors involved in stimulating the onset of puberty (see Chapter 8).

Orexins

The two orexins, orexin A and orexin B, were discovered around the turn of the millennium and are a pair of peptides that stimulate food intake when administered into the lateral hypothalamus in animals.

Orexin A appears to be the more active of the orexins, and is secreted by cells within the hypothalamus in response to metabolic changes and to other peripheral circulating gut hormones. Orexins A and B are also synthesised in the gut, but it is not clear whether circulating orexins have any role in the control of feeding (Kirchgessner, 2002).

Another possible role of the orexins is in the control of the sleeping–wakefulness cycle.

The combined effects of different gut hormones in patients with gastric bypass surgery

Gastric bypass surgery for the treatment of obesity has been remarkably successful, and it seems that the weight loss and metabolic changes that occur are not explained purely by the decreased absorptive effects of the surgery. Changes in some of the circulating gut hormones mentioned here that occur after gastric bypass surgery may be responsible for a change in appetite at the level of the hypothalamus. For example, bypass surgery results in an increase in GLP-1 and PYY production within days, and there is an associated fall in appetite. A large number of other possible gut peptides may also be responsible. Once these are discovered, and the pathways confirmed, it is hoped that therapy with gut hormones or analogues can be used to reduce weight instead of bypass surgery.

The combination of oxyntomodulin and PYY (3–36) administration has recently been shown to be additive in reducing food intake.

It certainly appears that determining the changes in circulating gut hormones and their specific effects on the brain could be an effective strategy for controlling obesity in the future.

References

Kirchgessner, A.L. (2002) Orexins in the brain-gut axis. *Endocrine Reviews*, 23, 1–15.

Tatemoto, K., Carlquist, M. & Mutt, V (1982) Neuropeptide Y: a novel brain peptide with structural similarities to peptide YY and pancreatic polypeptide. *Nature*, 296, 659–60.

Zhang, Y., Proenca, R., Maffei, M., Barone, M., Leopold, L. & Friedman, J.M. (1994) Positional cloning of the mouse *obese* gene and its human homologue. *Nature*, 372, 425–2.

Further Reading

Field, B.C., Wren, A.M., Peters, V., Baynes, K.C., Martin, N.M., Patterson, M. & Alsaraf, S. (2010) PYY3–36 and oxyntomodulin can be additive in their effect on food intake in overweight and obese humans. *Diabetes*, 59, 1635–9.

Stanley, S., Wynne, K., Mcgowan, B. & Bloom, S. (2005) Hormonal regulation of food intake. *Physiological Reviews*, 85, 1131–58.

CHAPTER 17

Hormones, Endocrine Tumours and the Gut

Introduction

The gastrointestinal (GI) tract is obviously important in the processes of digestion and absorption of nutrients. However, it also has an important endocrine role. We now know that the GI tract secretes a number of hormones that both regulate the secretion of digestive hormones and have an effect on the brain (the gut–brain axis). Neuroendocrine tumours of the GI tract can cause endocrine symptoms if they secrete excessive amounts of hormones.

Insulinomas

Tumours that secrete insulin will present with episodes of hypoglycaemia. In normal glucose homeostasis, insulin concentrations vary 10- to 100-fold between the fasting and fed state. Thus after a meal, very high insulin levels are a normal finding, and serve to drive glucose and other nutrients into cells. During fasting insulin levels are low, and it is then that the presence of a tumour continuing to make small amounts of insulin becomes significant. Patients thus have symptoms of hypoglycaemia that trouble them before meals or after an overnight fast (i.e. before breakfast).

The symptoms of hypoglycaemia include those of excessive sympathetic activation (tachycardia, palpitations and sweating) and those of neuroglycopaenia (essentially mental confusion). Patients may notice that eating resolves the effects of the sympathetic activation, and sometime put on a lot of weight as a consequence of trying to alleviate the symptoms of what are thought to be 'anxiety' effects. Insulinomas are rare (1 per million of population per year, similar to phaeochromocytomas). Patients initially have relatively non-specific symptoms and often seek help from psychiatrists and alternative practitioners until the diagnosis is recognised and the frequency of hypoglycaemia makes the diagnosis more plausible.

Once the diagnosis is suspected, patients should be referred to a specialist centre for further investigation. The main test is to provoke an episode of

Integrated Endocrinology, First Edition. John Laycock and Karim Meeran.

hypoglycaemia by fasting the patient for up to 72 hours. If hypoglycaemia does not occur in this period, the diagnosis is extremely unlikely. If hypoglycaemia is precipitated, it is important to check plasma levels of glucose, insulin and c-peptide, as well as take a simultaneous sample for a toxicology screen to ensure that there are no drugs being taken that can precipitate hypoglycaemia at that time. The diagnosis is confirmed if both insulin and c-peptide are raised at the same time as a glucose value of less than 2.2 mM. Clearly if any drugs that cause hypoglycaemia are found, then those are likely to be the cause.

Once the diagnosis is confirmed, the next stage is localisation of the tumour before arranging surgery. The tumours are usually at the tail of the pancreas but may be at the head. Imaging of the pancreas with ultrasound, computerised tomography (CT) scanning and magnetic resonance imaging (MRI) may reveal a lesion, but one cannot be certain that they are functional. More recently scanning with Gallium 68–labelled somatostatin analogues has shown up functional tumours. Angiography of the arterial supply of the pancreas may reveal a tumour blush. One of the most specific tests for localisation of an insulinoma is the calcium stimulation test, where calcium is injected into the four arteries that supply the pancreas, and systemic insulin levels are measured after each injection. When calcium is injected into the artery that supplies the insulinoma, a significant rise in plasma insulin occurs. The surgeon can then target the lesion supplied by that artery.

Patients with metastatic disease usually have liver metastases. Chemotherapy in the form of streptozotocin and 5-fluoro uracil has been used. Symptomatic control can be achieved with diazoxide, although this drug has many adverse effects. Local control of liver metastases can be achieved with a combination of hepatic arterial embolisation and surgery or radiofrequency ablation to remove the metastases.

Hepatic arterial embolisation involves an initial angiogram to determine the vascular extent of the liver metastases. These tumours and metastases are particularly vascular, which is why this technique is favoured. The normal liver has a dual blood supply, with every hepatocyte supplied by a branch of the hepatic artery as well as the hepatic portal vein. Metastases, in contrast, have only a single blood supply, from the hepatic artery, which provides angiogenic factors capable of stimulating the development of an increased local vasculature. Occlusion of the artery will thus cause selective death of the metastases, while preserving the hepatocytes which will survive on their hepatic portal vein supply. This technique has been refined to involve the injection of inert plastic microspheres into branches of the hepatic artery to selectively kill hepatic metastases.

Secretin

The first hormone ever described was secretin. In an experiment in the early 1900s, William Bayliss and Ernest Starling showed that duodenal secretion was stimulated by food even when the nerves to the duodenum were cut. This was the first proof of the existence of a circulating substance controlling the secretion of duodenal enzymes. They coined the term hormone to describe the circulating substance, and the first hormone ever described was thus secretin. Secretin increases the watery bicarbonate secretion from the pancreatic duct epithelium. Pancreatic acinar cells have secretin receptors in their plasma membrane. It promotes the growth and maintenance of the exocrine pancreas. Secretin also reduces gastrin release from normal stomach cells and hence reduces acid secretion in the stomach. Paradoxically, tumours that secrete gastrin are stimulated by secretin. Thus administration of secretin (the secretin test) can be used to make a diagnosis of a gastrinoma. In normal subjects the circulating gastrin level falls, whereas in patients with gastrinomas there is an increase in the plasma gastrin concentration.

Gastrin and gastrinomas

Gastrin is released from G cells in the antrum of the stomach in response to the ingestion of food or an increase in the pH of the stomach. It travels in the bloodstream to the parietal cells and enterochromaffin cells of the body of the stomach. The stimulated parietal cells secrete acid while the enterochromaffin cells secrete histamine which binds to H2 receptors on the parietal cells, further increasing acid secretion. Tumours that secrete gastrin (gastrinomas) will thus cause increased acid secretion, and patients with these tumours are said to have the Zollinger–Ellison syndrome. This was first described in patients having increased acid production, and ulceration of the GI tract because of the increased acid.

In patients with a raised circulating gastrin level, it is important to rule out secondary hypergastrinaemia, which may be due to lack of gastric acid, as may occur with achlorhydria, or in patients taking H2 antagonists or proton pump inhibitors. Both types of drug will suppress acid production, and the G-cells should respond normally by increasing gastrin in order to stimulate acid production. Thus the diagnosis of a gastrinoma should be made by the measurement of a combination of a low gastric pH (e.g. a pH of <2) with a simultaneously raised plasma gastrin concentration. If there is any doubt, a secretin test should be performed.

Localisation of gastrinomas

The primary gastrinoma tends to be located at the head of the pancreas or in the duodenal wall. Selective injection of secretin into the arterial supply of the pancreas will release gastrin from the primary tumour exactly as described for insulinomas in section 'Insulinomas'. Localisation of gastrinomas can also be identified by selective intra-arterial calcium injection. Thus the localisation and treatment of the primary gastrinoma are similar to those of an insulinoma. Gastrinomas are more likely to be malignant and have metastases, and they can be dealt with in the same way as metastatic insulinomas, with surgery, hepatic embolisation and chemotherapy. Somatostatin analogues seem to reduce the secretion of hormones from some of these tumours. Recently, radiolabelled treatment with lutetium (177)–labelled octreotide has shown promise in patients with metastatic disease of neuroendocrine tumours.

VIPomas

These are tumours that secrete vasoactive intestinal polypeptide (VIP). This peptide stimulates secretion of water and potassium into the colon. The massive amounts of VIP in turn cause profound and chronic watery diarrhoea and resultant dehydration, hypokalaemia, achlorhydria and acidosis (thus also known as WDHA syndrome, or pancreatic cholera syndrome). This is one of the rare causes of a hypokalaemic *acidosis,* since usually patients with hypokalaemia develop an alkalosis. Patients with VIPomas sometimes also get flushing and hypotension (vasodilatation), hypercalcaemia and hyperglycaemia. The hormone VIP is cleared by the kidneys. The dehydration can cause renal failure which in turn causes an increase in VIP levels, which in turn worsens the diarrhoea. Thus a crisis of diarrhoea and dehydration can occur. This was first described by John Verner and Ashton Morrison in 1958, and the condition was known as the Verner–Morrison syndrome. Treatment of a VIPoma involves aggressive rehydration and potassium replacement to improve renal function. If the pancreatic tumour can be found, surgery should be planned. VIPomas also respond to somatostatin analogues and chemotherapy. Metastatic disease to the liver can be treated as for gastrinomas (see section 'Gastrin and gastrinomas').

Glucagonoma

Glucagonomas are tumours that originate from the alpha cells of the islets of Langerhans. They are very rare, with an annual incidence of about 1 in 20 million of the population. This means that in the United Kingdom, there will be only three new cases per year. Apart from increasing plasma glucose levels, there are a number of unusual features that such patients express:

1. Hypercoagulability and increased risk of deep vein thrombosis, which occurs in about 10% of cases and may be responsible for up to 50% of deaths.
2. A necrolytic migratory erythematous rash.
3. Weight loss.
4. Stomatitis (inflammation of the mucous lining of the mouth), which may be due to multiple vitamin deficiencies.

Over 80% of patients have metastases at the time of presentation, with a majority of the primary tumours in the pancreas and almost all metastases in the liver. Hepatic embolisation is a useful mode of treatment.

Somatostatinomas

Somatostatinomas are rare tumours with an estimated incidence of about 1 in 40 million. About 1 in 10 of them is familial, the rest being described as sporadic. They occur with a similar distribution to gastrinomas. Those in the pancreas tend to be large tumours often associated with features of somatostatin excess, and duodenal tumours which are usually smaller. They may be associated with multiple endocrine neoplasia type 1 (MEN1) syndrome, neurofibromatosis type 1 or Von Hippel–Lindau syndrome.

The most common features of a somatostatinoma are the consequences of the somatostatin inhibiting other endocrine and exocrine secretions. Thus diabetes (due to low insulin levels), steatorrhoea (due to suppression of pancreatic enzyme synthesis) and cholelithiasis (gallstones, due to inhibition of gall bladder contraction) are common phenomena. Curiously, many of these symptoms respond to lanreotide and octreotide therapy. These two drugs are somatostatin analogues. Other treatments (e.g. for metastatic insulinomas) are used if required, including hepatic embolisation and surgery.

Further Reading

Turner, J.O., Wren, A.M., Jackson, J.E., Thakker, R.V. & Meeran, K. (2002) Localization of gastrinomas by selective intra-arterial calcium injection. Clinical Endocrinology, 57 (6), 821–5.

CHAPTER 18

The Parathyroids, the Endocrine Kidney and Calcium Regulation

Introduction

Calcium ions are the third most common cations in the body, after sodium and potassium ions, and they are essential because of the many different physiological roles that they play. Calcium not only is present in the blood and other extracellular fluids, but also is found in all the cells of the body. However, most of the calcium in the body (99%) is actually stored in bone in the form of complex hydrated calcium phosphate salts called hydroxyapatite crystals, which give bone its rigidity and strength. The calcium in bone provides approximately 1 kg to the average body weight. It is also found in the enamel of teeth.

In the blood, calcium (approximate concentration 2.5 mmol.L^{-1}) is present in three forms: as free ions (approximately 1.25 mmol.L^{-1}), dynamically bound to globulin and albumin fractions of the plasma proteins (approximately 1.15 mmol.L^{-1}) and as diffusible salts such as citrate and phosphate. The free calcium ions represent the component that is biologically active and is very precisely and finely controlled (see Figure 18.1).

Free (unbound) calcium and phosphate ions have a particularly important relationship with their salts, and the formation and deposition of calcium phosphate salts in tissues such as bone are closely regulated by the control exerted on the ion levels. When the concentration of calcium (or phosphate) ions in the blood falls, the dynamic equilibrium between the free ion and salt concentrations is disturbed such that dissociation of the salt to its constituent ions occurs, restoring the equilibrium. The calcium concentration in the blood is considerably higher (10^{-3} M) than that found in the cytoplasm of cells (10^{-7} M) so the control of calcium channels is important for regulating the diffusion of calcium down its concentration gradient, from either intracellular organelles (sites of calcium storage) or extracellular fluid into the cytoplasm. Calcium ions are also transported actively (e.g. from intracellular organelles into the cytoplasm

Integrated Endocrinology, First Edition. John Laycock and Karim Meeran.

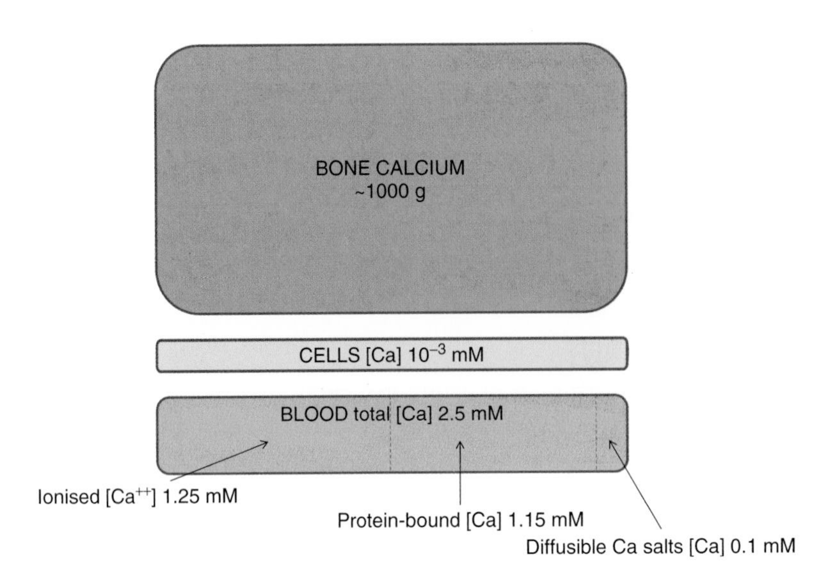

Figure 18.1 Distribution of calcium within the body.

PRINCIPAL FUNCTIONS OF CALCIUM

TISSUE	FUNCTIONS
• Bone	Strength, calcium reservoir
• Neuromuscular excitation channels	Regulation of sodium ion
• Muscle	Contraction
• Secretory (including nerve) cells (various tissues)	Exocytosis, stimulus–secretion coupling (e.g. transmitter release)
• Blood	Coagulation
• All cells	Co-enzyme, second messenger

Figure 18.2 The principal functions of calcium.

or from cytoplasm to the extracellular fluid) or by secondary transport mechanisms (e.g. sodium–calcium countertransport which is driven by the movement of sodium ions down their concentration gradient).

The ubiquitous distribution of calcium in the body is associated with the many functions of calcium (see Figure 18.2). For example, in bone, calcium in the form of hydroxyapatite provides rigidity and strength, and also acts as a large reservoir which can be broken down (resorbed) whenever the free ion concentration is reduced in the blood. In cells, calcium is stored in intracellular compartments such as the mitochondria, microsomes and endoplasmic reticulum (sarcoplasmic reticulum in skeletal muscle) providing rapidly accessible stores of the ion available for release into the cytoplasm upon appropriate stimulation. In the cytoplasm, calcium ions have many roles: they act as second messengers associated with the stimulation of intracellular enzymes, they act as co-enzymes, they are involved in the stimulus–secretion coupling which brings about the migration of intracellular granules towards the cell plasma membranes and the exocytosis of granule contents, and they bind to troponin or calmodulin proteins in skeletal and smooth (including cardiac) muscle respectively, allowing the formation of the actin–myosin sliding filaments which bring about contraction. In the blood, calcium ions also play a vital role, the main one being the influence they have on sodium channels in cell membranes, particularly in neuromuscular tissues. The calcium ions bind to membrane proteins associated with sodium channels, decreasing the number of open channels, thus regulating neuromuscular excitability. The tetany that can result from hypocalcaemia is caused by the increased excitation of neuromuscular cells, and can result in asphyxiation because of the contraction of muscles involved in breathing. This is the reason why the control of the blood calcium ion concentration is so exquisitely sensitive and keeps the circulating level within a very narrow normal range. In addition, calcium ions in the blood also act as a factor (Factor IV) involved in the coagulation cascade resulting in the formation of a blood clot.

Calcium enters the body in the diet, in meat and dairy products in particular but also in many green vegetables, seeds and nuts. It is mostly absorbed along the small intestinal tract, specifically the duodenum, mainly by transcellular transport involving specific binding proteins which are regulated. In addition, some calcium is absorbed passively by diffusion down its concentration gradient, and this paracellular transport (i.e. between cells) occurs all along the small intestine and becomes much more important when calcium content in the small intestine is high. Once in the blood, the calcium reaches all the cells in the body where it can be stored, or secreted out of the body in exocrine secretions from the glands of the gastrointestinal tract and elsewhere. Calcium is also lost continually in the cells of the dermis that are continually being shed, and in any blood loss from the body, this being of particular relevance in menstruating women. The main organs involved in the fine regulation of calcium are the kidneys; all the plasma calcium passes through the glomeruli into the filtrate and calcium reabsorption takes place particularly in the proximal tubules (70%) but also in the thin ascending limb of the loop of Henle (20%) and the distal convoluted tubule, the latter section of the nephron being the principal component of the tubular calcium which is regulated (see Figure 18.3).

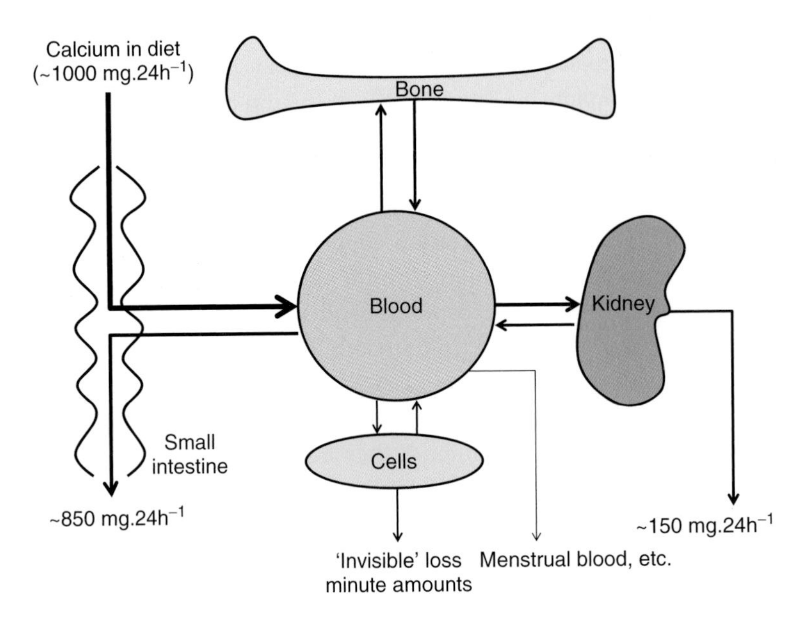

Figure 18.3 Diagram illustrating how the daily intake of calcium is handled by the main 'calcium-handling' tissues of the body.

Both calcium and phosphate ion concentrations in the blood are regulated by hormones, the main ones for calcium being parathormone, a vitamin D metabolite called calciferol and calcitonin. The first two hormones also regulate phosphate, as does fibroblast growth factor-23 (FGF-23).

Parathormone, or parathyroid hormone (PTH)

Synthesis, storage, release and transport

As its name implies, parathormone is synthesised and stored in cells of the parathyroid glands. Embryologically they are derived from the endoderm of the developing fourth pharyngeal pouch. Usually, there are four parathyroid glands (two superior and two inferior) situated at the back of the two main lobes of the thyroid gland in the neck, although the total number can occasionally vary in some individuals. Each parathyroid gland is encapsulated, and lies on the surface of the thyroidal tissue. It receives its main arterial blood supply from the inferior thyroid arteries which are branches of the thyrocervical trunk, with a small amount provided by the centrally located thyroid ima artery which usually arises from the brachiocephalic artery. The main venous outflow is via the inferior thyroid veins which join the larger brachiocephalic veins. The nerve supply to the parathyroids appears to be mainly sympathetic fibres from the superior cervical ganglion. Each parathyroid gland is comprised of two types of cell: a few oxyphilic cells the function of which is unknown, and the majority chief cells which produce parathormone. The chief cells are grouped together forming 'nests' which surround the blood vessels.

Parathormone (PTH) is a straight-chain polypeptide hormone of 84 amino acids which is cleaved from its larger 90–amino acid inactive precursor proparathormone molecule which itself is derived from its 115–amino acid pre-proparathormone precursor. Its gene is located on the short arm of chromosome 11. The pre-sequence is lost as the precursor molecule enters the endoplasmic reticulum. The proPTH molecule then enters the Golgi complex for further processing. Once in the secretory granules, some of the PTH is cleaved by proteases to carboxy-terminal fragments which are co-released with the PTH. Small amounts of amino terminal fragments are also produced. The release of PTH, and the C- and N- terminal fragments which represent up to 95% of the secretion, is from the secretory granules following the depolarisation of the cell membrane and a Ca^{++}-induced exocytotic process. Once in the blood, the half-life of PTH is 2–5 minutes, and it is metabolised in the liver and kidneys. The C-terminal fragments have a half-life up to 10 times longer.

A similar, related polypeptide called parathormone-related protein (PTHrP) is produced by various tumours, and accounts for the hypercalcaemia often seen in patients with them. In fact, there is a family of PTHrP-derived molecules which are now known to be produced by many different cells in the body. While there is some overlap between PTH and PTHrP (e.g. the hypercalcaemia), the various polypeptides also have differing effects such as the stimulation of cell proliferation, development, differentiation and death.

Receptors and mechanism of action

Parathormone has two receptors (PTHR1 and PTHR2) both of which also bind PTHrP. The PTHR1 binds PTH and some of the PTHrP derivatives. It is a member of the family of G protein receptors having seven transmembrane, an intracellular, and an extracellular, segments. Activation of this receptor is associated with increased adenyl cyclase activity and synthesis of the second messenger cAMP leading to activation of the protein kinase A pathway. These receptors are found particularly in bone and kidneys. PTHR2 mainly binds parathormone (not PTHrP), it is expressed in only a few tissues and it is also a G protein–coupled receptor associated with cAMP formation when activated. Its physiological role is presently unclear.

Actions of parathormone

The actions of PTH are primarily concerned with the regulation of calcium and phosphate ions in the blood. Its principal targets are therefore not

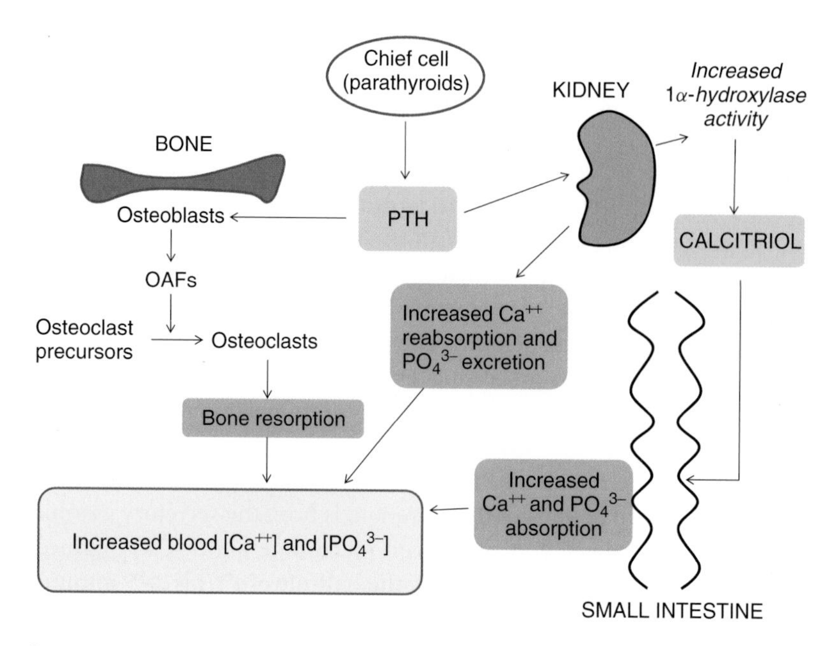

Figure 18.4 The main effects of parathormone (PTH) in the body.

surprisingly bone and the kidneys, with an important indirect effect on the small intestine (see Figure 18.4). Its overall effect is to increase the blood calcium ion concentration while maintaining a stable phosphate ion concentration.

Bone

Bone is generally comprised of a calcified matrix called osteoid, and three types of bone cell, the osteoblasts, osteoclasts and osteocytes which continually synthesise and remodel the matrix. The organic part of the osteoid is almost entirely made of a protein called type 1 collagen, with the remaining 5% called ground substance being composed of proteoglycans (carbohydrates linked to polypeptides). The complex hydroxyapatite crystals, which are comprised of hydrated calcium phosphate salts, form along the collagen fibres. This hydroxyapatite provides the long-term storage form of calcium and phosphate. However, in bone there is also a small readily exchangeable pool of calcium phosphate salts found mainly on the surface of newly formed bone which is in equilibrium with the calcium and phosphate ions in the extracellular fluid.

Osteoblasts are cells derived from mesenchymal stem cells in bone marrow. They lie along the surface of bone, and are concerned with the synthesis of osteoid. They are linked to adjacent cells (other osteoblasts on the surface, or osteocytes deep in the recently formed bone) by long cytoplasmic processes which extend between them. They have an abundant endoplasmic reticulum and numerous secretory granules which release their contents to the exterior of the cell by exocytosis. The released molecules are enzymes involved in the synthesis of collagen, and other components of the osteoid. The osteoid is synthesised around the osteoblasts, and as it becomes mineralised the osteoblasts become surrounded and their activity decreases. At this stage, they are called osteocytes; they no longer synthesise osteoid, but maintain contact with surface osteoblasts by means of the cytoplasmic processes which connect them, and which become channels, or canaliculi. It is believed that the osteocytes can actually break down the bone matrix surrounding them, allowing for its reconstruction according to the mechanical loads the bone is currently subjected to. The third type of bone cell is the osteoclast which is derived from quiescent precursor myeloid cells produced in bone marrow which enter the circulation and subsequently become attached to bone surfaces where they are activated. They are large multinucleated cells, similar to macrophages and monocytes, which are activated by osteoclast-activating factors (OAFs). These include two cytokines released from osteoblasts and their precursors: receptor activator of nuclear kappa B ligand (RANKL) and macrophage colony-stimulating factor (M-CSF). Another cytokine, interleukin 1B (IL1B), also stimulates differentiation of the osteoclasts from its progenitor cells. In contrast, osteoprotegerin (OPG), another molecule, decreases RANKL binding to its RANK receptor reducing osteoclast formation and activation. Once activated, the osteoclasts release acid and proteolytic enzymes into nearby osteoid, stimulating bone matrix resorption which releases calcium (and phosphate) into the extracellular fluid and, ultimately, the blood (see Figure 18.4).

Although parathormone increases the resorption of bone by stimulating osteoclast activity, there are no PTH receptors on these cells. Instead, it binds to its PTHR1 receptors on osteoblasts, stimulating the release of OAFs (RANKL and M-CSF) and decreasing inhibitory OPG synthesis, resulting in the activation of the osteoclasts. Consequently the osteoid is broken down and Ca^{++} and PO_4^{3-} are released into the general circulation. PTH also promotes the development of osteoblasts and osteoclasts from their precursor cells. The bone-resorptive action of PTH is therefore dependent on the stimulation of osteoblasts directly, and osteoclasts indirectly. By releasing calcium and phosphate ions from the bone matrix into the circulation, these ions could readily associate with each other to form salts which could then be deposited spontaneously in the soft tissues. This likelihood is normally prevented by various mechanisms, including the PTH-induced excretion of phosphate by the kidneys, maintaining the dissociation constant between the ions and their salts.

For example:

$$3[Ca^{++}] \times 2[PO_4^{\ 3-}] \leftrightarrow [(Ca^{++})_3(PO_4^{\ 3-})_2]$$

[calcium ions] × [phosphate ions] in equilibrium with

[calcium phosphate salts]

It should be noted that while the stimulation of bone resorption is the principal action of PTH, this hormone can also increase bone synthesis to a lesser extent; the overall effect seems to be dependent on circulating levels.

Kidneys

PTH has various effects on the kidneys all of which maintain or increase the plasma Ca^{++} concentration, acting through its PTHR1 receptor and cAMP generation.

Calcium and phosphate reabsorption: PTH stimulates calcium reabsorption along various sections of the nephron, specifically in the proximal convoluted tubule (approximately 65%), in the thick ascending limb of the loop of Henle (20%) and in the distal convoluted and connecting tubules (15%). In contrast, PTH promotes the excretion of phosphate ions by decreasing their reabsorption along the proximal (approximately 80%) and distal tubules. Phosphate reabsorption is dependent on sodium–phosphate co-transporters, and their activity is down-regulated by PTH. Proximal tubular sodium reabsorption by these transporters is also down-regulated, so PTH has some natriuretic activity. Apart from being part of the phosphate regulatory system, the increased excretion of phosphate by PTH is important because while the plasma calcium concentration is increased (by its action on bone) the phosphate concentration is maintained, thereby decreasing the likelihood that calcium phosphate salts will be deposited in the soft tissues.

Calcitriol synthesis: Between the cells of the proximal tubules are endocrine cells which are the source of a steroid hormone which is a metabolite of vitamin D_3 called 1,25 dihydroxycholecalciferol, or calcitriol. PTH acts on these cells to stimulate calcitriol synthesis by acting as a transcription factor for the 1α-hydroxylase enzyme. Calcitriol is also involved in the regulation of calcium (and phosphate) metabolism, and is considered in more detail later.

Small intestine

PTH has little or no direct effect on calcium or phosphate ion absorption through the wall of the small intestine, but does exert an important indirect influence through its action on stimulating calcitriol synthesis (see earlier section). Calcium and phosphate absorption are both enhanced by PTH through this indirect action.

Control of synthesis and release

The main stimulus for parathormone release from the chief cells of the parathyroid glands is a decrease in the circulating calcium ion concentration. The chief cells respond to changes in circulating calcium ions by means of a calcium-sensing receptor (CaR) in the plasma cell membrane, which is also expressed in cells throughout the body, including the parafollicular cells of the thyroid, the renal tubules and brain cells. The receptor belongs to the large family of receptors which are G protein linked and which have extracellular, seven-transmembrane and intracellular domains. Activation of the receptor by Ca^{++} results in the inhibition of the adenyl cyclase–cAMP pathway and stimulation of the phospholipase (PLC) C-IP_3/DAG pathways. Since the stimulus for PTH synthesis and release is a fall in circulating Ca^{++} levels, the reduction in ligand binding to the CaR would be associated with the dis-inhibition of the adenyl cyclase–cAMP pathway and a decreased activation of the PLC pathway. In the presence of normal circulating Ca^{++} levels, the synthesis and release of PTH are minimal, but as levels decrease there is a sharp increase in both these activities. Calcitriol inhibits, and phosphate stimulates, PTH gene expression, both acting mainly indirectly by affecting the plasma Ca^{++} concentration, but also directly through their own receptors and intracellular mechanisms in their target tissues (see Figure 18.5).

Calcitriol (1,25 $(OH)_2$-cholecalciferol)

Calcitriol is a steroid hormone derivative of vitamin D_3. It plays an important role in maintaining a stable Ca^{++} concentration in the blood, mainly by its action on calcium and phosphate absorption along the small intestine.

Synthesis, storage, release and transport

Calcitriol is a bioactive metabolite of the steroid vitamin D_3 (cholecalciferol). This vitamin is either ingested in the diet (e.g. present in various fish oils, milk, eggs, butter and some cereals) or actually synthesised in the skin from its early precursor, 7-dehydrocholesterol. Another, related vitamin called ergocalciferol (vitamin D_2) which is found in certain algae and fungi can also be utilised.

In the epidermis and dermis layers of the skin, 7-dehydrocholesterol is converted to vitamin D_3 by ultraviolet B (UVB) light with a wavelength between 285 and 315 nm. The skin pigment melanin, glass, clothing, the

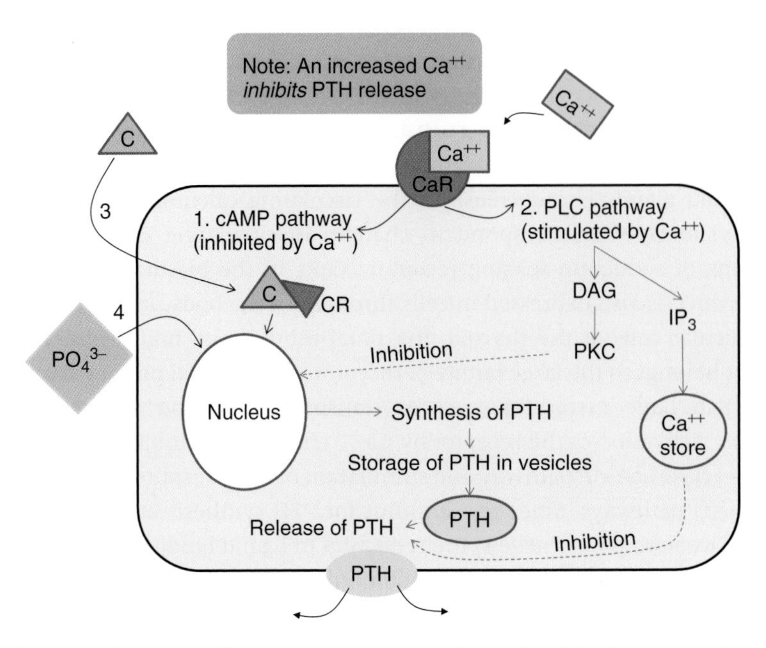

Figure 18.5 Diagram illustrating a parathyroid chief cell and the factors associated with the synthesis and release of PTH. CaR = calcium receptor; C = calcitriol; CR = calciferol receptor; DAG = diacylglycerol; IP3 = inositol triphosphate; PLC = phospholipase C and PKC = protein kinase C. For further details, see section 'Control of synthesis and release'.

low sun of winter and grey skies are all factors which decrease the amount of cholecalciferol synthesised in the skin. Cholecalciferol once synthesised is transported in the circulation bound to a binding protein (cholecalciferol-binding protein (CBP), also known as vitamin D–binding protein (VDBP)). It reaches the liver where it gets rapidly hydroxylated in various positions, the main hydroxylation being catalysed by a 25-hydroxylase enzyme, forming 25-hydroxycholecalciferol (25 OH-D_3). This precursor of calcitriol is stored in the liver which normally contains at least 3 months' supply of it, and which accounts for the liver being a good source of vitamin D_3. The 25 OH-D_3 also circulates in the blood bound to the binding protein CBP. Having very little biological activity of its own with respect to calcium and phosphate metabolism, 25 OH-D_3 is then 1α-hydroxylated in various tissues, but mainly in epithelial cells of the renal proximal convoluted tubules. The 1α-hydroxylase activity is the determining factor which produces the principal bioactive steroid derivative of vitamin D_3, 1,25 $(OH)_2$ cholecalciferol, also known as calcitriol (see Figure 18.6).

Calcitriol, like its precursors, is a steroid hence it is lipophilic. Consequently, it is not stored to any significant extent in the cells where it is produced as with other steroid hormones, but is essentially synthesised

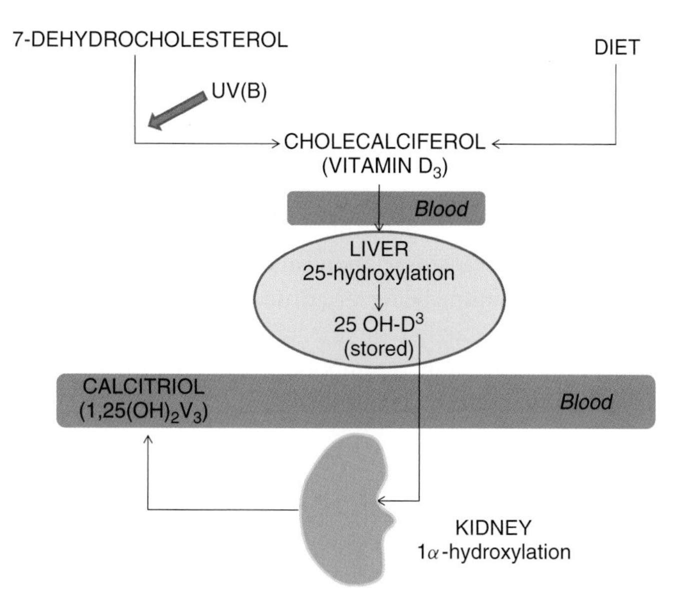

Figure 18.6 Diagram illustrating the synthesis pathways for calcitriol.

only when the proximal tubular cells are appropriately stimulated. Once in the circulation, it is bound mainly to CBP (85%), but some is also bound to albumin, with less than 0.5% present in the free, biologically active state. Its half-life is quite long, between 3 and 6 hours, and it is metabolised by the liver and kidneys to various products which are generally inactive and which are excreted partly by the kidneys but mainly by the liver via the bile duct into the duodenum. There is also evidence for some enterohepatic recycling.

A number of other metabolites of 25 OH-D_3 are produced in the body, including 24,25$(OH)_2D_3$ and 1,24,25$(OH)_3D_3$ which are synthesised by the action of 24-hydroxylase found in many tissues including bone, kidney and intestine. Their physiological roles are unclear at present. Indeed, these metabolites are probably inactive and the 24-hydroxylation pathway may be an important inactivating mechanism for the removal of calcitriol.

Receptors and mechanism of action

Calcitriol binds to specific intracellular receptors (vitamin D_3 receptors (VD_3R)) which also bind related ligands such as the 25 OH-D_3 metabolite. These receptors are part of the retinoic acid family of receptors.

The mechanism of action is believed to be mainly genomic with effects only apparent after a few hours. Once bound to its receptor, it acts as a transcription factor modulating gene activity resulting in new protein synthesis, but it does so only after forming a complex with another receptor called retinoid X receptor. However, there are more rapid effects, such as the opening of calcium channels and the stimulation of the phospholipase C in cell membranes, so non-genomic mechanisms of action are also quite likely but need to be defined.

Actions

Calcitriol exerts effects on all three of the principal tissues involved in calcium metabolism: the intestinal tract, bone and kidneys, with the effect on the duodenum and jejunum being of particular importance. It also has an important regulatory action on the production of PTH by the chief cells of the parathyroid glands, an important influence on phosphate metabolism as well as other effects (Figure 18.7).

Small intestine

Calcium absorption in the gut represents approximately one-third of the normal daily calcium intake, which can be of the order of 1000 mg. However, some calcium is also lost in intestinal secretions, so the actual overall uptake will be of the order of 200 mg.24h^{-1}. Calcium is believed to cross the intestinal wall by three mechanisms: transcellular and vesicular transport across the intestinal cells, and paracellular transport (i.e. between the cells). There is good evidence that the first two mechanisms are dependent on the activity of calcitriol and it is also likely that the paracellular route, which involves specific calcium channels between the cells, is stimulated

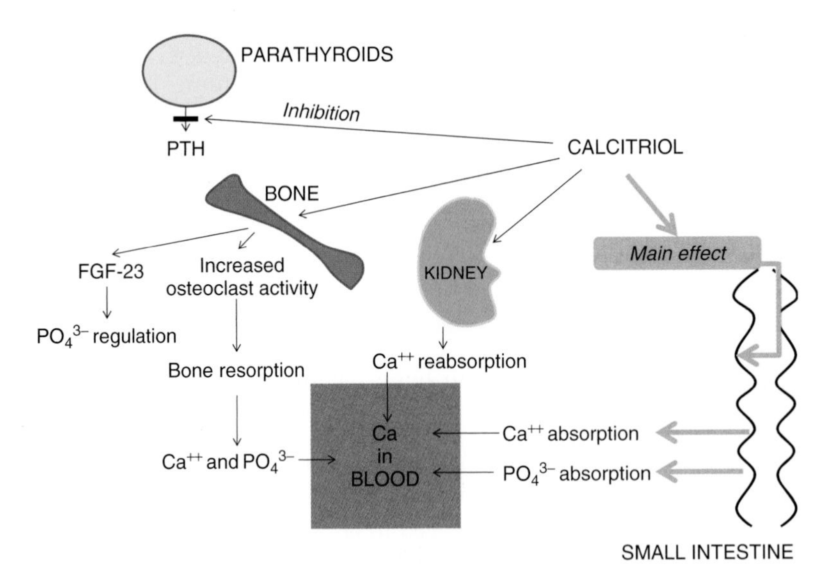

Figure 18.7 Diagram illustrating the effects of calcitriol on calcium and phosphate regulation.

by it. Calcitriol also stimulates intestinal phosphate absorption through a separate mechanism.

With respect to calcium absorption, this involves the activity of transport proteins within the intestinal cells, or enterocytes. These proteins are located at the luminal brush borders, within the cell cytoplasm and at the serosal membranes, and they are all calcitriol sensitive. Calcium in the diet reaches the small intestine, particularly in the duodenum and jejunum, where its transport into the enterocytes is associated with a calcium-binding protein, alkaline phosphatase and a Ca^{++} Mg^{++}–ATPase all of which are induced by calcitriol. However, there are also at least two calcitriol-sensitive calcium channels present in these sections of the small intestine (as well as elsewhere, such as in the kidneys), but their importance regarding intestinal calcium transport is unclear. Once the calcium ions have entered the cells, they bind to intracellular proteins such as calmodulin and calbindin which are directed to the brush border by calcitriol. Calbindin in particular has a far greater affinity for Ca^{++} than do the brush border proteins involved in the initial transfer process, and it is probably the main mechanism for transporting the Ca^{++} from the luminal brush border to the serosal membranes, as it associates with the intracellular microtubular network. The calcium ions are finally secreted out of the cells by a calcitriol-inducible, ATPase-dependent, calcium (and magnesium) pump which has an even greater affinity for Ca^{++} than calbindin.

Kidney

In the kidney, calcitriol stimulates the synthesis of proximal tubular calcium channels and promotes the reabsorption of Ca^{++} through them. In the distal tubules, calbindin protein synthesis is stimulated and this enhances the reabsorption of calcium along this part of the nephron. Furthermore, calcitriol inhibits phosphate reabsorption indirectly, mainly in the proximal tubules, by stimulating the synthesis of fibroblast growth factor-23 (FGB23), a molecule (hormone) which appears to have a profound inhibitory control on phosphate metabolism.

Bone

Calcitriol has various effects on bone but they are physiologically less important than its actions on stimulating Ca^{++} absorption in the small intestine. It stimulates osteoblasts which synthesise osteoclast-activating factors such as RANKL, and another protein osteocalcin the function of which is still unclear, while the synthesis of type 1 collagen is inhibited. Calcitriol stimulates osteoclast activity and the formation of osteoclasts from its precursor macrophage stem cells. Thus, to some extent, calcitriol will tend to increase bone matrix resorption.

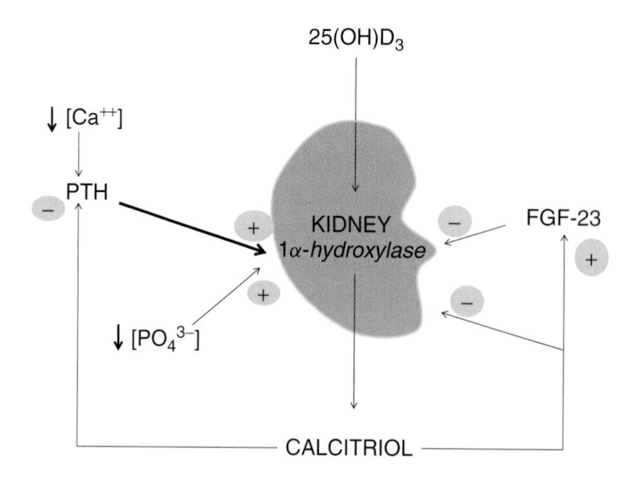

Figure 18.8 Diagram illustrating the principal controlling factors on calcitriol synthesis and release.

PTH synthesis

Cacitriol has clear genomic effects on the chief cells of the parathyroid glands, including the inhibition of cell proliferation and the synthesis of PTH. It thus acts as a direct negative feedback mechanism between renal calcitriol production and its main stimulus PTH (see Figure 18.8).

Other effects

Patients suffering from vitamin D deficiency often exhibit severe muscle weakness due to proximal myopathy. It is unclear precisely how calcitriol affects muscle physiology but certainly muscle cells synthesise VDRs, there is evidence to suggest that it increases amino acid transport into the cells, and it stimulates the synthesis of troponin C (an intracellular protein involved in the regulation of actin–myosin sliding filaments).

In addition, and as has been mentioned, calcitriol has direct and indirect effects on phosphate metabolism. It stimulates the absorption of dietary phosphate along the small intestine directly through a mechanism or mechanisms which are poorly understood. It also stimulates the synthesis of FGF-23 which is now believed to be an important endocrine factor which regulates sodium-dependent phosphate absorption and reabsorption, by inhibiting sodium phosphate co-transporters in the brush borders of enterocytes and proximal tubular cells respectively.

Control of calcitriol synthesis and release

Calcitriol synthesis is regulated mainly by PTH, but is also influenced by FGF-23, phosphate, calcium and calcitriol itself. PTH stimulates the synthesis and release of calcitriol through its stimulatory action on 1α-hydroxylase. Calciferol (vitamin D) thus cannot be converted to calcitriol in the absence of PTH. Hypophosphataemia is also associated with an increased calcitriol synthesis, but it is still unclear how the 1α-hydroxylase-containing proximal tubular cells detect changes in circulating phosphate levels. Increased calcium ion levels are associated with an inhibition of 1α-hydroxylase activity, and there is a direct inhibition of this enzyme by increased circulating calcitriol levels by a short negative feedback loop. FGF-23 has an inhibitory effect on calcitriol synthesis, this also being a classic negative feedback effect (see Figure 18.8).

Calcitonin

The third endocrine control mechanism regulating the circulating calcium ion concentration involves a hormone synthesised by the parafollicular cells (sometimes called the C-cells) located within the thyroid gland. This hormone, unlike PTH and calcitriol, decreases the circulating calcium ion concentration through effects on bone and kidney. In humans, unlike in other vertebrates, its effects are relatively weak and short-lived, and consequently it is believed to play only a minor role in the regulation of calcium metabolism.

Synthesis, storage, release and transport

Calcitonin (occasionally called thyrocalcitonin) is a 32–amino acid polypeptide which is initially synthesised as a pre-procalcitonin molecule in the parafollicular cells of the thyroid gland. These cells are embryologically derived from the neural crest, migrating to the pharyngeal pouches and becoming incorporated into the developing thyroid. The parafollicular cells lie in small clumps, between the follicles in the main lobes of the thyroid. As with other proteins, the signal peptide is lost as the molecule enters the endoplasmic reticulum and the 136–amino acid procalcitonin is cleaved within the vesicles arising from the Golgi complex to produce calcitonin. The human calcitonin gene is located on the short arm of chromosome 11. The precursor procalcitonin can also be spliced to form another calcitonin-like molecule called calcitonin gene–related peptide (CGRP), this alternative splicing occurring in various tissues including nerve cells. Both calcitonin and CGRP can be produced by tumours, and can therefore serve as biochemical markers. The main stimulus for calcitonin synthesis and release is an increased circulating calcium ion concentration.

Calcitonin circulates in the blood, principally unbound and bioactive. It is metabolised mainly in the kidneys, but also in the liver.

Receptor and mechanism of action

Calcitonin is one molecule belonging to a whole family of related molecules which include CGRP, amylin and adrenomedullin. It binds to its G protein–coupled calcitonin receptor (CR) in target cell membranes in tissues such as kidney, bone and brain. Both calcitonin and CGRP also act as ligands for another receptor called calcitonin receptor–like receptor (CRLR). The functional calcitonin receptor (CR) is coupled with different G proteins which are associated with either the activation of adenyl cyclase and increased protein kinase A (PKA) activity, or phospholipase C associated with inositol triphosphate (IP_3) and diacylglycerol (DAG) as well as protein kinase C (PKC) activity. The specific mechanism of action therefore depends on which enzyme is stimulated, resulting in the activation of either the cAMP or the IP_3 and DAG and PKC pathways within different target cells. Indeed the CRLR is actually associated with the activation of phospholipase C in the target cell membranes and this is associated with the formation of PKC.

Actions of calcitonin

While the physiological importance of calcitonin remains unclear, particularly in the long term, it clearly has effects on the kidney and bone. In the kidney, the receptors are located mainly in the proximal tubules where calcitonin decreases reabsorption of calcium and sodium. Indeed, its natriuretic effect is quite pronounced. In bone, calcitonin inhibits osteoclast activity and stimulates osteoblast proliferation and activity, resulting in an increase in bone osteoid formation. It has been postulated that calcitonin is protective of bone, for instance in situations when excessive bone resorption might be potentially harmful such as during pregnancy and lactation. Other postulated roles include sperm capacitation and fetal development.

Control of synthesis and release

The synthesis of calcitonin in the parafollicular cells is stimulated mainly by an increase in circulating calcium ion levels. Other factors which stimulate its release include certain gastrointestinal hormones such as gastrin, pentagastrin and pancreozymin, as well as β-receptor agonists such as adrenaline, and glucocorticoids. Inhibitors include somatostatin and calcitriol (see Figure 18.9).

Phosphate regulation and fibroblast growth factor-23

While much less is known about the factors and mechanisms involved in regulating phosphate, it will already be appreciated from earlier sections

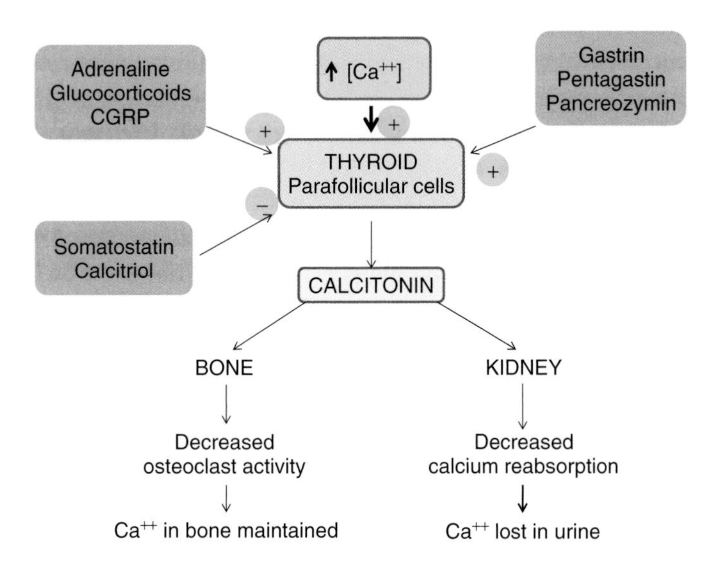

Figure 18.9 Diagram illustrating the controlling factors of calcitonin synthesis and release, and its principal actions on calcium metabolism.

in this chapter that PTH and calcitriol both influence absorption and reabsorption of this anion in the small intestine and kidneys respectively.

PTH stimulates phosphate absorption indirectly, through the stimulation of renal 1α-hydroxylase and calcitriol synthesis, and also directly inhibits renal phosphate reabsorption. Thus this hormone increases phosphate loss from the body by its renal effect, but indirectly restores the phosphate by increasing calcitriol production thereby stimulating intestinal phosphate absorption. Thus it is calcitriol which has the important effect on maintaining the circulating phosphate concentration.

A recently identified polypeptide hormone which also plays a role in regulating phosphate levels is fibroblast growth factor-23 (FGF-23), originally called phosphatonin. Its gene is on chromosome 12, and transcription resulting in the production of the molecule occurs in osteocytes, these being cells derived from osteoblasts. Overexpression or administration of FGF-23 to experimental animals is associated with the development of hypophosphataemia, so this molecule induces decreased uptake or increased loss of phosphate from the circulation. Evidence to date indicates that this molecule impairs sodium–phosphate co-transport in the brush borders of both enterocytes and proximal tubular cells. These inhibitory actions on the phosphate transport system account for the decreased uptake across the small intestine and the decreased reabsorption in the proximal tubules. Production of FGF-23 is stimulated by phosphate in the diet and in the circulation, as well as by calcitriol. As mentioned earlier, there is a direct negative feedback loop between calcitriol (which stimulates FGF-23 production) and FGF-23 (which inhibits 1α-hydroxylase activity, hence calcitriol production).

Diseases of the parathyroids, the endocrine kidney and calcium disorders

One of the most fundamental survival principles is the maintenance of a stable level of ionised calcium in the blood. This is why the exquisitely sensitive system of calcium regulation exists. If this fails, nerve and muscle transmission will become spontaneous, resulting in death of the organism. Therefore, if the calcium concentration in the blood falls, it becomes essential to maintain plasma levels at the expense of bone (Figure 18.4). Two regulatory hormones exist (PTH and 1,25 dihydroxycholecalciferol) and they act to control the level of calcium in the blood by regulating the movement of calcium from gut to bone via the blood, and also have a role in the regulation of plasma phosphate. The system has thus evolved to enable regulation of both calcium and phosphate ions, and also allows both to be stored in bone creating the skeleton which provides the strong frame for our bodies.

When blood calcium levels are low, the parathyroid glands will normally increase their secretion of PTH which, if prolonged, is called secondary (appropriate) hyperparathyroidism. PTH stimulates osteoclast activity resulting in increased bone resorption. Unchecked, this would lead to osteoporosis. In addition, the high level of PTH will activate the renal enzyme 1-α hydroxylase, and increase the synthesis of the bioactive vitamin D product, calcitriol. This in turn will increase the absorption of calcium from the gut, hopefully leading to normalisation of calcium.

Calcitriol is essential for normal calcium absorption, so a low vitamin D status can also precipitate secondary hyperparathyroidism. The combination of low vitamin D together with secondary hyperparathyroidism causes significant reduction in calcium absorption from the gut as well as bone loss, and that combination causes demineralisation of bone and hence osteomalacia rather than osteoporosis which essentially is the loss of osteoid. The key differences between osteomalacia and osteoporosis are described in Figure 18.10, where osteoporosis is the loss of total bone with a normal ratio of osteoid to calcium. On the other hand, osteomalacia is a demineralisation of bone, so that there is a relative excess of osteoid.

Osteoporosis

The human body is very efficient at removing anything that is not being used. Thus in the same way that muscle will atrophy if not used, bone atrophies if it is not subject to the stresses and strains of everyday living.

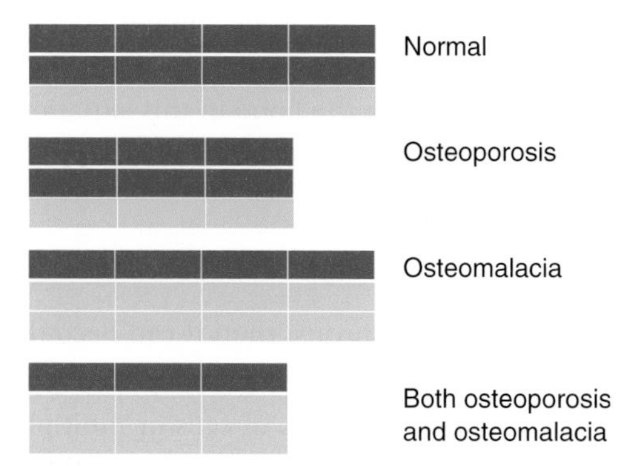

Figure 18.10 The boxes represent units of bone. When the bone is mineralised, it is represented in green. Decalcified osteoid is represented in yellow. Thus normal bone here has 12 units, of which eight are calcified. In osteoporosis, three of those are lost, so only nine units remain, of which six are calcified. In osteomalacia, there are 12 units of bone, but only four are calcified. If you are really unlucky, you might have both osteoporosis and osteomalacia, and in that circumstance, you will have only nine units of bone of which three are calcified.

Thus weightless individuals (e.g. when travelling in space) lose quite a lot of bone, and this results in osteoporosis. However, the same things occur on Earth with disuse, as in bedridden individuals. Bone density can be increased with regular exercise, especially weight-bearing exercise. Osteoporosis can also be caused when bone turnover is increased, as with hyperthyroidism and Cushing's syndrome, both conditions associated with increased breakdown of bone matrix. Lack of sex steroids, especially in women after the menopause, also causes osteoporosis because of the loss of the protective effect of oestrogens. In addition, osteoporosis is a feature of normal aging.

Osteomalacia (in children presenting as Rickets) occurs due to lack of vitamin D and the effects of its active derivative calcitriol. Thus it commonly occurs when vitamin D is scarce, and this is becoming an increasingly common problem in countries far from the equator. Osteomalacia also occurs when individuals are on drugs that increase the breakdown of vitamin D, such as phenytoin and other anticonvulsants, or are on an inadequate diet.

Renal osteodystrophy

Because the kidneys are the site of the enzyme that hydroxylates 25 hydroxycholecalciferol to produce calcitriol, renal impairment can result in a deficiency of this enzyme. This will result in low levels of calcitriol, which in turn will cause secondary hyperparathyroidism. The renal impairment in addition means that there is a decreased excretion of phosphate, so that a low blood level of calcium with a high level of phosphate can result. The consequent effects of very high levels of PTH on the bone in this situation mean that significant loss of bone (osteoporosis) can result. The pattern is known as renal osteodystrophy.

Primary hyperparathyroidism

If one of the parathyroid glands becomes tumourous, then it can become overactive. It then secretes unregulated (or sometimes partially dysregulated) parathormone. Essentially, either a single clone of cells or multiple parts of the parathyroid gland can secrete excess PTH (primary hyperparathyroidism), and in that circumstance the calcium levels in the blood increase. The high calcium levels are filtered at the renal glomeruli, and often overwhelm the tubules' ability to reabsorb calcium, resulting in high concentrations of urinary calcium which often precipitates causing very painful renal stones.

Over time, the excess loss of bone causes osteoporosis. In the case of primary hyperparathyroidism causing bone loss, it is the density at the wrist that is most significantly affected.

The hypercalcaemia can cause depression, or a feeling of general malaise. It is often said that patients with hypercalcaemia are more likely to moan than others. In addition, hypercalcaemia also can cause abdominal pain and constipation. Some students remember the list of problems that patients with hyperparathyroidism have as '(Fractured) Bones, (Renal) Stones, (Psychic) Moans and (Abdominal) Groans' (abdominal pain).

Sporadic hyperparathyroidism is the most common cause of hypercalcaemia in the community. Hyperparathyroidism can also occur in genetically predisposed individuals, and in these people the presentation can be different. Because all the cells contain a mutation that predisposes to hyperparathyroidism, it is not uncommon for more than one clone to proliferate, and sometimes multifocal or multigland primary hyperparathyroidism can be seen.

Genetic causes of primary hyperparathyroidism include conditions that cause multiple endocrine neoplasia, and are thus known as MEN. There are two common variants, known as MEN1 and MEN2.

MEN1

This is an autosomal dominant condition that presents with hyperparathyroidism as well as other neoplasia, especially in the pituitary and pancreas, with associated problems. Patients who carry the MEN1 gene thus develop primary hyperparathyroidism at a young age compared to those who develop the sporadic variety. Families who carry the MEN1 gene need to have regular screening of their blood calcium levels.

MEN2

This is an autosomal disorder that presents with medullary thyroid cancer most commonly, but also with adrenal phaeochromocytomas and primary hyperparathyroidism. A similar regular screening programme needs to be undertaken for these patients.

More details of MEN1 and MEN2 are given in Chapter 19.

Treatment of primary hyperparathyroidism

Surgery is the mainstay of treatment, but it is important to prove the diagnosis, and localise the disease first, and also to find a surgeon who is very experienced in the management of primary hyperparathyroidism. Initial investigation of a patient who presents with hypercalcaemia is to also measure the circulating PTH and vitamin D levels, and if the PTH is raised in the face of hypercalcaemia, that confirms primary hyperparathyroidism. The next stage is to exclude one rare variant cause of hypercalcaemia, namely, familial hypocalciuric hypercalcaemia (FHH). This is done by measuring the urinary calcium; essentially, if the urine calcium to creatinine ratio is high, then the diagnosis of FHH is excluded.

Once the diagnosis of primary hyperparathyroidism is confirmed, the next stage is to localise the disease, and try to determine which glands are affected. This is now done with a combination of ultrasound scan of the neck, and a functional nuclear medicine scan which is particularly good at localising active parathyroid tissue (sesta-mibi scanning). If the two scans are in agreement as to which gland is enlarged, then that gland can be removed, usually resulting in cure of the patient. If the scans are not concordant, then the surgeon needs to very carefully identify all four parathyroid glands in order to determine which one is overactive, and then that one can be removed.

In patients with MEN1, because of the risk of recurrent disease in any gland left behind, it is important to minimise that risk. Because recurrent surgery is more difficult, it is important to optimise the time of the first operation. One way to do this is to wait until the patient has significant hypercalcaemia, and then remove most of the parathyroid glands, leaving behind only half of one normal gland. This will mean that there is sufficient parathyroid tissue present to prevent the patient becoming hypocalcaemic, but with only a small risk of recurrence (see Chapter 19).

Sarcoidosis

Sarcoidosis is a disease of the lung where giant cells express ectopic 1α hydroxylase, which is not under the control of PTH. In some patients with sarcoid, therefore, vitamin D is activated when it should not be, and this in turn can result in hypercalcaemia. Treatment of the sarcoidosis with steroids or other standard treatment will suppress the giant cell expression of the enzyme. Thus the diagnosis of hypercalcaemia in sarcoidosis is successfully followed by treatment with a course of steroids.

Hypoparathyroidism

The commonest cause of hypoparathyroidism is the inadvertent removal of the glands during a thyroidectomy. Sometimes it is not possible to preserve them in patients who are having a thyroid cancer removed. A rarer cause is autoimmune hypoparathyroidism.

If all four glands have been removed, significant hypocalcaemia will result, and this can be managed only by a combination of calcium replacement and treatment with calcitriol. Titrating the doses of these two drugs is critical, and needs to be carried out by an expert. In the absence of PTH, normalising the calcium with calcitriol will result in a high level of phosphate in the blood, so that one might give excess oral calcium in order to allow the dose of calcitriol to be lowered. It is important to realise that hypoparathyroidism cannot be managed with calcium and calciferol alone. Calcitriol is essential.

Treatment of secondary hyperparathyroidism and osteomalacia

Vitamin D deficiency, causing secondary hyperparathyroidism and consequently osteomalacia and osteoporosis, is now remarkably common. This is especially the case in those who are not exposed to adequate sunlight (e.g. the housebound elderly), and also those with dark skin who are in a temperate environment. Other causes include poor dietary intake.

Treatment is simply with vitamin D. Cholecalciferol is available orally in a dose of between 1000 and 2200 units to be taken daily. Vitamin D levels should be checked, and patients advised to start with 2200 units daily. This preparation is not licenced as a drug, but is available from many health food shops as a supplement. Many patients will need to stay on the vitamin D for a prolonged period, and possibly for life. The vitamin D should be continued at least until the circulating PTH level is no longer elevated.

Treatment of osteoporosis

The treatment of choice is to find and treat the cause. Thus hyperthyroidism and Cushing's syndrome should be treated, and the bone density measured. Women are at risk of postmenopausal osteoporosis. Treatment is fully discussed in Chapter 7.

Appendix: Clinical Scenario

Clinical Scenario 18.1

Mr Barnes has been well all his life, and develops depression for no apparent reason. He has just been promoted at work, and his family seem supportive. However, the depression becomes severe enough for him to see his GP about it.

The patient had hoped he could see a councillor, but the waiting list is for several months. His GP has offered him Prozac, but he would prefer an alternative. He thus takes St John's wort as a treatment, and feels 'reasonably well' on this medication. He was unlucky enough to fall on the ice the following winter.

An X ray of his wrist is shown in Figure 18.11.

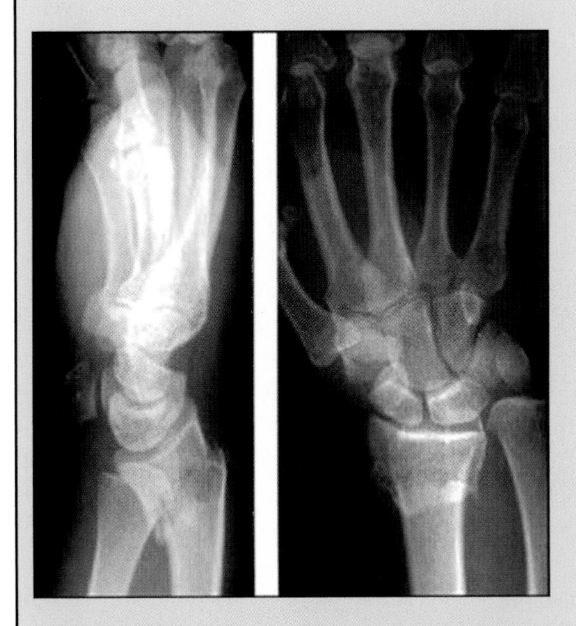

Figure 18.11 X rays of Mr Barnes' wrist after the fall.

His depression worsened. He recovers from his fracture over the next year or so, and was then stable for a few months but then admitted with severe abdominal pain. His symptoms improved when he passed a kidney stone.

Questions

Q1. What does this plain X ray of the wrist show?
Q2. Give three possible diagnoses.
Q3. Outline an investigation and management plan.

Answers to Clinical Scenario 18.1

A1. What does this plain X ray of the wrist show?

This patient has fallen on his wrist, but instead of having his hand outstretched, which causes the common Colles fracture, this X ray shows that he must have fallen with his wrist flexed, causing a Smith's fracture. This is because the hand is placed forwards rather than backwards.

A2. Give three possible diagnoses.

He has now had a fracture (bones), abdominal pain (groans), kidney stones and depression (psychic moans). Thus as suggested in this chapter, the combination of 'Bones, Stones, Moans and Groans' would suggest primary hyperparathyroidism. This would result in hypercalcaemia.

The differential diagnosis of the hypercalcaemia includes hypercalcaemia of malignancy, and hypercalcaemia of sarcoidosis.

A3. Outline an investigation and management plan.

Measurement of blood calcium and PTH levels is crucial. In the presence of hypercalcaemia, the PTH should be suppressed, and a diagnosis of primary hyperparathyroidism is suggested by a non-suppressed PTH. In the case of a suppressed PTH, investigation for malignancy and sarcoidosis should be undertaken. A chest radiograph might show bilateral hilar lymphadenopathy, suggesting sarcoidosis, and hypercalcaemia of sarcoidosis will respond to a short course of steroids.

If the hypercalcaemia is severe, rehydration is the main stay of treatment. Most patients are dehydrated at presentation, because the high circulating calcium level results in hypercalciuria, which in turn causes an osmotic diuresis, which presents as nephrogenic diabetes insipidus. Almost all patients will respond to rehydration. Occasionally frusemide is used to enable large volumes of saline to be administered, which may be required to treat the hypercalcaemia.

Bisphosphonates should be avoided in suspected hyperparathyroidism, as they make subsequent investigations impossible. Bisphosphonates may be useful in the hypercalcaemia of malignancy.

Once the calcium balance has improved, patients can then be investigated to find the source of the PTH, and if a single parathyroid adenoma is found (using sesta-mibi scanning and an ultrasound scan of the neck), a parathyroidectomy is required.

In young patients who present with primary hyperparathyroidism, it is important to consider heritable causes such as MEN1. Details of this can be seen in Chapter 19.

Further Reading

Pondel, M. (2000) Calcitonin and calcitonin receptors: bone and beyond. *International Journal of Experimental Pathology*, 81 (6), 405–22.

Villa, I., Dal Fiume, C., Maestroni, A., Rubinacci, A., Ravasi, http://ajpendo.physiology.org/content/284/3/E627.full - aff-3 F. & Guidobono, F. (2003) Human osteoblast-like cell proliferation induced by calcitonin-related peptides involves PKC activity. *American Journal of Physiology*, 284, E627–33.

Warde, N. (2011) Bone: the odyssey of osteoclast precursors. *Nature Reviews Rheumatology*, 7, 557.

CHAPTER 19

The Genetics of Endocrine Tumours

Multiple endocrine neoplasia types 1 and 2, and Von Hippel Lindau disease

Multiple endocrine neoplasia types 1 and 2 (MEN1 and MEN2), and Von Hippel-Lindau disease, are all dominantly inherited conditions that put patients at risk of particular endocrine disease. The alleles responsible for most variants are now known, although the actual mechanism for the cause of the tumours remains an area of active research.

Multiple endocrine neoplasia type 1 (MEN1)

Patients with MEN1 are at risk of primary hyperparathyroidism, pituitary tumours and pancreatic hormone–secreting neoplasia. The condition should be suspected in anyone who presents with two or more endocrine neoplasia, or anyone with a family history of MEN1 and one of the features characteristic of an endocrine disorder. In addition, some patients also present with adrenocortical tumours which may be non-functioning, carcinoid tumours; facial angiofibromas; collagenomas or lipomatous tumours.

Primary hyperparathyroidism is the most common manifestation of MEN1. Patients present with typical features of hypercalcaemia, but at a much younger age than in sporadic primary hyperparathyroidism. Thus they are at risk of 'bones, stones, moans and groans' as described in Chapter 18. Interestingly the gender ratio of primary hyperparathyroidism in MEN1 is 1:1, as one might expect from a dominant condition, although this is at variance with sporadic disease which occurs more commonly in females (3:1 female-to-male ratio).

The hypercalcaemia is usually mild, and in patients with a family history will present asymptomatically with biochemical hypercalcaemia, initially with a normal or raised parathyroid hormone (PTH) concentration.

Unlike sporadic disease, the histology is usually one of parathyroid hyperplasia, (rather than a single adenoma) and the surgical management of such patients needs careful consideration, especially since patients may require neck surgery more than once. Finding a recurrence of primary

---

Integrated Endocrinology, First Edition. John Laycock and Karim Meeran.

hyperparathyroidism is much more difficult than operating on someone who has never had surgery, because finding a small parathyroid adenoma in a patient with a healed neck can be extremely difficult. These factors need to be taken into account when planning surgery for such patients. One has to plan for future parathyroid surgery that may be required at a later stage.

Some centres advocate the removal of all four parathyroid glands, but obviously this will result in hypoparathyroidism and dependence of patients on calcium and one-alpha calcidol for the rest of their life. Calciferol (vitamin D) that is not one-alpha hydroxylated will not work in the absence of PTH. Other centres suggest that each adenoma is removed if and when they occur, but the disadvantage of this is that patients might need several operations in their lifetime, each one more difficult than the last. Maybe the best preference is for a full discussion with the patient as to these options, including the option of removal of all but half of a parathyroid gland. Thus at the time of surgery, the enlarged gland(s) are removed, and one normal gland is partly removed. The remaining half-gland will hopefully supply enough PTH to prevent hypocalcaemia and the risk of recurrent primary hyperparathyroidism is reduced.

Pancreatic tumours

The incidence of pancreatic tumours varies in published series, but is between 30% and 80% of MEN1 patients, depending on their age and the expertise of the centre. Most of these tumours produce excessive amounts of hormone, for example gastrin, insulin, glucagon or vasoactive intestinal polypeptide (VIP), and are associated with distinct clinical syndromes. A blood sample for the measurement of fasting gut hormones is taken for diagnosis.

Gastrinoma

These gastrin-secreting tumours represent over 50% of all pancreatic islet cell tumours in MEN1. Conversely, approximately 20% of patients with gastrinomas will have MEN1. Gastrinomas are the major cause of morbidity and mortality in MEN1 patients. They may develop multiple ulcers that can perforate, and gastrinomas in these patients have metastatic potential. This association of recurrent peptic ulceration, marked gastric acid production and non-islet β cell tumours of the pancreas is referred to as the Zollinger-Ellison syndrome. Additional prominent clinical features of this syndrome include diarrhoea and steatorrhoea. The diagnosis is established by demonstration of a raised fasting serum gastrin concentration in association with an increased basal gastric acid secretion. Medical treatment of

MEN1 patients with the Zollinger-Ellison syndrome is directed to reducing basal acid output and this may be achieved by the parietal cell proton pump inhibitors. The ideal treatment for a non-metastatic gastrinoma is surgical excision. However, in patients with MEN1 the gastrinomas are frequently multiple or extra-pancreatic and the role of surgery has been controversial, with some centres advocating aggressive recurrent operations, and others observing patients and treating them with proton pump inhibitors for many years. For example, in one study only 16% of MEN1 patients were free of disease immediately after surgery, and at 5 years this had declined to 6%; the respective outcomes in non-MEN1 patients were better, at 45% and 40%.

The treatment of disseminated gastrinomas is difficult, and hormonal therapy with octreotide, which is a human somatostatin analogue; chemotherapy with streptozotocin and 5-fluoroaracil; hepatic artery embolization and removal of all resectable tumours have all occasionally been successful.

Insulinoma

These islet β cell tumours secreting insulin represent one-third of all pancreatic tumours in MEN1 patients. Insulinomas also occur in association with gastrinomas in 10% of MEN1 patients, and the two tumours may arise at different times. Insulinomas occur more often in MEN1 patients who are younger than 40 years, and many of these arise in individuals younger than 20 years, whereas in non-MEN1 patients insulinomas generally occur past the age of 40 years. An insulinoma may be the first manifestation of MEN1 in 10% of patients, and approximately 4% of patients presenting with insulinoma will have MEN1. Patients with an insulinoma present with hypoglycaemic symptoms which develop after a fast or exertion, and improve after glucose intake. Biochemical investigations reveal raised plasma insulin concentrations in association with hypoglycaemia. Circulating concentrations of C-peptide and proinsulin, which are also raised, may be useful in establishing the diagnosis, as may an insulin (c-peptide) suppression test. Further details on the localisation of insulinomas are given in Chapter 17. Surgical treatment, which ranges from enucleation of a single tumour to a distal pancreatectomy or partial pancreatectomy, has been curative in some patients. Chemotherapy, which consists of streptozotocin or octreotide, is used for metastatic disease. Medical treatment includes regular carbohydrate meals and diazoxide.

Glucagonoma

These islet α cell, glucagon-secreting pancreatic tumours occur in fewer than 3% of MEN1 patients. The characteristic clinical manifestations of a skin rash (necrolytic migratory erythyema), weight loss, anaemia and stomatitis may be absent and the presence of the tumour is indicated only by glucose intolerance and hyperglucagonaemia. The tail of the pancreas is the most frequent site for glucagonomas and surgical removal of this region is the treatment of choice. However, treatment may be difficult as 50% of patients have metastases at the time of diagnosis. Medical treatment with octreotide or with streptozotocin has been successful in some patients.

VIPoma

Patients with VIPomas, which are vasoactive intestinal peptide (VIP) secreting pancreatic tumours, develop watery diarrhoea, hypokalaemia and achlorhydria, referred to as the WDHA syndrome (see also Chapter 17). This clinical syndrome has also been referred to as the Verner-Morrison syndrome or the VIPoma syndrome. VIPomas have been reported in only a few MEN1 patients and the diagnosis is established by documenting a markedly raised plasma VIP concentration. Hypokalaemia and dehydration occurring acutely may be fatal, and careful and aggressive rehydration is required initially. Since VIP is cleared by the kidneys, the dehydration can cause acute renal failure, with a consequent sharp increase in VIP concentrations. Thus rehydration can help the severe diarrhoea, by helping reduce the VIP level. Surgical management of VIPomas, which are mostly located in the tail of the pancreas, has been curative. However, in patients with unresectable tumours, treatment with streptozotocin, octreotide and corticosteroids has proved beneficial.

PPoma

These tumours, which secrete pancreatic polypeptide (PP), are found in a large number of patients with MEN1. No pathological sequelae of excessive PP secretion are apparent and the clinical significance of PP is unknown, although the use of serum PP measurements has been suggested for the detection of pancreatic tumours in MEN1 patients.

Pituitary tumours

The incidence of pituitary tumours in series varies between 15% and as high as 90% of patients with MEN1, depending of the aggressiveness of the screening programme.

The commonest of these is a prolactinoma, which accounts for about 60% of pituitary tumours in MEN1. About 20% cause acromegaly and 5% cause Cushing's disease, with the rest being 'non functioning'. Of those who present with a functioning pituitary tumour, less that 3% will have MEN1, so that 97% of pituitary tumours at least occur sporadically.

Patients with MEN1 should have screening for pituitary tumours, including regular measurements of prolactin.

The management of these tumours is exactly the same as in sporadic disease, with surgery, radiotherapy and medical treatment, and is covered elsewhere in this book.

Other tumours include carcinoid tumours that occur in about 3% of MEN1 patients, facial angiomas in close to 90% and adrenal tumours that are usually non functioning in up to 40%. There are a small number of MEN1 patients who have functioning adrenal tumours that can cause Cushing's or Conn's syndrome.

Genetics of MEN1

The gene causing MEN1 was localised to chromosome 11 by genetic mapping studies that investigated MEN1 associated tumours for loss of heterozygosity. This is where one allele carries the mutation and one does not, hence the somatic cells of an MEN1 patient are heterozygous. That hererozygosity is lost in cells that become tumours. Segregation studies in MEN1 families also contributed. The results of these studies, which were consistent with Knudson's model for tumour development, indicated that the MEN1 gene represented a tumour suppressor gene. The protein product of that gene has been called 'MENIN' and is made up of 610 amino acids. It seems to be expressed in many tissues.

Studies of protein-protein interactions have revealed that MENIN interacts with several proteins involved in transcriptional regulation, genome stability, cell division and proliferation.

Multiple Endocrine Neoplasia type 2 (MEN2)

Patients with MEN2 are at risk of Medullary Thyroid Cancer, parathyroid hyperplasia and phaeochromocytoma. The risk of Medullary Thyroid Cancer is close to 100%, so this is one condition where genetic screening of family members has truly changed the lives of many people. Children who inherit the MEN2 gene are advised to have a total thyroidectomy, and can continue lifelong screening for parathyroid disease with regular calcium measurements, and for phaeochromocytomas with regular catecholamine measurements.

Without screening, patients with MEN2 are at risk of death by the age of 50, either from medullary thyroid carcinoma and the cardiovascular effects of undiagnosed phaeochromocytoma.

The syndrome was first described by John Sipple in 1961. There are three major variants of MEN2, and it is now known that there are several different mutations of the RET proto-oncogene that are associated with different presentations. The three major variants of MEN2 are MEN2a, MEN2b and familial medullary thyroid carcinoma (FMTC).

MEN2a

Patients who inherit the gene for MEN2a are at risk of medullary thyroid carcinoma, phaeochromocytomas and parathyroid hyperplasia. Almost 100% of patients with MEN2a develop thyroid cancer by the age of 10 years, so prophylactic thyroidectomy is suggested. The risk of phaeochromocytoma in these patients is in the region of 50% while only 20% develop parathyroid disease. Thus it is usual to try and preserve the parathyroids during prophylactic thyroidectomy, as removal of the parathyroids results in lifelong reliance of the patient on calcium and one alpha calcidol, to control calcium concentrations, which can be difficult.

MEN2b

These patients have similar risks to MEN2a patients with regard to medullary thyroid cancer and phaeochromocytoma, but no risk of parathyroid disease. Instead they present with bowel ganglioneuromata, and mucosal neuromas that can be seen by looking into the mouth. They also have skeletal abnormalities, including kyphoscoliosis, and a facial appearance similar to Marfan's syndrome (a so called 'Marfanoid habitus'). However they do not have any features of true Marfan's syndrome.

FMTC

These patients only present with thyroid cancer as the name suggests. It is important to watch carefully for phaeochromocytomas, in case a family is labeled with FMTC, but in fact have MEN2a.

The gene for MEN2 is on chromosome 10, and is the RET proto-oncogene. Unlike MEN1, strong genotype-phenotype correlations exist within MEN2 such that there are clear associations between mutations at specific codons and MEN2 subtypes. Mutations clustered in the cysteine-rich extracellular domain (codons 609, 611, 618, 620, 630 and 634) are the primary causative factor in approximately 98% of cases of MEN2A. These highly conserved cysteines are important for receptor dimerization and mutations result in ligand-independent dimerization and activation of the RET receptor complex. Mutations in the intracellular tyrosine kinase domain (codons 768, 790 and 804) are less common, traditionally associated with FMTC, and rarely associated with other MEN2A-related tumours. Ninety-five percent of MEN2B cases involve a single point mutation leading to the substitution of methionine 918 for a threonine altering the substrate recognition pocket of the catalytic core of the receptor.

MEN1

Parathyroid hyperplasia
Pituitary adenomata, including Prolactinomas, Cushing's disease and Acromegaly
Pancreas (including insulinomas, gastrinomas, glucagonomas and VIPomas)

MEN2a

Medullary carcinoma of the thyroid in almost all cases
Phaeochromocytoma
Parathyroid hyperplasia

MEN2b

Medullary carcinoma of the thyroid in almost all cases
Phaeochromocytoma
Mucosal Neuromata and a 'Marfanoid habitus'

Further Reading

Brandi, M.L., Gagel, R.F., Angeli, A. et al. (2001) Guidelines for diagnosis and therapy of MEN type 1 and type 2. *J Clin Endocrinol Metab.* 86(12), 5658–71.

Norton, J.A., Fraker, D.L., Alexander, R. et al. (1999) Surgery to cure the Zollinger-Ellison syndrome. *New England Journal of Medicine*, 341, 635–44.

Sipple, J.H. (1961) The association of phaeochromocytoma with carcinoma of the thyroid gland. *American Journal of Medicine* 31, 163–166.

Trump, D., Farren, B., Wooding, C., Pang, J.T., Besser, G.M., Buchanan, K.D., Edwards, C.R., Heath, D.A., Jackson, C.E., Jansen, S., Lips, K., Monson, J., O'Halloran, D., Sampson, J., Shalet, S.M., Wheeler, M.H., Zink, A. & Thakker, R.V. (1996) Clinical studies of multiple endocrine neoplasia type 1 (MEN1) in 220 patients. *Quarterly Journal of Medicine*, 89, 653–69.

Turner, J.J., Wren, A.M., Jackson, J.E., Thakker, R.V. & Meeran, K. (2002) Localization of gastrinomas by selective intra-arterial calcium injection. *Clinical Endocrinology*, 57 (6), 821–5.

CHAPTER 20

Future Prospects

Introduction

What does the future hold for endocrinology? It is impossible to predict the long-term developments likely to take place, but in the shorter term there are various areas of endocrinology which are likely to have reached an interesting stage. Clearly there will be steady advances made in our better understanding of how the body functions, as our continually expanding knowledge of cellular and molecular biology becomes translated into physiological actions and regulatory systems. Some developments will inevitably be associated with ethical issues which will require national and international discussion, and ultimately approval, before they can be utilised for the benefit of humankind. Various areas of development are already clearly identified.

Neuroendocrinology

One such area is the ever-continuing progress being made in the field of neuroendocrinology, as the effects of hormones on behaviour become increasingly unravelled. For example, the more widespread distribution of receptors for a particular hormone such as oxytocin or vasopressin does not match the relatively sparse projections of the relevant neurones (oxytocinergic and vasopressinergic). The discovery that hormones can be released from dendrites directly into the third ventricle provides a useful explanation for the transport of hormones to specific regions of the brain where the receptors, but not the neurone terminals, are found. Thus there is much interest currently in potential roles for hormones, including neuropeptides and neurosteroids, in regulating various behaviours in humans, such as parenting, social recognition, trust and maybe even mating. Already there is interest in identifying the role of such hormones in clinical conditions such as autism. Another area of interest, awaiting further research, is the importance of hormones in establishing neuronal and other cellular networks and responsiveness to various stimuli such as to stressors, *in utero* or in the early post-natal period.

Integrated Endocrinology, First Edition. John Laycock and Karim Meeran.

The genome

Identifying the structure of the human genome has already been established and many advances in our understanding of the human body have been made, including the discovery of many genes, hence proteins, some of which were initially labelled as orphan molecules. Many of these molecules have already been, or will in future be, identified to be receptors, enzymes or hormones for example. Following on from the identification of biologically relevant molecules and their structures will hopefully follow the development of new drugs which can target receptors specifically, and thus provide new treatments for a variety of disorders. Then of course there is the exciting possibility of replacing defective genes, something which is already beginning to become a real possibility.

Stem cells

Stem cells are another development with tremendous potential in the future. The creation of whole organs and tissues from stem cells has already begun, and surely it will not be long before endocrine tissues such as pancreas, thyroid and adrenal will be available for implant into humans. Will ova and spermatozoa also become available through stem cell research? Ethical issues will have to be resolved when considering the use of embryonic cells, but already other sources of stem cells are being studied.

Organ replacement

There has always been a demand for organs and tissues for transplantation, whether to replace a defective or failing heart or kidney or for a damaged limb or even badly scarred skin. The grafting of tissue from one individual to another, including from an animal to a human, has been known about for over a century, and over the last 50 years various attempts have been made in this field, including transplanting hearts and kidneys from primates into humans. In the early days, this form of organ replacement was unsuccessful mainly because of donor tissue rejection by the recipient. This form of transplantation, of animal tissues into humans, is called xenotransplantation. Because of the continued demand for replacement organs and tissues, there has been much research into developing appropriate means (e.g. immunological) by which the procedure can be improved. This essentially involves animal organs which are immunologically adapted for transplantation into humans. One example would be genetically modified porcine endocrine pancreatic tissue. However, there are dangers, for instance the risk of cross-species transmission of disease.

While many patients in urgent need of an organ or tissue would be prepared to accept xenotransplantation, there are clearly ethical, and other, issues which need to be considered and resolved.

Technical advances

Advances in surgical techniques are likely to continue, particularly as computers and their accompanying software become ever more sophisticated. Thus already accepted as commonplace is keyhole surgery, and computerised surgery 'from afar' (linked by advanced telecommunications systems) with the expert specialist surgeon directing surgery at a distance (maybe in another country) from the actual operating theatre. This also applies to diagnostic equipment. Scanning devices have advanced a long way from simple X ray machines, and already the latest devices such as functional magnetic resonance imaging (fMRI) scanners are providing ever-increasing insight into body function by linking activity (e.g. in the brain) with sensory, or other, input. And what about nanotechnology, the development of materials on a 'nano' (i.e. molecular, at 10^{-9}) scale? This is almost at the level of science fiction, and yet rapid advances are already being made in engineering such microscopic molecular machines, including motors, which can be utilised as biological devices, for example for targeting drugs to specific cells (e.g. cancer cells).

More expert opinions are now already obtained using multidisciplinary meetings across multiple sites, and even internationally. Such meetings are attended by radiologists, surgeons of various specialities, radiotherapists and of course endocrinologists. A patient with Cushing's disease might thus initially be referred for the investigation of Cushing's syndrome. Once the diagnosis is confirmed, the case will be presented at a meeting which can be transmitted to several hospitals, and surgeons of all types can see the radiology and hear the discussion. The patient needs to be reviewed by both pituitary and adrenal surgeons, with an overview from endocrinologists, to decide which is the best option. The images are transmitted live for all sites to review. Web-based interactivity is likely to speed up communications, as has already happened with the picture-archiving and communication system (PACS) in radiology, where images of radiographs can be rapidly reviewed in any NHS hospital. The benefits for teaching and training are clear.

Patients having more say in their care relies on them being well informed, and there are many examples in endocrinology where this is the case. Patients with Graves' disease are increasingly able to choose between drugs, radioiodine and surgery, but this relies on them being informed about the risks and benefits of each. The web is already a good source of information (and misinformation!), but as medical knowledge increases, we should be able to impart the best individualised management plan to every patient.

These, and other developments, are undoubtedly going to expand our knowledge and increase our ability to manipulate the human body to our advantage either by enhancing already sound performance, or by replacing defective organs and tissues including endocrine glands. While research into expanding our longevity also carries on apace (e.g. the relevance of telomeres, genes, diet and mentality), the need to understand more fully how the endocrine system functions and interacts with all other systems in the body is essential for our future.

Index

Note: figures and tables are indicated by *italic page numbers*

Integrated Endocrinology, First Edition. John Laycock and Karim Meeran.